Springer Series in Biophysics 2

Editor: P. M. Bayley, London

Springer Series in Biophysics

Volume 1 **Structure, Dynamics and Funktion of Biomolecules**
Edited by A. Ehrenberg, R. Rigler, A. Gräslund and L. Nilsson

Volume 2 **Biophysics and Synchrotron Radiation**
Edited by A. Bianconi and A. Congiu Castellano

Biophysics and Synchrotron Radiation

Edited by
A. Bianconi and A. Congiu Castellano

With 192 Figures

Springer-Verlag Berlin Heidelberg New York
London Paris Tokyo

Professor Dr. Antonio Bianconi
Professor Dr. Agostina Congiu Castellano
Department of Physics
University of Rome
"La Sapienza"
Piazzale Aldo Moro, 2
I-00185 Rome, Italy

ISBN-13: 978-3-642-71492-4 e-ISBN-13: 978-3-642-71490-0
DOI: 10.1007/978-3-642-71490-0

2131/3130-543210

Preface

This book the second volume in the "Springer Series in Biophysics" collects together contributions to the conference on "Biophysics and Synchrotron Radiation" held in July 86 at Frascati. This meeting addressed the advances on the structure of biological molecules obtained by using synchrotron radiation. In fact it was most timely to review the results of the research in biophysics which is rapidly developing at synchrotron radiation facilities. Moreover, there was interest to discuss the new perspectives opened up by the future high brilliance synchrotron radiation sources.

With the use of synchrotron radiation, x-ray spectroscopy of biological molecules is firmly established in the techniques of EXAFS and XANES. Contributions to the detailed knowledge of local structure of active sites of metalloproteins by this approach are presented in this volume, together with a number of studies of the interaction of metal ions with other important biological macromolecular systems.

Structural determination of very large biological systems at high resolution, including a protein and its substrate, are reported. The experimental advances in protein crystallography presented here reduce the time for solving protein structures, thus satisfying a major requirement of the rapidly-expanding field of protein engineering.

Great scientific interest is now concentrated on detecting the dynamics of the structure of biological molecules, related to their biological function. In this volume new experimental approaches to this problem are presented, showing the enormous potential offered by synchrotron radiation in introducing the dimension of time into x-ray diffraction, x-ray scattering and x-ray spectroscopy.

The success of this meeting was already ensured by the large and enthusiastic participation of the leading scientists in the field, and prolonged discussions in a friendly atmosphere. We hope that these contributions to this exciting area of biophysical research will now be enjoyed by a wider audience.

A. BIANCONI A. CONGIU CASTELLANO

Contents

X-Ray Diffraction

X-Ray Spectroscopy

X-Ray Scattering

X-Ray Microscopy

Time-Resolved Fluorescence

List of Contributors

You will find the addresses at the beginning of the respective contributions

X-RAY DIFFRACTION

The Impact of Synchrotron Radiation on Protein Crystallography

Guy Dodson
Chemistry Department
University of York, YO1 5DD

Introduction

Protein crystallography since the 1950's has developed into one of the major arms of biology owing to its capacity to reveal the 3 dimensional structure in atomic detail of proteins and other macromolecules. Because of this it has become an essential platform for understanding many biological processes. The success of protein X-ray analysis rests principally on the power of the experimental methods for determining the phase. In this technique heavy atoms which scatter X-rays strongly are introduced into the crystal - usually by soaking. Providing the crystal organisation is unaffected by this reaction the changes in the X-ray diffraction pattern can be used to locate the heavy atoms, then to calculate the phase angle and hence obtain the electron density from which the structure is interpreted.

The problems faced by the protein crystallographer mostly relate to getting suitable crystals and preparing satisfactory heavy atoms derivatives. First there is the fragility of protein crystals. The protein molecules are supported in the lattice by only occasional contacts and are mostly surrounded by solvent, which forms a large proportion of the crystal volume. Thus, there is often a problem of poor diffraction which limits the quality or resolution of the X-ray image. Associated with this is the tendency of protein crystals to deteriorate while being irradiated with X-rays; this behaviour is usually more pronounced in crystals containing bound heavy atom. These experimental hurdles have generally been overcome by replacing crystals during data collection -which of course can raise problems if there is any crystal variability. In the case of heavy atom derivaties, variation in the pattern of substitution and isomorphism are relatively common problems. A second difficulty, too often experienced, is making enough satisfactory heavy atom substitutions. In principle one isomorphous derivative is sufficient to determine phasing if anomolous scattering data are also measured accurately. Usually however two or even three derivatives are found to be needed greatly increasing the experimental burden in the analysis.

Protein molecules are large, usually containing between 1000 - 5000 atoms and as biochemical and biological research expands more and larger proteins and macromolecular complexes are coming available for study. To determine their structure accurately large data sets are needed and thus methods for obtaining tens of thousands of X-ray reflections are required. As a result measurement and processing of protein crystal data has become a major committment.

The problem of crystal fragility and the size of the data sets are bottlenecks in the subject. When the protein crystals being analysed are scarce and/or very precious - a fairly common situation - these difficulties are magnified and their effective use is of paramount importance. The intensity and tuneability of SR however have provided solutions to some of the problems faced by protein crystallographers. Consequently over the last few years its useage has increased rapidly and it is clear that the potential of synchrotron sources in protein crystallography is being realised.

In this brief review the role of SR in the analysis of protein crystals will be illustrated by reference to some selected studies going on within the UK which have benefitted from the radiation's intensity and/or tuneability. A very much more detailed survey of synchrotron x-radiation and protein crystallography is available which should be read by all interested in the field. (Helliwell, 1984).

|SR = synchrotron radiation|

INTENSITY RELATED ADVANTAGES

Data Collection Rates and Crystal Lifetime

The rate of data collection from synchrotron sources is always going to be dictated by the characteristics of the crystals but nonetheless the rate of collection generally is greatly increased with synchrotron x-radiation. For comparable data sets collection is of the order of 10 - 20 times faster.

Usually the high intensity does not cause an equivalent deterioration in the crystal. Indeed with synchrotron x-radiation good diffraction continues long enough to collect many times more reflections - and especially high angle reflections - per crystal owing to its much more rapid rate of data production.

The rate of data collection is very greatly increased by the use of Laue diffraction methods which exploit the white beam characteristics of the synchrotron source. The geometry of the Laue method allows about 15° of reciprocal space to be recorded simultaneously which offers immense potential for rapid data collection. Recent experiments on Lectin (Helliwell et al.1985) and phosphorylase b (Hadju et al., 1986) have already been carried out. The Laue method appears to be ideal for time resolved analysis of protein mechanisms and considerable effort is being applied to its development in experiments on phosphorylase b by Dr Louise Johnson and her colleagues in Oxford and Daresbury.

Accuracy

The diffraction data collected with synchrotron x-radiation is generally of good quality with internal consistency now as good or even better than that from data collected by diffractometers or other conventional methods. To some extent this can be attributed to the superior collimation which reduced the spread of the reflections. So far in protein studies diffraction data have been collected on films. Typically agreement factors for symmetry related reflections have values between 4 - 10 per cent in F^2 for data extending to 3.0Å spacing; by protein crystal standards this is very acceptable. Data collected at shorter wavelengths ($\lambda \sim 1$Å) following the arrival of the wiggler at Daresbury is distinctly more accurate owing no doubt to the reduction of absorption effects and possibly in some cases,of radiation damage as well.

Large Unit Cells

The very large unit cells that occur in proteins and especially in viruses requires very tight collimation to resolve adjacent reflexions. This has been achieved at the various synchrotron centres and not surprisingly SR has been successfully applied to a number of virus studies. These also benefit from the shorter exposure time needed to record data. The determination of the cold virus by Michael Rossmann and colleagues at Purdue (1985) is an excellent example of the successful application of SR to a large unit cell crystal (Rossmann et al., 1986).

Extension of Resolution

A result of using synchrotron x-radiation has been the general improvement in resolution of the data sets. The availability of intense radiation at short wavelengths has further extended the resolution of some data sets. For example in the haemoglobin studies at York a data set collected at ~ 1.0Å extends to 1.5Å spacing - previous data sets were about 2.0Å spacing in resolution (Liddington, 1986).

These large data sets are essential for accurate analysis of protein crystals. Their processing has become a time consuming operation while their manipulation in protein crystallographic refinement calculations has become an important aspect - discussed in the paper by Dodson and Liddington in this volume.

Examples of Protein Structure Studies using SR

SR brings obvious benefits to protein crystallographic research. These are perhaps best illustrated by reference to some selected examples of studies being undertaken in the UK - mostly using the facilities at the SRS at Daresbury and sometimes those at LURE.

Glyceraldehyde phosphate dehydrogenase

The enzyme which exhibits cooperative activity is part of the glycolytic pathway; it is a tetramer of molecular weight 145,000 with 4 identical subunits. The catalytic mechanism involves NAD^+ which binds to the enzyme in a negatively cooperative way (Leslie and Wonnacott, 1984). X-ray structural studies have demonstrated that there are ligand induced sequential conformational changes within the enzyme. An accurate description of the structural environment of the NAD and its contacts to the protein is therefore a necessary step to understanding the basis of the enzymes cooperative behaviour. The 1.8Å data from the apo enzyme has been refined to an agreement factor $R = \Sigma||Fo| - |Fc|| / \Sigma|Fo|$ of .189. This has led to a detailed and accurate structure of the enzyme and attached cofactor - the rms difference in the NAD atoms in the symmetry related subunits is .1Å, that of the protein atoms is .2Å. In addition some 680 water molecules have been identified. In the electron density the protein and cofactor atoms are sometimes individually resolved and such features as aromatic rings with central holes appear. (A Wonnacott, private communication).

Glycogen phosphoraylase b

The enzyme glycogen phosphoraylase is the key enzyme in controlling glycogen breakdown. It is a very large tetramer, again of 4 identical subunits, of total molecular weight about 400,000. It exhibits complex allosteric catalysis with binding sites activating sites and regulatory sites. Clearly the task of identifying the structural and chemical behaviour of the enzyme is an enormous one. However the first steps towards this have been taken in the refinement of the native enzyme with 1.9Å data collected from the synchrotron source at Daresbury. In this study the agreement factor R, is .198 on some 72,000 reflections (90% > 3σ) whose internal consistancy is of the order of 7 per cent in F^2, an impressively low value. The refinement is a colossal undertaking, the structure is still being analysed and details of sequence, conformation, H bonding patterns and solvent structure are still being sorted out. (D Stuart, private communication).

The accurate phases produced by the refinement of the enzyme are an important ingredient in the pioneering use of SR to determine the structures on intermediate states in the catalytic mechanism by Dr Louise Johnson and her colleagues in Oxford. In these experiments the substrate heptinitol was used since when it is phorphorylated it remains bound to the enzyme in the active site. The rate of phosphorylation by phorphorylase b can be controlled by regulating the levels of enzyme activator. Thus the unreactive substrate can be studied and then the enzyme switched on by adding activator. The intermediate reaction states are obtained by rapid data collection (45 - 60 minutes) after short soaks with activator. Some 100,000 reflections (reducing to 26,000 unique terms) were recorded in the data collection, yielding data sets of about 3Å resolution. (By comparison an equivalent data set would take at least 1 week on a conventional X-ray source). Electron density maps of the substrate : enzyme complex, some intermediate stages and the enzyme product complex have been obtained. These reveal the arrival on the heptenitol of the phosphate group which is in a position that would represent the attacking position in a reaction mediated by 5'-phosphate of the cofactor through a water molecule. (Hadju et al., 1986). The details of any conformational changes in the catalytic residues and their contacts with the substrate can only be described however when the crystallographic refinements of the enzyme substrate complexes are completed - a substantial undertaking.

The use of the Laue technique to record the diffraction data of intermediate states is a most exciting development. A 3.0Å data set (some 5000 unique reflections) can be recorded from these crystals in 250 milliseconds.

T state haemoglobin

The allosteric mechanism of haemoglobin has been studied intensively for many years, particularly by Dr Max Perutz and his colleagues at Cambridge (Perutz, 1970). The object of this research is to identify the structural tension which tips the molecule from one allosteric state to the other. In the studies at York, carried out by Dr Derewenda and Dr Liddington, a series of liganded T state haemoglobins have been investigated using SR at LURE and the SRS. Their refinement is descussed in a later paper. (Dodson and Liddington - this volume). SR has enabled high resolution (∿ 2.0Å) data to be collected relatively quickly so that the 3 structures could be determined and refined to yield coordinates accurate enough (∿ .2Å) for detailed comparison.

The 3 structures consisted of T state deoxy Hb, T state semi oxy Hb and T state met Hb in which both α and β subunits were liganded. Their comparison showed that as the α subunit is liganded there is concerted movements of the haem and the EF corner with which it is in contact. This movement is in response to the stereochemical changes generated when the Fe switches from 5 to 6 fold coordinate which requires the Fe to move more into the haem plane and to attain a more octahedral geometry. Now the H bond between the F4 carbonyl and the F7 imidazole NDI is extended by the movement of the Fe and the establishment of octahedral geometry. In the T state met Hb where the movement to the octahedral structure is most pronouned this H bond is stretched to 3.2Å. We conclude that this resistance to the development of an octahedral environment for the Fe is an important factor in reducing the oxygen affinity at the α subunit in T state Hb. (Liddington, 1986)

β lactamase I (Class A)

The structure of the β lactamase enzyme has been recently determined in Oxford in the laboratory of Professor David Phillips (B Samraoni et al, 1986). The β lactamases are largely responsible for bacterial resistance to β lactam antibiotics and therefore the structure of this enzyme is important for understanding the stereochemistry of antibiotic specificity. The crystals of lactamase I are small having dimensions .3 x .1 x .03 mm^3 and consequently satisfactory data could not be collected from conventional sources. With synchrotron radiation native and derivative data were collected (at the SRS) out to about 2.5Å spacing. The interpretable map was obtained in which the secondary structure and much of the main chain was identified. The crystal structure is now being refined, the current agreement factor R, is .25, and it is clear that the detials of the enzyme's catalytic and specificity sites will be established with accuracy.

Diffuse Scattering from Protein Crystals

The occurrence of pronounced diffuse scattering in diffraction patterns of protein crystals is sometimes bery clear when the intense SR sources are used. It has been speculated by Helliwell and colleagues (1986) that the marked diffuse scattering in the enzyme β-phosphogluconate dehydrogenase originates from the molecule's flexing in the crystal. The enzyme is V shaped and a hinge bending mode of vibration is plausible. It may be that the diffuse scattering may be a new route to identifying the dynamic motions in proteins in crystals and could provide experimental support for molecular modelling of protein motion.

THE USE OF TUNEABILITY

Phase determination

The obvious application of tuneability of SR is to optimise anomolous dispersion effects and hence get information for determining the phases. So far this approach has been limited to strengthening the conventional isomorphous phasing methods. With careful discrimination it should be possible to determine phases by multi wavelength methods from the anomolous dispersion associated with one heavy atom containing crystal alone. Experiments are under way on several metal containing proteins to derive phases; the method will benefit from the greater accuracy inherent in area detector technology.

The study on cytochrome C4 by Sawyer and colleagues (1983) illustrates the use of anomolous scattering by, in this case, the Fe atoms in the haems of the protein. At 1.739Å f" ∿ 5$\bar{e}$ (of ∿ 3$\bar{e}$ at Cukα). Careful data collection, R_{sym} ∿ 5% in F^2, gave a data set with detectable Freidel differences. The Patterson with coefficients $||F^+| - |F^-||^2$ yielded clear peaks for the Fe - Fe vectors. The phased determination from the Fe anomolous differences were sufficiently good to identify enough atom portions in a difference Fourier map. (Helliwell 1984).

A very valuable application of SR tuneability can be seen in the analysis of the storage protein ferritin being studied at Sheffield by Professor Pauline Harrison's laboratory. This protein is essentially a spherical shell of icosohedral symmetry (60 subunits) about 125Å in diameter. A typical preparation of ferritin contains about 2000 iron atoms per molecule on average. Data collection with wavelengths below Fe absorption edge (λ = 1.739Å) was prevented by severe radiation damage (Baker et al. 1985). When the wavelength was altered to values >1.739Å absoption was greatly reduced and data collection was achieved straight forwardly to a resolution of 2.5Å.

Conclusions

The main experimental benefits of SR are:

1. Increased resolution in many data sets and the successful analysis of small and weakly diffracting crystals.
2. Improved accuracy in the data sets.
3. Very good collimation with acceptable intensity particularly necessary for collecting data from crystals with very large unit cells.
4. Rapid rates of data collection, which makes it possible to carry out time resolved studies on protein mechanisms.
5. The wide range of available wavelengths has allowed experimenters to optimise anomolous scattering and absorption effects.

The development and application of SR techniques to protein crystallography has taken place at a most opportune time. There are parallel developments in computing which can handle the increased size of the data sets and greater complexity of the structures under study. In particular the crystallographic refinement methods now available ensure that the full advantage of the data can be used in positioning the atoms accurately and in defining with reasonable reliability their thermal motion. In addition the growth of molecular biology has brought new methods for identifying proteins and producing them for study. The technique of site directed mutagenesis in which specific amino acids can be altered within the protein is an immensely important development. It will allow the relation between sequence and structure in proteins to be analysed rigorously and it will, in time, give biochemists and crystallographers the capacity to design new molecules with new properties. Experiments of this kind will inevitably rely on knowing the protein's structure and the detailed environment of altered residues. This puts fresh importance on the rapid determination of protein crystal structures and this in turn, emphasizes the growing role of synchrotron radiation in protein crystallography

Acknowledgment

I should like to acknowledge the helpful discussions with the protein crystallographic research groups in the UK some of whose work I have presented here.

References

1. Baker, P.J. et al., Synchrotron Radiation : Appendix to the Daresbury Report 1984-1985 p.109.
2. Brzozowski, A., et al., Nature, 307, 74, 1984.
3. Hadju, J. et al., Biochem. Soc. Trans. In press.
4. Helliwell, J.R., Rep. Prog. Phys, 47, 1403, 1984.
5. Helliwell, J.R., et al., Synchrotron Radiation : Appendix to Daresbury Annual Report 1984-1985 p.115.
6. Helliwell, J.R., et al., Biochem Soc. Trans. In press.
7. Leslie, A.G.W. ,and Wonacott, A.J., J. Mol. Biol. 178, 743, 1984.
8. Liddington, R., D.Phil Thesis, York University, 1986
9. Perutz, M.F., Nature, 237, 495, 1977
10. Rossmann, M.G., et al., Nature, 317, 145, 1985.
11. Samraoui, B., et al., Nature, 320, 378, 1986.

Crystallography of the Complex tRNA ^{Asp}-Aspartyl-tRNA Synthetase from Yeast

J. CAVARELLI, R. GIEGE, A. PODJARNY, B. REES, M. ROTH*, J.C. THIERRY and D. MORAS

Institut de Biologie Moléculaire et Cellulaire
15, rue Descartes 67084 STRASBOURG-CEDEX, FRANCE.
*Institut Laue Langevin 156X, 38042 GRENOBLE-CEDEX, FRANCE.

Transfer RNAs and aminoacyl-tRNA synthetases are involved in a molecular recognition step of crucial importance for the fidelity of the translation of the genetic information during protein synthesis (1,2). Aminoacyl-tRNA synthetases catalyse the attachment of the correct aminoacid to its cognate tRNA, in a highly specific two step reaction (3).

$$\text{aa + ATP + aaRS} \xrightleftharpoons{Mg^{++}} \text{(aa AMP)aaRS + PPi} \qquad (1)$$

$$\text{(aa AMP)aaRS + tRNA} \rightleftharpoons \text{aa-tRNA + AMP + aaRS} \qquad (2)$$

The first step corresponds to the aminoacid activation and the second to the transfer to tRNA. For each amino acid, if many isoacceptor tRNAs can exist, only one synthetase is available which can charge all isoacceptors. This class of enzyme exhibits a puzzling structural diversity. Polypeptide chain lenghts range from about 350 amino acids like in **E. coli** tryptophanyl-tRNA synthetase to three times more like in yeast valine-tRNA synthetase. Moreover the oligomeric state of the proteins can vary from monomers to (α_4 or $\alpha_2\beta_2$) tetramers. Table 1 summarizes some informations on the oligomeric state and molecular weight of these molecules.

The best structural informations presently available on tRNA-synthetases interaction are given by chemical labelling experiments in solution (4). The first results confirm the diversity of situations from one system to another.

TABLE 1 : OLIGOMERIC STRUCTURES OF AMINO-ACYL tRNA SYNTHETASES

TYPE	AMINO ACID SPECIFICITY	MOLECULAR MASS OF THE SUBUNIT (x 10^3)	ACTIVE SITES	SPECIES
SHORT MONOMERS	Cys,Glu,Gln, Arg	55 - 75	1	E.Coli,B.Stearo
LONG MONOMERS	Val,Ile,Leu	100 - 130	1	E.Coli,B.Stearo
SHORT DIMERS α_2	Trp,Tyr,His, Ser, Pro,Thr, Lys,Asp,Asn	37 - 65	1/ α	E.Coli,B.Stearo, Yeast, Beef Pancreas.
LONG DIMERS α_2	Met	80	1/ α	E.Coli,Yeast.
TETRAMERS α_4	Ala	95	1/ α	E.Coli.
TETRAMERS $\alpha_2\beta_2$	Phe	$\alpha \approx 70$ $\beta \approx 60$	1/ $\alpha\beta$	Yeast.
	Phe,Gly	$\alpha \approx 90$ $\beta \approx 40$	1/ $\alpha\beta$	E.Coli.

The aspartic acid system from yeast

Aspartyl-tRNA synthetase (AspRS), a dimer of Mr ≃ 125000, was purified from **Saccharomyces cerevisiae** (baker's yeast) (5). $tRNA^{Asp}$ (Mr = 24160) was prepared from crude brewer's yeast tRNA (Boehringer Mannheim, FRG). Crystals were grown using the sitting drop technic, with ammonium sulfate as precipitating agent (6). The fact that AspRS and its cognate tRNA were crystallized using ammonium sulfate suggested the possibility of using the same precipitating agent for cocrystallization experiments (7). A summary of the conditions and preliminary crystallographic data of the caracterized crystal forms are given in table 2 (8).

Table 2: Crystallographic data of different crystal forms in the yeast aspartic acid system (from ref. 8).

Crystal Form	Space group	Unit cell dimension a (Å)	b (Å)	c (Å)	V ($Å^3$)	Molecules per asymmetric unit	Precipitant
tRNA							
I	$P6_322$	98	98	300	2.57×10^6	2	spermine
II	$P6_222$	98	98	150	1.28×10^6	1	spermine
III	$C222_1$	171	98	150	2.51×10^6	4	spermine+ isopropanol
IVa	$C222_1$	61.7	67.5	149.5	0.61×10^6	1	spermine+A.S 62%
IVb	$C222_1$	60.3	68	149.5	0.61×10^6	1	spermine+A.S 62%
AspRS							
I	$P4_12_12$	92	92	185	1.57×10^6	0.5	A.S. (54%)
II	$P4_12_12$	89	89	480	3.81×10^6	1	PEG 6000(8%)
Complex							
I	I432	354	354	354	4.44×10^7	1	A.S. (50%)
II	$P2_12_12_1$	209	228	138	6.58×10^6	2	A.S. (50%)

Crystals of complex contain two tRNA molecules for each dimeric enzyme. In the cubic form the asymmetric unit contains one synthetase and two $tRNA^{Asp}$, the orthorhombic asymmetric unit contains two enzymes and four tRNAs. The strategy of the structure determination took advantage of the high solvent content of the cubic crystals (78%) and the existence of non crystallographic symmetry. The methods of contrast variation and density modifications have an optimal efficiency in these conditions.

Data collection

Neutron diffraction data were measured at the Institut Laue Langevin, Grenoble (France) on the small-angle scattering camera D17, adapted for single crystal work. The wavelength used was 11.2 Å. The 2D Position Sensitive Detector ($64 \times 64 cm^2$) was at a distance of 81.5 cm and rotated 22° away from the position orthogonal to the primary beam. Five different D_2O/H_2O contrasts were used and scaled together (9-11).

X-Ray data were collected at LURE (Orsay) on the D12 experiment using an Arndt-Wonacott oscillation camera from Nonius or at CHESS (Cornell) using a supper oscillation camera. The main results are summarized in table 3 . At LURE the setting was similar to that used for the data collection from crystals of cowpea mosaic virus (CpMV) (12). Life time of the complex crystals could be increased by lowering the temperature of the experimental set-up to 4°C. Film scanning was performed on a Joyce-Loebl rotating drum scanner. Some films were scanned on an Optronic film scanner at the EMBL laboratory in Heidelberg. Data processing was done using the OSC program developped by Rossmann and modified to run on a Vax computer (13). Table 3 presents a summary of the presently available data on the cubic form.

TABLE 3 : DATA PROCESSING OF THE CUBIC FORM AT 7 Å RESOLUTION

DERIVATIVE	NC	NF	RSYM	NR. T	NR. I	NP	RMERG
NATIVE LURE*							
batch 1	7	16	5 - 9	14478	3826	63.	11.8
batch 2	4	13	4 - 8	13950	4503	74.3	9.2
batch 1 + 2	11	29	4 - 9	28428	4744	78.3	11.7
NATIVE CHESS*	10	30	5 - 8	33481	5143	85.	11.8
HG LURE *	2	13	5 - 8	15239	4238	70.	8.8

Crystals in batches 1 and 2 have different morphological shapes. Batch 2 crystals diffract slightly better.
NC = Number of crystals, NF = Number of films

NR = Number of reflections (T = total ,I = independent)
NP = 100 x (nobs / ncalc) ncalc = number of theoretical reflections.
nobs = number of measured reflections.
N number of unique reflections present in the data set.

$$R = 100 \times \sum_{r}^{N} \left(\sum_{i}^{n(r)} (| I(i,r) - I(r) |) \right) / \left(\Sigma\, n(r)\, I(r) \right).$$

I(i,r) = i-th individual measurements of the r-th reflections.
n(r) = number of the multiple measurements for the r-th reflections

$$I(r) = \Sigma\, I(i,r)/n(r)$$

RSYM = agreement of multiple measurements within each film.
RMERG = " overall " R-Factor for the complete data set .

* **LURE** : wavelength= 1.412 Å , crystal to film distance = 170 mm. exposure time : 5000 - 7000 seconds / degree.
* **CHESS** : wavelength= 1.558 Å , crystal to film distance = 100mm. exposure time : 40 seconds / degree.

Structure determination

For the cubic form the crystal structure was solved to low resolution using the method of contrast variation (14). The molecule of synthetase was located using data at the match point of the tRNA (65% D_2O). A systematic R factor search in the assymetric unit was performed against data to 40 Å resolution (50 reflexions). The best fit gives an R factor of 40% and is satisfactory with respect to packing considerations. A model of the enzyme formed by two identical ellipsoids related by a local two-fold axis was then refined to 32 Å resolution (94 reflexions, 13 parameters) giving an agreement factor of 41%.

tRNA molecules were located using a combination of informations from three sources : a systematic R-factor search against 33% D_2O data (enzyme matched), map phased by contrast transfer from the 65% D_2O data structure factors (centroid map) and difference map at 0% D_2O phased by the protein model.

After location of the molecules in the unit cell the molecular envelop was improved by the following procedure : least-spaces refinement of pseudo-atoms (gaussian spheres) representing the molecules, density modification, readjustment of the model by optical fit on a graphic display. The density modification step includes solvent levelling outside the molecular envelope and negative and positive attenuation within it (15). The procedure was usually repeated few times at each resolution step.

Additional informations about the orientation of the tRNA molecules were obtained from rotation function studies, which enabled to locate the non crystallographic two-fold axis (figure 1). Finally a model dependent of 97 parameters was refined by least-squares methods to an R factor of 31% for 1251 reflections (62% of the data to 10 Å resolution). 80 parameters correspond to the 20 gaussian spheres representing the synthetase dimer, 8 to each of the tRNA molecules (6 rigid body parameters for the tRNA backbones, 1 occupation parameter and 1 temperature factor). The tRNA backbone used in the refinement was obtained from the position of the phosphate groups in the structure of the free $tRNA^{Asp}$ (16). An omit map ($2F_o$-F_c, α_c) computed after removal of one tRNA molecule from F_c calculation showed clear density for the tRNA backbone.

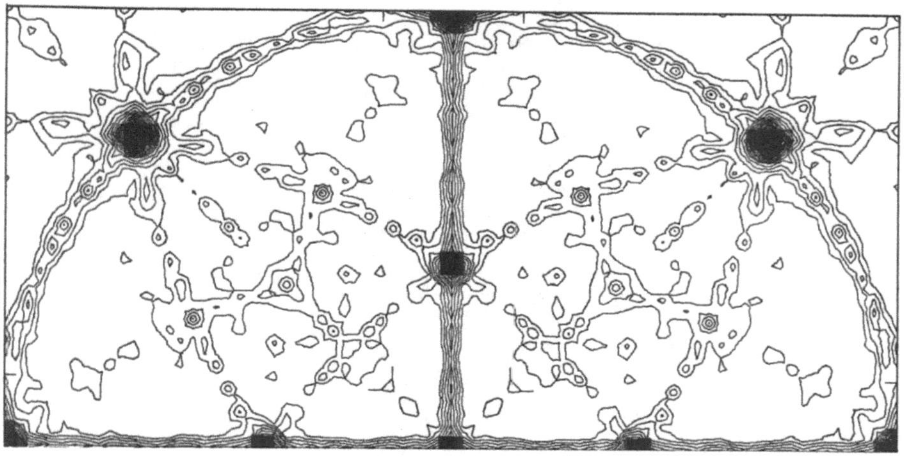

FIGURE 1 : Stereogram of the self rotation function of tRNA^{asp}-AspRS complex in the cubic form (χ = 180 °, 0≤ ψ ≤ 90°, 0≤ φ ≤ 180 °) Radius of integration = 150 Å ,resolution 7Å-13Å, 500 Reflections.

Molecular and crystal structure

The tRNA^{Asp}-AspRS complex is shown on figure 2. The molecule has an elongated shape of about 155 Å long, a dimension in good agreement with that found for the free enzyme (17). One side of tRNA, the variable loop one, is in contact with the synthetase. The contact area matches very well the one derived from chemical labelling experiments (4). The anticodon loop is close to the molecular two-fold axis whereas the CCA-end points toward the extremity of the enzyme. The overall shape of the tRNA is close to that of the free molecule. The backbone of two methionyl-tRNA synthetases (tryptic fragments of about 500 amino acids) could be grossly fitted in the AspRS density and account for a large part of it (18).

The packing is dominated by two main contacts around two-fold symmetry axes. The first one involves RNA-RNA interactions from the D-loop side, the other associates the protein part of the molecules. This arrangement is only possible because of the peculiar interaction between tRNA^{Asp} and AspRS which enables a sandwich type of association without heterologous RNA-protein interactions.

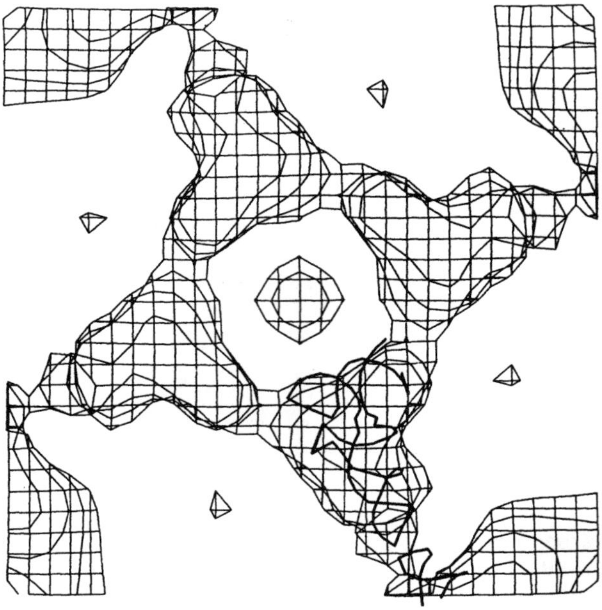

FIGURE 2a : Electron density at 25 Å resolution (neutron data) showing the characteristic L-shape of the tRNA molecule.

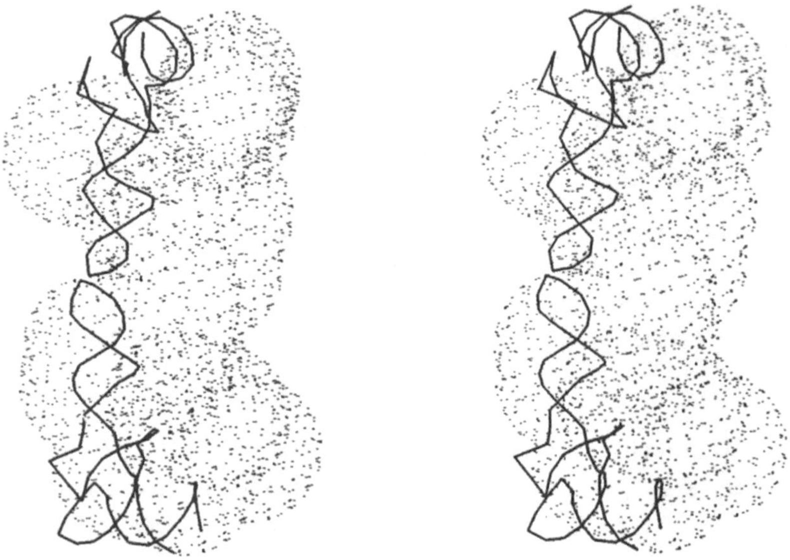

FIGURE 2b : Stereo view of the low - resolution model of the complex . The CCA ends are omitted from the tRNA backbone.

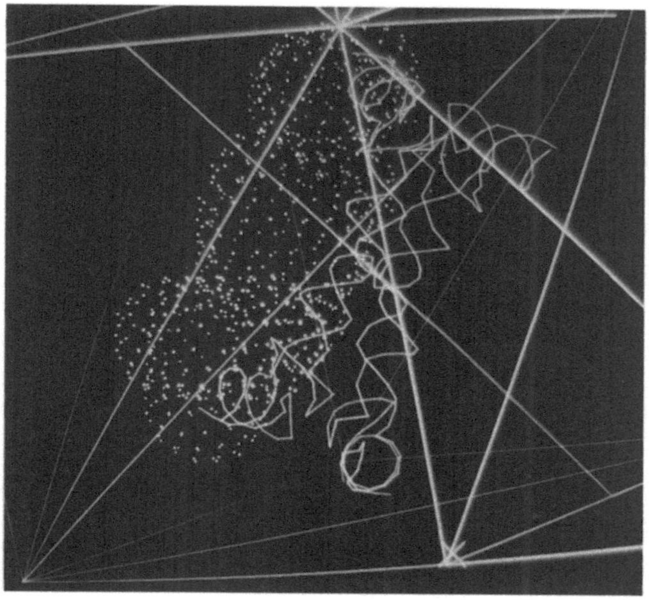

FIGURE 2c : Location of the model in the cubic unit cell ,showing the nucleic acid contacts.

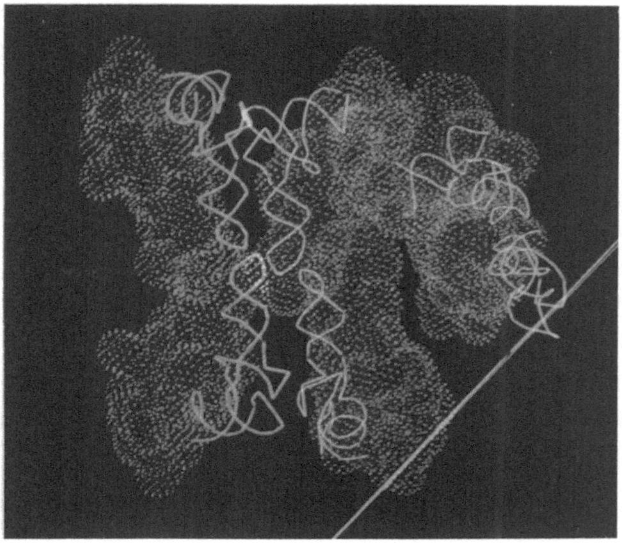

FIGURE 2d : Packing of three complex molecules emphasing the protein-protein and nucleic acid - nucleic acid interactions.

References

(1)Schimmel, P.R., Söll, D. and Abelson, J.N. (Eds.) Transfer RNA: Structure, Properties and Recognition, Cold Spring Harbor Monogr. Ser. **9A**, New-York (1979) 579 pp.

(2)Ofengand, J. (1982) In "Protein Biosynthesis in Eukaryotes" (Ed. Perez-Bercoff) pp. 1-67.

(3)Söll, D. and Schimmel, P.R. (1974) In "The enzymes", P.D. Boyer Ed., Acad. Press, NY, **10**, p. 489-538.

(4)Romby, P., Moras, D., Bergdoll, M., Dumas, P., Vlassov, V., Westhof, E., Ebel, J.P. and Giegé, R. (1985) J. Mol. Biol. **184**, 455.

(5)Lorber,B., Kern, D., Dietrich, A., Gangloff, J., Ebel, J.P. and Giegé, R. (1983) Biochem. Biophys. Res. Commun. **117**, 259-267.

(6)Dock, A.C., Lorber, B., Moras, D., Pixa, G., Thierry, J.C. and Giegé, R. (1984) Biochimie **66**, 179-201.

(7)Lorber, B., Giegé, R., Ebel, J.P., Berthet, C., Thierry, J.C. and Moras, D. (1983) J. Biol. Chem. **258**, 8429-8435.

(8)Moras, D., Dumas, P., Thierry, J.C., Westhof, E., Lorber, B., EBEL, J.P. and Giegé, R. (1984) In "Natural Products Chemistry", Zalewski and Skolik Eds., Elsevier Sciences Publisher, pp. 407-419.

(9)Moras, D., Lorber, B., Romby, P., Ebel, J.P., Giegé, R., Bentley-Lewit, A. and Roth, M. (1983) J. Biomol. Struct. Dyn. **1**, 209-223.

(10)Roth, M. and Lewit-Bentley, A. (1982) Acta Cryst. **A38**, 670-679.

(11)Roth, M., Lewit-Bentley, A. and Bentley, G.A. (1984). J. Appl. Cryst. **17**, 77-84.

(12)Usha, R., Johnson, J.E., Moras, D., Thierry, J.C., Fourme, R. and Kahn, R. (1984) J. Appl. Cryst. **17**, 147-153.

(13)Rossmann, M.G. (1979) J. Appl. Cryst. **12**, 223-238.

(13a)Schmid, M.F., Weaver, L.H., Holmes, M.A., Grutter, M.G., Ohlendorf, D.H., Reynolds, R.A., Remington, S.J. and Matthews, B.W. (1981) Acta Cryst., **A37**, 701-710.

(14)Jacrot, B., (1976) Rep. Prog. Phys. **39**, 911-953.

(15)Bentley, G.A., Lewit-Bentley, A., Finch, J.T., Podjarny, A.D. and Roth, M. (1984) J. Mol. Biol. **176**, 55-75.

(16)Westhof, E., Dumas, P. and Moras, D. (1985). J. Mol. Biol. **184**, 119-145.

(17)Dietrich, A., Giegé, R., Comarmond, M.B., Thierry, J.C. and Moras, D. (1980) J. Mol. Biol. **138**, 129-135.

(18)Zelwer, C., Risler, J.L. and Brunie, S. (1982) J.Mol.Biol. **155**, 63-81.

Time-Resolved X-Ray Fibre Diffraction Studies of Structural Transitions in the DNA Double-Helix Using the Daresbury SRS

W.Fuller, V.T.Forsyth, A.Mahendrasingam, R.J.Greenall, W.J.Pigram and K.Bellamy

Department of Physics, University of Keele, Staffs., ST5 5BG, U.K.
and
SERC Daresbury Laboratory, Daresbury, Warrington, WA4 4AD, U.K.

INTRODUCTION

X-ray fibre diffraction has played a crucial role in the determination of the three-dimensional stereochemistry of deoxyribonucleic acid (DNA) (Watson and Crick, 1953; Wilkins, Stokes and Wilson, 1953; Franklin and Gosling, 1953). Early studies showed that the DNA molecule consists of two polynucleotide strands which can assume a variety of highly regular helical conformations. Of particular significance was the observation that the conformations assumed were essentially independent of the biological origin of the DNA and hence of the particular genetic information coded in the sequence of adenine, thymine, guanine and cytosine bases along each polynucleotide strand (Hamilton et al., 1959). The double-helical model for DNA proposed by Watson and Crick resolved in a very elegant way the paradox of how a molecule with the very high degree of regularity indicated by the x-ray fibre diffraction data could also code within its structure a great variety of genetic information. In this model the repeating unit is an average nucleotide with the sugar-phosphate chain which links together successive bases having the same conformation irrespective of the base to which it is attached. Since the sugar-phosphate chains lie on the outside of the molecule its overall appearance and hence the way neighbouring molecules can pack together in crystalline fibres is highly regular.

In the 33 years since the Watson and Crick model was proposed a wide variety of naturally occurring and synthetic DNAs have been intensively studied by x-ray fibre diffraction. These studies have led to the identification of five principal conformations of the DNA double-helix which have been designated as A, B, C, D and S. During the past few years considerable progress has been made in identifying the factors which favour the assumption by DNA of these various conformations. Three factors have been identified as being of particular significance. The first is the degree of hydration of the DNA molecule. The second is the nature and concentration of ions in the vicinity of the DNA. The third is the occurrence of certain highly repetitive base-sequences along the DNA. The elucidation of the stereochemical basis for the role of water, ions and nucleotide sequence in determining DNA conformation represents one of the most important aspects of nucleic acid structure currently under investigation. While fibre diffraction studies of polynucleotides remain as the most powerful technique available for determining detailed stereochemical information on the DNA double-helix, important contributions are also being made by a variety of other physical and chemical techniques and also from studies of oligonucleotides by x-ray diffraction analysis of single crystals and by NMR studies in solution. The x-ray fibre diffraction studies of polynucleotides and single crystal studies of oligonucleotides should be regarded as largely complementary with fibre diffraction tending to emphasise regularity in helical conformations and single crystal studies providing information on conformational variation along an oligonucleotide chain as a function of nucleotide sequence. NMR studies of oligonucleotides are of particular importance because of the opportunity they provide to compare the structure of an oligonucleotide in a single crystal with that in solution (Patel et al., 1982; Clore et al., 1985). Such studies can be used in evaluating the extent to which conformational variation along an oligonucleotide chain in a single crystal should be attributed to its nucleotide sequence rather than to "end effects" due to the shortness of the oligonucleotide or to distortion due to crystal packing forces. The evidence available to date indicates that whilst there is variation in the nucleotide conformation along an oligonucleotide double-helix in solution this variation is generally rather less than that observed in oligonucleotide single crystals.

The relationship between stereochemical information obtained from single crystal determinations of oligonucleotides and that obtained from fibre diffraction studies of polynucleotides can be expected to be extremely valuable from the point of view of developing an understanding of the relationship between regularity and variation in the structure of the DNA double-helix. From the single crystal structures which have been determined to date it would appear that the results of such studies are most meaningfully discussed in the context of the helical structures determined from fibre diffraction studies of polynucleotides. In pursuing such an approach it is important to recognise that the helical regularity observed in the structures determined by fibre diffraction analysis is not simply a consequence of interactions between neighbouring DNA molecules within the fibre. In fact there is clear evidence from a variety of physical and chemical techniques that regular helical conformations of the DNA duplex occur in solution. General indications of such order are provided by the highly cooperative behaviour of DNA during melting studies. Further the band shift (Wang, 1979; Peek and Wang, 1981) and enzyme degradation (Rhodes and Klug, 1981) methods indicate that in solution naturally occurring DNAs have a regular helical structure with the number of nucleotide-pairs per helix pitch similar to that observed for the B form of DNA in oriented fibres. In fact there is a small, but nevertheless reproducible difference of about 5% in the number of residues per turn determined from these two solution techniques and that from fibre diffraction analysis. The origin of this difference is not clear, but its further investigation is clearly of very great interest as a contribution to an understanding of the affect of variation in the polynucleotide environment on the conformation of the DNA double-helix. Finally the fact that naturally occurring DNAs have been observed to assume the B conformation in a wide variety of crystalline and semi-crystalline packing arrangements provides further evidence that the conformation assumed by the DNA double-helix is not simply an artefact of fibre formation.

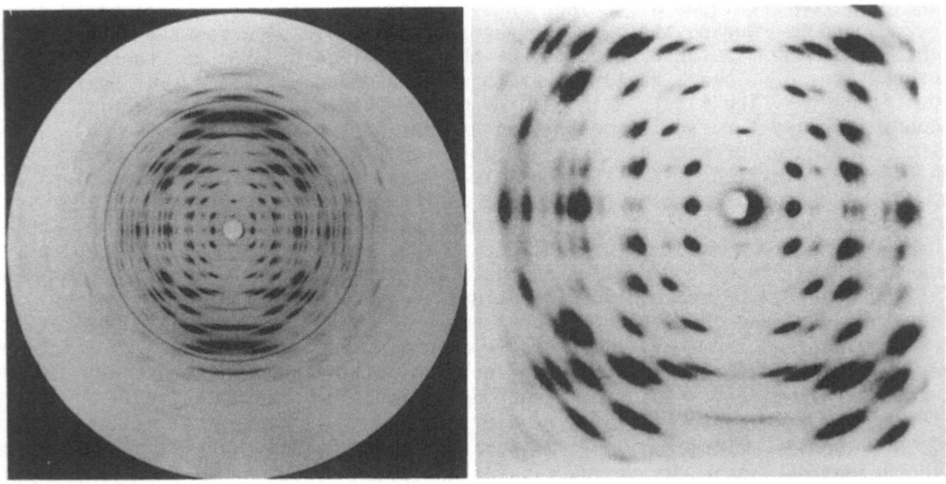

FIGURE 1 Comparison of diffraction patterns from a fibre of lithium poly (d(A-T)).poly (d(A-T)) recorded on (left) photographic film and (right) the Rutherford detector (Bateman and Connell, 1980).

From the above discussion it is clear that structural information derived in fibre diffraction analyses of DNA can be expected to be relevant to the conformation of the molecule outside the fibre. In making this point it is also important to emphasise that in the in vivo environment the concentration of DNA is typically closer to the situation within a fibre than to that in the dilute solutions in which many physical and chemical studies are conducted. Further in favourable circumstances the accuracy of the structural information obtained from fibre diffraction analyses of polynucleotides can be comparable to that attainable in structural analyses of oligonucleotide single crystals and single crystal analyses of globular macromolecules i.e. a few tenths of an angstrom in the position of the more strongly scattering atoms such as phosphorus. At first sight this result might seem surprising. However it stems from the fact that if the covalent bond lengths and angles along the polynucleotide chain are assumed to have similar values to those observed for similar atomic groupings in single crystals of nucleic acid bases, nucleosides and nucleotides the number of parameters required to define the complete helical structure is typically reduced to about 10. In a highly crystalline diffraction pattern like that illustrated in Figure 1 for the D-conformation of the synthetic polynucleotide poly (d(A-T)).poly (d(A-T)) double-helix it is possible to determine the intensity of a few hundred independent peaks in the diffraction pattern. Because a fibre diagram corresponds to a single crystal rotation pattern there is systematic overlap within each diffraction peak of symmetry related reflections. There is also, depending on the particular values of the lattice parameters, some accidental overlap of diffraction intensities. Both systematic and accidental overlap occurs because a crystalline fibre consists of a large number of crystallites oriented so that their c-axes are closely parallel while remaining in random orientation about this axis. It is however quite incorrect to state that data from fibre diagrams is necessarily cylindrically averaged. While this statement is correct for TMV gels and semi-crystalline and non-crystalline fibres, it is certainly not the situation for crystalline fibres. Despite some overlap of reflections the data from a crystalline nucleic acid fibre largely retains its three-dimensional character allowing full three-dimensional models of the nucleic acid double-helix to be determined and also the orientation of the double-helix about its own axis with respect to the unit cell axes.

Detailed stereochemical information on the conformations which can be assumed by the DNA double-helix and the identification of the factors which are important in inducing transitions between them can be expected to be of central importance for an understanding of the biological activity of DNA. Indeed it would be surprising if the possibilities for structural transitions in the DNA double-helix which have been observed in polynucleotide fibres had not been exploited in the evolution of the biochemical processes in which DNA is involved. While some conformational transitions in the DNA double-helix have been studied in solution (Pohl and Jovin, 1972), fibre diffraction studies give by far the most complete opportunity for investigating such transitions. Indeed a number of conformations have to date only been observed in oriented fibres. Furthermore no other physical or chemical technique allows as complete a monitoring of conformational change as can be achieved using fibre diffraction. The advent of synchrotron radiation sources with a gain of typically a factor of 100 in intensity over conventional sources in fibre diffraction analysis has dramatically increased the power of the fibre diffraction method for studying conformational transitions in the DNA double-helix. It should be stressed that in addition to allowing conformational transitions to be analysed, the application of synchrotron radiation also allows fibre diffraction data to be recorded which is of much higher quality than it is practical to collect using conventional sources. This is allowing more detailed refinement to be undertaken of the conformations which form the initial and final points of the transitions being studied and hence represent a further important contribution of synchrotron radiation studies to the characterisation of structural transitions in the DNA double-helix.

EXPERIMENTAL PROCEDURE

As was noted in the previous section the conformation assumed by a DNA double-helix depends on its degree of hydration and the ionic content of its environment. Recently we have made a great deal of progress in establishing procedures for preparing fibres which behave in a highly predictable way with regard to the conformation assumed by the DNA at a particular relative humidity of the fibre environment (Mahendrasingam et al. 1983a and b). We are thus able to induce structural transitions in the DNA molecules in an oriented fibre with a very high degree of reproducibility. Further since these transitions are generally reversible we are able to undertake systematic and controlled studies of these transitions. The fibres used in these studies are typically 0.1 mm in diameter and a few millimetres in length. X-ray fibre diffraction patterns were recorded on either line 7.2 or line 9.6 at the Daresbury SRS. The beam line optics and fibre diffraction camera have been described (Nave et al, 1985). Diffraction data was recorded with monochromatic radiation at various wavelengths between 0.8 and 1.6 A with pinhole collimation and using either photographic film or a multiwire two-dimensional detector. The water content of the fibre and hence the degree of hydration of the DNA molecules was controlled by varying the water content of helium gas which continuously passed through the camera.

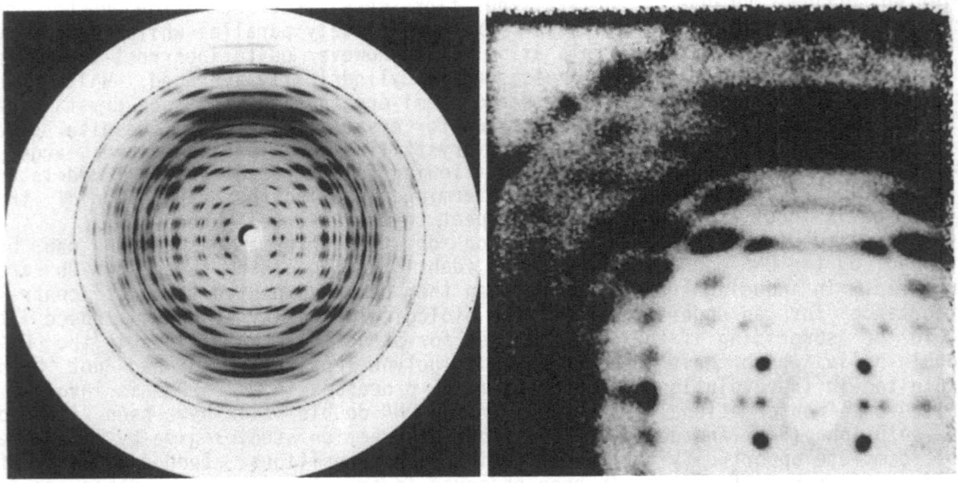

FIGURE 2 Comparison of diffraction patterns from a fibre of rubidium poly (d(A-T)).poly (d(A-T)) recorded on (left) photographic film and (right) the Gabriel detector.

Experiments were conducted with two types of area detector and diffraction patterns recorded with them are compared in Figures 1 and 2 with patterns recorded from the same specimens using photographic film. In fairness to the designers of these detectors it should be emphasised that neither was operating at its full specification when these studies were made. Nevertheless it should also be emphasised that the studies which were done represent by far the most successful and extensive application of area detectors in the acquisition of high angle fibre diffraction data yet achieved at the SRS and clearly represent a major advance in the observation of structural transitions in a macromolecular structure. The nominal resolution of the Rutherford detector is 0.5 mm by 1.5 mm compared to 1.5 mm x 1.5 mm for the Gabriel detector. While the point to point resolution in the patterns recorded using both these area detectors is clearly much inferior to that in the patterns recorded using photographic film, it is still adequate in the pattern recorded from the Rutherford detector for a large fraction of the diffraction peaks distinguished on photographic film to still be resolved. Diffraction patterns with this degree of resolution are certainly adequate for many types of fibre diffraction analysis of nucleic acids although if the resolution obtained in these studies cannot be improved then one would expect it would be necessary to continue to complement data collected using area detectors with parallel data collection using film. Superficial examination of Figure 2 might suggest that the performance of the Gabriel detector is much inferior to that of the Rutherford detector. However, as Professor J. Bordas has pointed out to us, the rather different experimental arrangements used in the studies with the two detectors tend to favour the Rutherford detector. First the range of 2θ over which the data is collected is greater for the Gabriel detector because the main beam position is closer to one corner of the detector than was the case with the Rutherford detector. This has the effect of increasing the loss of definition due to parallax effects which while rather greater for the Gabriel are in fact a significant problem with both detectors. Further while both experiments recorded data from a highly ordered D-form fibre of DNA, the Gabriel was used to collect data from a rubidium salt of the polynucleotide while the Rutherford detector was used with a lithium salt. The presence of crystals of excess rubidium iodide in amorphous regions of the specimen result in additional strong diffraction rings one of which tends to increase the apparent background level in Figure 2.

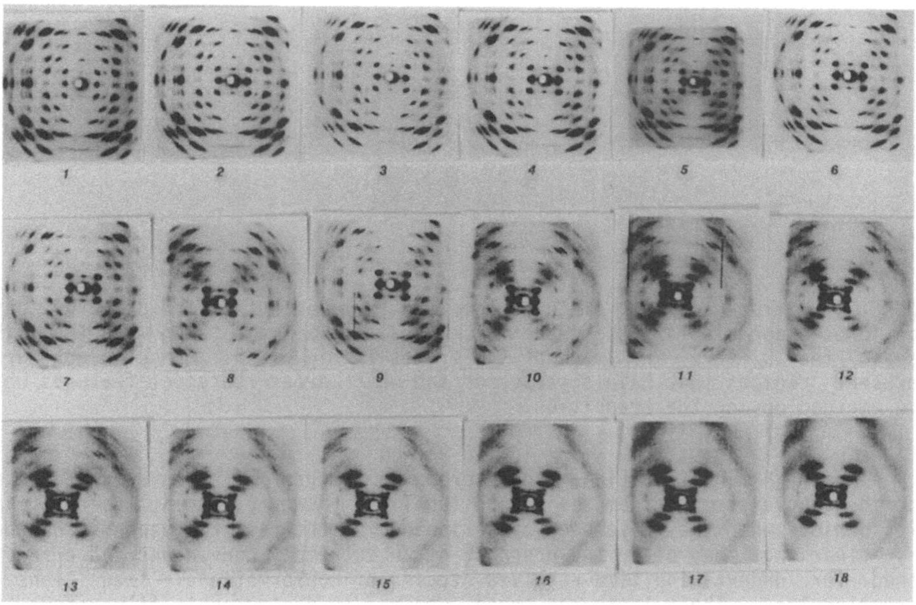

FIGURE 3 Diffraction patterns recorded using the Rutherford detector from a fibre of lithium poly (d(A-T)).poly (d(A-T)) undergoing the D→B conformational transition.

Both area detectors gave a gain in exposure time compared with photograph film of better than a factor of 5 so that, depending on the beam current in the SRS, exposure times were of the order of one minute. The efficiency of the Gabriel detector was very high with more than 90% of the incident photons counted. The comparable figure for the Rutherford detector was about 60%. The diffraction pattern in Figure 2 was recorded on the Gabriel detector with a total count rate of about 100,000 per second. The detector saturates at a total count rate of 300,000 per second or a local count rate of 30,000 per second per millimetre of anode wire. The corresponding figures for the Rutherford detector are probably about 100,000 counts per second total and less than 10,000 per second per millimetre local.

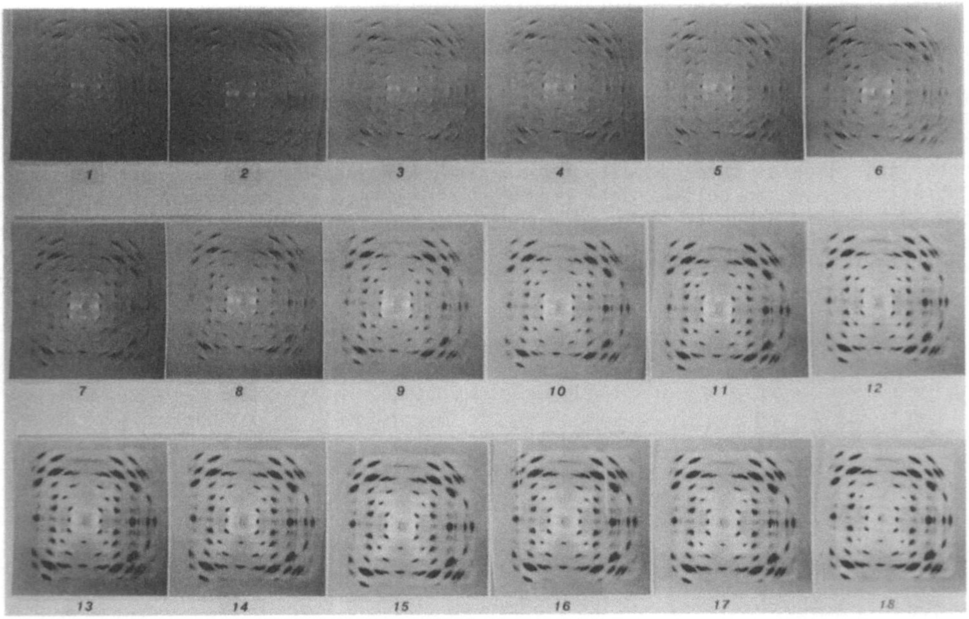

FIGURE 4 Difference x-ray pictures computed from the Rutherford detector data. Each photograph displays the result of subtracting a frame taken when the fibre was in the D form from each of the frames in Figure 3. Hence black areas represent regions that have disappeared and white areas represent features which have emerged during the transition.

The major difficulties encountered in the use of the area detectors described above stemmed from their unreliability. However, during the relatively brief periods when they have been operating during the past two years it has been very clear that they offer the prospect of an extremely convenient and effective method for undertaking time-resolved studies of conformational transitions in macromolecules. Their increased speed compared to photographic film offers the prospect of a gain of at least a factor of 5 in time resolution. Furthermore successive time slices can be "end to end" whereas with photographic film there is currently at the Daresbury facility a dead-time of 15 to 20 seconds between successive exposures even using an automatic carousel. However, perhaps the most important advantage of using area detectors in studies of this kind is that the experiment can be conducted in a genuinely real time mode. The systems we are studying are complex and even though we can now induce structural transitions in the DNA double-helix in a very reproducible way there are still uncertainties associated with the synchronisation of the relative humidity of the fibre environment with the initiation of the transition. The ability to continuously monitor the x-ray fibre diffraction pattern using an area detector allows very much greater control over the experiment and can be expected to lead to the observation of effects which might well be missed with film recording.

The power of a synchrotron radiation facility equipped with an area detector is well illustrated by the sequence of 18 diffraction patterns in Figure 3. These were selected from a series of diffraction patterns recorded as 64 one minute time slices during the D→B transition in an oriented fibre of the lithium salt of the synthetic polynucleotide poly (d(A-T)).poly (d(A-T)). The very extensive data in these patterns presents formidable problems in data handling, data reduction and also of course in the development of new approaches for the analysis of the data itself. During the past two years a great deal of progress has been made by the computer support group in the Daresbury Laboratory in implementing existing programmes such as GUCKMAL which allow data reduction to proceed interactively and by developing new packages (e.g. "OTOKO" and "BSL") to enable time-resolved data to be manipulated and displayed. Figure 4 illustrates the use of the "BSL" program (written by G. Mant, SERC Daresbury Laboratory) to display changes in the diffraction pattern occurring during a D→B transition. Each pattern in Figure 4 displays the result of subtracting frame 1 of Figure 3 from subsequent patterns in the series illustrated in that figure. Hence black areas represent diffraction features that have disappeared and white areas represent new features which have emerged during the transition.

THE D→B CONFORMATIONAL TRANSITION IN THE DNA DOUBLE HELIX

The D-conformation of the DNA double-helix is of particular interest for a number of reasons. First the diffraction data recorded from it contains well-defined maxima extending to d spacings of less than 1.5 A indicating the existence of a highly regular structure. Second the D conformation has only been observed from a number of synthetic DNAs with highly repetitive nucleotide sequences, e.g. from poly (d(A-T)).poly d(A-T)) in which there is an alternating sequence of adenine and thymine residues along each polynucleotide chain (Davies and Baldwin, 1963). The conformation has not been observed for naturally occurring DNAs with no particular pattern in their nucleotide sequence. Nor has it been observed for synthetic polynucleotides in which there is an alternating sequence of guanine and cytosine residues along each chain, although it is observed for polynucleotides in which inosine, an analogue of guanine, alternates with cytosine (Mitsui et al., 1970). Because of the very limited number of sequences for which this conformation has been observed it has been suggested that such sequences in DNA might have a highly specific role through their capacity to form the D conformation at a crucial stage in the macromolecule's biological activity. The third reason for particular interest in the structure of the D conformation of DNA is the fact that it can be observed in a variety of isomorphous forms in which the alkali metal ions neutralising the DNA phosphates are successively lithium, sodium, potassium and rubidium.

If the D conformation does indeed have a highly specific biological role then transitions between it and the B conformation which is generally regarded as the form normally assumed in vivo are of very great interest. We have established conditions for reproducibly inducing a reversible transition between the D and B forms on a timescale which allows the associated conformational changes to be monitored using synchrotron radiation (Mahendrasingam et al. 1986; Forsyth et al, 1986). Figure 5 illustrates the progress of such an experiment where over a period of approximately seventeen hours a fibre of poly (d(A-T)).poly (d(A-T)) underwent a series of four successive reversible D→B→D transitions. These were induced by varying the relative humidity of the fibre environment. As can be seen from Figure 5 it was possible to vary the rate at which the D→B→D transformations took place by varying the rate of change of relative humidity.

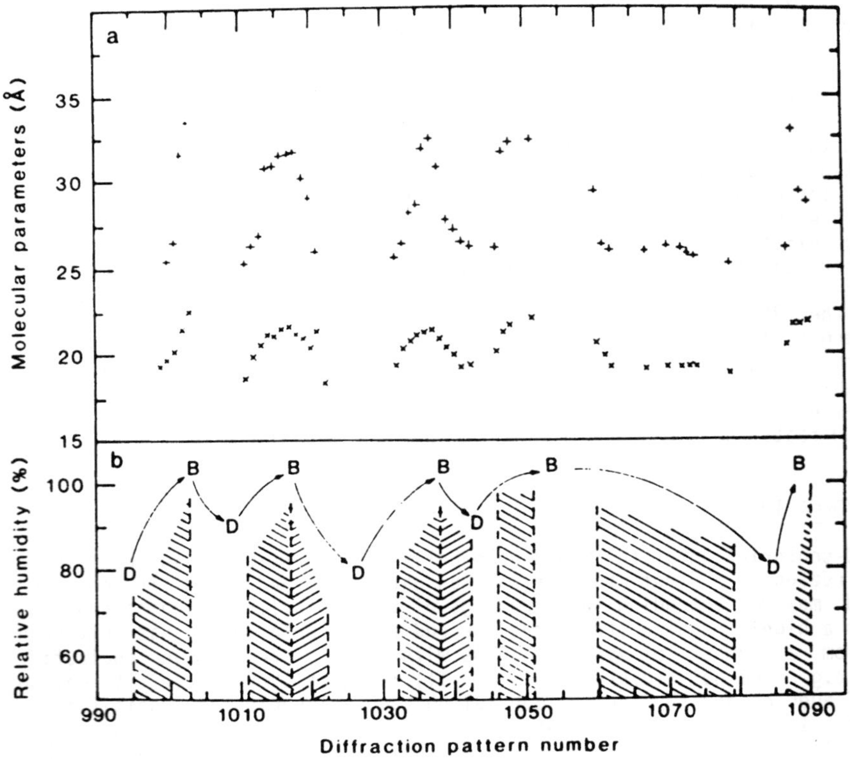

FIGURE 5 (a) Variation in the lateral spacing (x) and helical pitch (+) of the intermediate semi-crystalline form observed in the D→B conformational transition.
(b) Variation in the relative humidity of the fibre environment and molecular conformation during the experiment. The gaps between the shaded regions represent periods when the storage ring was being refilled.

Because the point to point resolution of data collected using photographic film is currently significantly better than that collected using area detectors the changes observed during this experiment are illustrated in Figure 6 using data from photographic film. However similar effects were observed in parallel experiments using area detectors.

The sequence of diffraction patterns in Figure 6 was selected from patterns taken during the D→B transition. Figure 6(a) illustrates a typical D-type diffraction pattern which represents the starting point of the transition. As the relative humidity is raised a number of changes in the diffraction pattern are observed including in particular the appearance of a satellite reflection identified as IV in Figure 6(c). With increasing relative humidity this reflection moves nearer to the centre of the pattern and increases in intensity. The appearance of this reflection and the change in its position and intensity with relative humidity can be attributed to the occurrence of an intermediate conformation through which the D→B transition takes place. In 6(f), the final pattern in the series, the D pattern has completely disappeared having been replaced by a semi-crystalline B pattern. These observations are most readily interpreted in terms of a mechanism whereby with increasing relative humidity an increasing number of molecules which were originally in the D conformation undergo a gradual change to a conformation with a larger pitch. Reflection IV whose layer line spacing corresponds to this pitch length is sufficiently sharp to indicate that there is relatively little variation in this pitch length through the structure. As the relative humidity increases still further the pitch of the molecules in this intermediate form gradually increases and the number of molecules with this conformation also increases at the expense of the number of molecules remaining in the D conformation. This process continues until the pitch of the intermediate form reaches the value of 34A characteristic of the semi-crystalline B conformation and all the molecules have changed into this form. Although this discussion has focussed on the reflection labelled IV in Figure 6(c) there are many other changes in the diffraction pattern associated with the D→B transition e.g. the reflection labelled III in 6(c) which corresponds to the side-by-side packing of the molecules. The variation in the pitch and side-by-side spacing of the intermediate form throughout the four D→B→D transitions is illustrated in Figure 5.

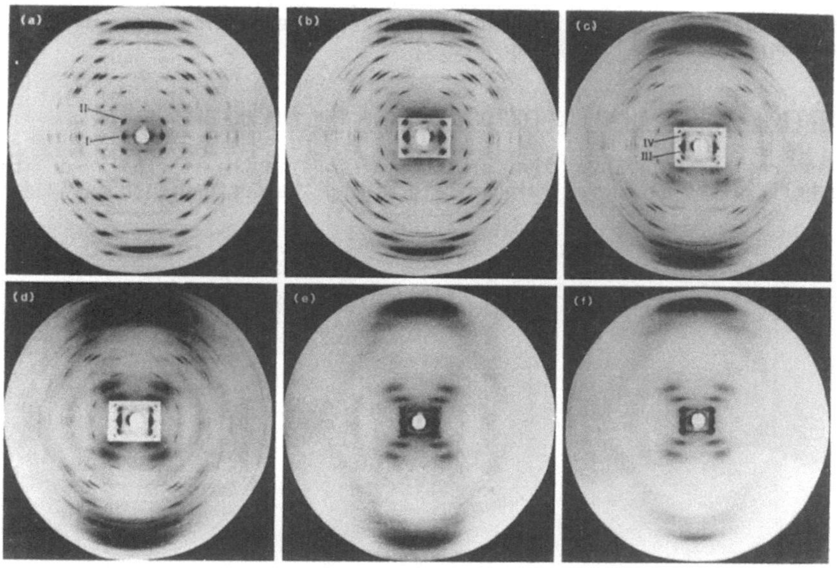

FIGURE 6 Diffraction patterns recorded on photographic film illustrating typical stages in the D→B transition. In (a) the reflection marked I is related to the lateral intermolecular separation and that marked II to the helical pitch of molecules in the D form. In (c) the reflection marked III is related to the lateral intermolecular separation and that marked IV to the helical pitch of molecules in the semicrystalline intermediate form. In the reproduction of patterns (b), (c) and (d) diffraction in the central region of the pattern has been attenuated so that neighbouring reflections are more easily resolved.

The results described above have immediate implications for the resolution of a controversy with regard to the handedness of the D conformation of the DNA double-helix. While the B conformation of DNA is generally regarded as being right-handed both left and right-handed models have been proposed for the D conformation. The most recent of the left-handed models (Drew and Dickerson, 1982) attracted particular attention because of the possible biological importance of a number of features on which it was based e.g. the base-pairing pattern observed by Hoogsteen (1959) rather than Watson-Crick pairing between the adenine and thymine bases. While the diffraction calculated from this left-handed model is in less satisfactory agreement with the observed diffraction than that reported for the best right-handed models (Arnott et al, 1983), the left-handed model has nevertheless continued to be canvassed as a serious possibility for the D conformation. However, the observations described above demonstrate that the D→B transition involves a gradual change in the helix geometry in which the pitch is observed to change essentially continuously between the 24 A characteristic of the D form and the 34 A characteristic of the the B form. This precludes the possiblity of a conformational change as dramatic as one involving a change in helix sense. These studies therefore, allow the rejection of not only the left-handed model proposed by Drew and Dickerson but also more generally all left handed models for the D conformation of DNA.

It should be emphasised that to date only a small fraction of the data recorded during the D→B transition has been analysed. In addition to the appearance of the sharp reflections III and IV discussed above there is during the transition substantial variation in the continuous diffraction in patterns 6(a) to 6(f). We are currently undertaking the analysis of this more extensive diffraction data in an effort to determine the stereochemical pathway followed in the D→B conformational transition.

THE LOCATION OF IONS IN THE D-CONFORMATION OF THE DNA DOUBLE-HELIX

As was noted above the presence of ions in the vicinity of the DNA double-helix plays a crucial role in determining the conformation assumed by the

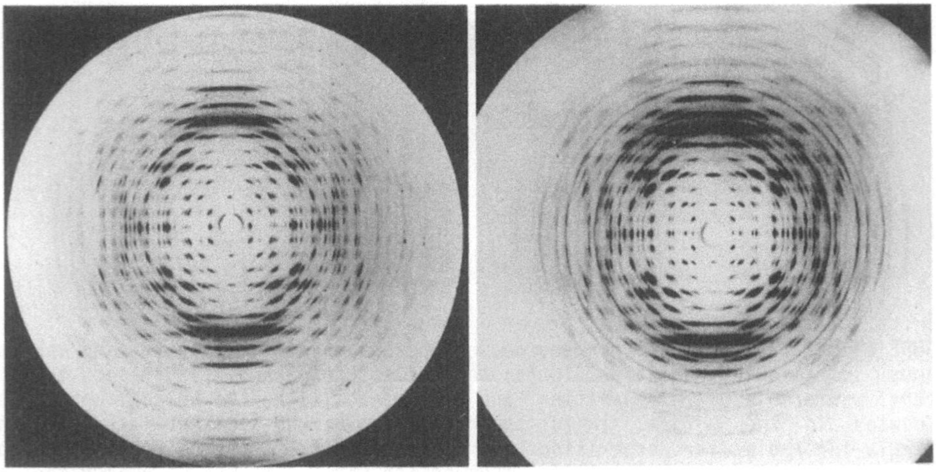

FIGURE 7 Comparison of the potassium (left) and rubidium (right) isomorphous forms of the D-conformation of poly (d(A-T)).poly (d(A-T)). Intensity differences of corresponding reflections are large enough to be seen by eye (e.g. on $\ell = 1$) and extend to high angle reflections (i.e. d less than 4A).

polynucleotide. The D form is a particularly striking example of this, being observed only when the fibre contains approximately one alkali metal cation per nucleotide-pair in addition to the number of ions required to neutralise the DNA phosphates together with an equivalent number of additional halide anions. It would appear that to obtain the D form it is necessary to saturate the solvent surrounding the DNA in the fibre unit cell. An important step towards understanding the stereochemical role of ions in stabilising particular DNA conformations is the determination of the location of solvent and ions around the DNA. Single crystal studies of oligonucleotides have yielded important information on the arrangement of solvent molecules around DNA and fibre diffraction studies using synchrotron radiation have been used to study the arrangement of caesium ions around the B form of the DNA double-helix (Bartenev et al., 1983). Our recent observation of an isomorphous series for the D form of DNA in which the alkali metal ion is successively lithium, sodium, potassium and rubidium offers the possibility of using fibre diffraction data to determine the positions of ions around a polynucleotide. The intensity differences between isomorphous rubidium and lithium salts of the D form of poly (d(A-T)).poly (d(A-T)) are very marked as can be seen from inspection of the diffraction patterns in Figure 7.

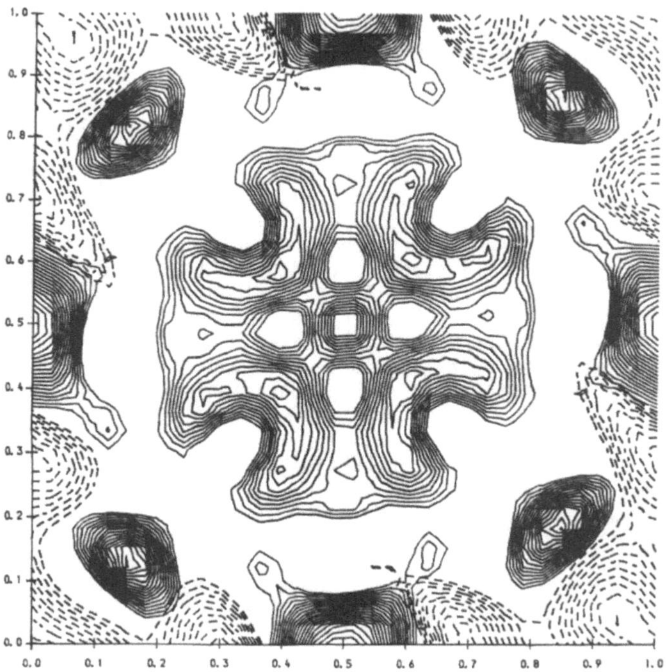

FIGURE 8 Fourier difference c-axis projection (shown dashed) calculated by combining the amplitude differences between the rubidium and lithium D forms of poly (d(A-T)).poly (d(A-T)) and phases calculated from the model of the D conformation refined against the lithium data set. The full line represents the corresponding projection calculated using both amplitudes and phases from the refined D model.

We are currently calculating a three-dimensional difference synthesis in which amplitude differences derived from differences between isomorphous patterns are combined with phases calculated from the model of the D conformation refined against the data from the lithium salt of the D form. It should be stressed that the Fourier maps obtained at this stage are still very preliminary, but the nature of the results can be indicated by the c-axis projection of the map in Figure 8. The peaks observed in this difference projection all occur in the three-dimensional map in regions between DNA molecules. Many of these peaks form an approximately helical array, but others occupy positions which only have the symmetry of the space group. This is consistent with the occurrence of meridional diffraction in regions which should be zero if the structure consisted solely of an array of perfectly regular helices. If all or some of these peaks do indeed correspond to ions, the information these Fourier syntheses will provide on the ionic environment of the D conformation will be of great interest for an understanding of the factors which are important in stabilising the D form of the DNA double-helix. These studies are being pursued in parallel with EXAFS studies of ions associated with the D conformation of DNA.

CONCLUSIONS AND FUTURE POSSIBILITIES

Recent advances in our capability to reproducibly observe particular conformations of the DNA double-helix and to induce transitions between these conformations in a controlled way have together with the advent of synchrotron radiation dramatically increased the possibilities for obtaining detailed stereochemical information on the structure of the DNA double-helix. We can expect to see more detailed refinements of the principal conformations of DNA and characterisation of the stereochemical pathways associated with transitions between them. We should also hope to obtain precise information on the location of water molecules and ions around the DNA and to relate changes in these positions to the relative conformational energies of different structures. Taken in conjunction with parallel studies using EXAFS and neutron scattering and information from single crystal studies of oligonucleotides it should be possible to identify the interactions which are important in stabilising polynucleotide structures and to allow predictions to be made of the conformational changes to be expected as a function of changes in the environment of the polynucleotide.

We are grateful to the Science and Engineering Research Council for support, to Professor J. Bordas, Dr. C. Nave, Mr. J.S. Worgan and Dr. R. Lewis for help with the use of the area detectors and to Mrs. H.E. Moors, Mr. M. Daniels and Mr. M. Cheney for help with preparation of the manuscript.

REFERENCES

Arnott, S., Chandrasekaran, R., Puigjaner, L.C., Walker, J.K., Hall, I.H., Birdsall, D.L. and Ratliff, R.L., Nucleic Acids Res., 11, 1457 (1983).

Bartenev, V.N., Golovamov, I.Eu., Kapitonova, K., Mokulskii, M.A., Volkova, L.I. and Skuratovskii, I., J.Mol.Biol., 169, 217 (1983).

Bateman, J.E. and Connell, J.E. (1980) Nucl. Instrum. Methods, 173, 525, (1980).

Clore, G.M., Gronenborn, A.M., Moss, D.S. and Tickle, I.J., J.Mol.Biol., 185, 219 (1985).

Davies, D.R. and Baldwin, R.L., J.Mol.Biol., 6, 251 (1963).

Drew, H.R. and Dickerson, R.E., EMBO J., 1, 663 (1982).

Forsyth, V.T., Greenall, R.J., Hussain, R., Mahendrasingam, A., Nave, C., Pigram, W.J. and Fuller, W., Biochemical Society Transactions, 14, 553 (1986).

Franklin, R.E. and Gosling, R.G., Nature, 171 740, (1953).

Hamilton, L.D., Barclay, R.K., Wilkins, M.H.F., Brown, G.L., Wilson, H.R., Marvin, D.A., Ephrussi-Taylor, H. and Simmons, N.S., J.Biophys.Biochem. Cytol.. 5, 397 (1959).

Hoogsten, K., Acta.Cryst., 12, 822 (1959).

Mahendrasingam, A., Rhodes, N.J., Goodwin, D.C., Nave, C., Pigram, W.J., Fuller, W., Brahms, J. and Vergne, J., Nature 301, 535 (1983a).

Mahendrasingam, A., Pigram, W.J., Fuller, W., Brahms, J. and Vergne, J., J. Mol. Biol., 168, 897, (1983b).

Mahendrasingam, A., Forsyth, V.T., Hussain, R., Greenall, W.J., Pigram, W.J. and Fuller, W., Science, 233, 195 (1986).

Mitsui, Y., Langrdige, R., Shortle, B.E., Cantor, C.R., Grant, R.C., Kodama, M. and Wells, R.D., Nature, 228, 1166 (1970).

Nave, C., Helliwell, J.R., Moore, P.R.,Worgan, J., Thompson, A.W., Greenall, R.J., Miller, A., Burley, S.K., Bradshaw, J.P., Pigram, W.J., Fuller, W., Siddons, D.P., Deutsch, M. and Tregear, R.T., J. Appl. Cryst, 18, 396 (1985).

Patel, D.J. Kozawski, S.A., Rice, J.A., Marky, L.A., Breslauer, K.J., Broka, C. and Itakura, K., Topics in Nucleic Acid Structures (Neidle, S. ed.), Part 2, 287 Macmillan Press, London (1982).

Peck, L.J. and Wang, J.C., Nature, 292, 375 (1981).

Pohl, F.M. and Jovin, T.M., J. Mol. Biol., 67, 375, (1972).

Rhodes, D. and Klug, A., Nature, 292, 378 (1981).

Wang, J.C. (1979) Proc. Natl. Acad. Sci. USA 76, 200 (1979).

Watson, J.D. and Crick, F.H.C. Nature, 171, 737 (1953).

Wilkins, M.H.F., Stokes, A.R. and Wilson, H.R., Nature, 171, 737 (1953).

Multiwavelength Anomalous Diffraction in Protein Crystallography: Phase Calculations, Experimental Procedures and Results

R. Fourme, M. Chiadmi & R. Kahn
LURE (CNRS, MEN & CEA) Université de Paris-Sud, B209D, F91405 Orsay Cedex

I- Introduction

Once suitable crystals have been grown, the phase problem- i.e. the determination of structure factor phases from Bragg reflection intensity data - remains the crucial difficulty of the ab-initio elucidation of any three-dimensional macromolecular structure by X-ray diffraction methods. The techniques which have been most commonly used to circumvent this problem have involved the special diffraction characteristics associated with the presence of a small number of heavy atoms which are either naturally present or purposely introduced in the parent structure of interest.

(i) The first technique is <u>multiple isomorphous replacement</u> (MIR). For a given Bragg reflection, let us call $\mathbf{F}_{ph}$, $\mathbf{F}_p$ and $\mathbf{F}_h$ (with modulus F_{ph}, F_p and F_h and phases φ_{ph}, φ_p and φ_h) the structure amplitudes from respectively a particular derivative, the light-atom part and the heavy atom part of this derivative. The following relation holds:

$$\mathbf{F}_p = \mathbf{F}_{ph} - \mathbf{F}_h$$

If one considers a triangle built in the complex plane on these three vectors (fig. 1), known quantities are the modulus of the light-atom structure amplitude (obtained from the native crystal data set), the modulus of the derivative structure amplitude and the modulus and phase of the heavy atom structure amplitude (assuming that, prior to the phase determination, the location, occupancy and thermal parameters of heavy atoms have been determined). It is clear that the information from a single triangle is not sufficient to determine unambiguously φ_p, because $\mathbf{F}_p$ may be placed on either side of $-\mathbf{F}_h$. The ambiguity can be solved by constructing another similar triangle, which requires measurements on another derivative, and by choosing the solution for $\mathbf{F}_p$ which is common to both diagrams. Such geometric or analytic solutions of the phases assume that the structure factors are perfectly measured and that there is perfect isomorphism between the parent and the derivative structures. In reality, experimental and model errors are present. A treatment of these errors has been first given by Blow and Crick (1) , given certain assumptions regarding the randomness of errors and various simplifications . The principle will be recalled briefly:

For a particular derivative and assuming some value φ for the protein phase angle, the calculated value of $F_{ph}{}^2$ is

$$F_{ph}{}^2(\mathrm{calc}) = F_p{}^2 + F_h{}^2 - 2F_pF_h\cos(\varphi-\varphi_h)$$

with the corresponding 'lack-of-closure' of the phase triangle

$$x = |F_{ph}(calc) - F_{ph}(obs)|$$

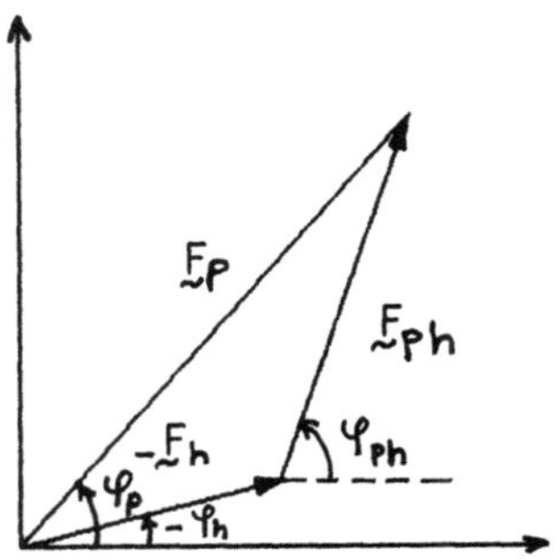

Fig. 1 Phase triangle for a single isomorphous derivative

The probability distribution of φ is given by the equation

$$p(\varphi) = N\exp(-x^2/2\sigma^2)$$

where N is a normalization factor and σ is an estimate of the cumulate error. Allowing φ to vary by increments between 0 and 2π, a probability distribution is obtained for the parent/derivative pair. This process is repeated for the n available pairs ($n \geqslant 2$). Finally, the joint probability distribution $P(\varphi)$ is calculated

$$P(\varphi) = \Pi p(\varphi).$$

Blow and Crick have shown that the 'best' Fourier - defined as the Fourier transform which is expected to have the minimum mean square deviation difference $\langle(\delta\rho)^2\rangle$ from the Fourier transform of the true structure factors when averaged over the whole unit-cell - has Fourier coefficients which are given, for each reflection, by the equation

$$F_{best} = F_p \int_0^{2\pi} P(\varphi)\exp(i\varphi)d\varphi / \int_0^{2\pi} P(\varphi)d\varphi \quad \text{(eq. 1)}$$

(ii) The second phasing technique is based on anomalous dispersion. With conventional X-ray sources, anomalous dispersion has been in most cases confined to give auxiliary information to that obtained from MIR. But the situation is changing rapidly, essentially because intense and tuneable synchrotron X-ray sources and electronic area detectors are now available to perform optimized multiple wavelength anomalous diffraction (MAD) experiments; this progress in experimental procedures has also stimulated theoretical developments, which indicate that MAD methods have considerable untapped potential for future application. In MIR methods, the variations (both in phase and modulus) of the reference vector $\mathbf{F}_h$ in each phase triangle come from different locations of heavy atoms. In MAD methods, these variations result from the fact that the atomic scattering factor of the anomalous scatterers in a single sample undergo large changes at X-ray wavelengths that occur in the vicinity of their absorption edges. Simply by changing the X-ray wavelength, one generates from one and the same crystal as many pseudo-derivatives as wanted. In fact, as the (+) mate of a Bijvoet pair has generally not the same modulus and the same phase as the complex conjugate of the (-)

mate, as shown in fig.2a, measurements of Bijvoet pair intensities at a particular wavelength where anomalous dispersion is significant brings about as many information as two different isomorphous derivative data sets measured at a wavelength where anomalous dispersion is negligible.

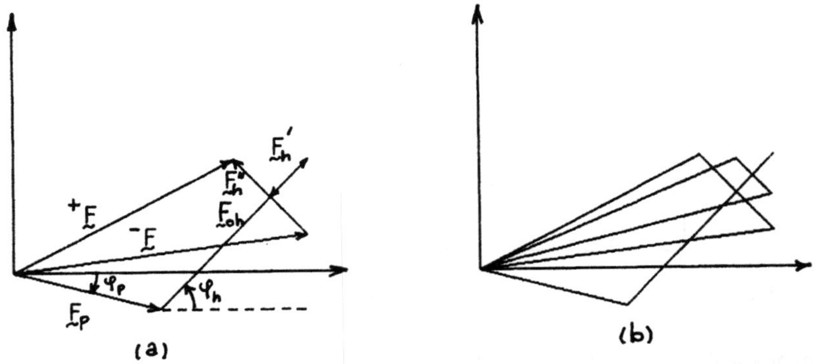

Fig.2 Vector diagram for a Bragg reflection and the complex conjugate of its Bijvoet mate a) at one wavelength b) at two wavelengths

At first glance, MAD has important advantages with respect to MIR, because relying on a single sample is a major simplification which also eliminates all problems connected with lack of isomorphism. Nevertheless, a more careful inspection reveals, as discussed later in this paper, various theoretical and practical problems which prevent the full exploitation of their considerable potential. We present briefly the principle of phase calculations and experimental procedures which led to the determination of an unknown protein structure from measurements at three wavelengths with strong anomalous dispersion, using synchrotron radiation and an electronic area detector.

II- MAD phase calculations

The following assumptions , simplifications and notations will be used. $\mathbf{F}_p$ with phase φ_p is the structure amplitude of the parent structure (with all atoms exhibiting an anomalous dispersion which is negligible when compared to that of the anomalous scatterers). Each parent molecule is labelled with a few heavy atoms which have a significant and variable anomalous dispersion in the selected X-ray wavelength. For the sake of simplicity and because it is a common case, it is assumed that all heavy atoms are of the same type; the scattering factor of these atoms is the complex number $f_\lambda = f_0 + f'_\lambda + if''_\lambda$ where f_0 is Bragg angle-dependent and wavelength-independent whereas f'_λ and f''_λ are Bragg angle-dependent and wavelength-dependent. The measured magnitude of the structure factor of a particular Bragg reflection (**H**) is $^+F_\lambda$ (intensity $^+I_\lambda$); for this reflection, the 'normal' components of the structure amplitude of the heavy atoms and of the total structure are respectively $\mathbf{F}_{0h}$ with phase φ_h and $\mathbf{F}_{0ph}$ with phase φ_{ph}; the real and imaginary 'anomalous' components are respectively $\mathbf{F}'_{\lambda h}$ and $\mathbf{F}''_{\lambda h}$ with phases φ_h and $\varphi_h + \pi/2$. The measured magnitude of the structure factor of the Bijvoet mate is $^-F_\lambda$

(intensity $^{-}I_{\lambda}$). In the very narrow wavelength range in the vicinity of one of the absorption edges of the heavy atoms, F_p and F_{0h} are essentially constant while $F'_{\lambda h}$ and $F''_{\lambda h}$ will change substantially. Fig.2a shows the vector diagram for the reflexion (**H**) and the complex conjugate of its Bijvoet mate. Fig.2b shows the effect of a change of wavelength on this diagram.

It is possible to derive phases by an <u>algebraic</u> analysis which has been given by Karle (2) in a general form with no approximations for any number and type of anomalous scatterers. The resulting relations are largely linear with appropriate selection of unknown quantities. The quantities of interest are expressed only in terms of functions of the normal parts of the atomic scattering factors and are separate for each type of anomalous scatterers. This analysis has been used in the MAD study of lamprey hemoglobin, using data collected at SSRL, Stanford (3).

Another approach is based on <u>probabilistic</u> methods. We have developed such methods in two steps

(i) The starting point for the first step was a paper by Phillips and Hodgson (4) suggesting a direct extension of the error analysis given by Blow and Crick for the MIR method. One of the (+) or (-) data sets at a particular wavelength, which we call 'reference' set, is selected to play the same role as the native data set in MIR methods; the other sets are treated as isomorphous derivatives . A probability distribution is calculated for each possible pairing of the reference set with the other sets and the joint probability distribution is calculated. To fix ideas, let us assume that Bijvoet pairs have been measured at three wavelengths and that the (+) set at λ_1 has been chosen as the reference; probability distributions are obtained for the five pairs (+1)/(-1), (+1)/(+2), (+1)/(-2), (+1)/(3), (+1)/(-3) and the joint probability distribution is calculated. From the centroid phase of the joint probability distribution, the 'best' value of the reference phase is derived and the 'best' reference electron density is computed using equation (1). The method is relatively simple and requires little computer time; it has a few drawbacks

- the contribution of the reference set is obviously overweighted.
- in contrast to the phases, the modulus of the Fourier coefficients are not optimised taking into account the whole experimental information.
- prior to the computing of the 'best' electron density map, the modulus and the phases of the Fourier coefficients must be corrected for the imaginary component of the anomalous scattering of the heavy atoms in order to obtain a real map.

(ii) The above drawbacks are removed by an improved procedure which leads to the 'best' amplitude and phase of the wavelength-independent part of each structure factor, given measurements at various wavelengths which are treated on an equal basis (5).

Let us define wavelenght-independent quantities X and Y as

$$X = F_{oph} \cos(\varphi_{ph} - \varphi_p)$$

$$Y = F_{oph} \sin(\varphi_{ph} - \varphi_p)$$

and known wavelength-dependent quantities x_λ and y_λ as

$$x_\lambda = F_{oh} * f'_\lambda / f_o$$

$$y_\lambda = F_{oh} * f''_\lambda / f_o$$

It is straightforward to show that

$$^{+}F_{\lambda}^{2} = {}^{+}I_{\lambda} = (X + x_{\lambda})^{2} + (Y + y_{\lambda})^{2}$$
$$^{-}F_{\lambda}^{2} = {}^{-}I_{\lambda} = (X + x_{\lambda})^{2} + (Y - y_{\lambda})^{2}$$

The probability distribution is now a function of X and Y. A separate distribution may be be defined for each set of (+) and (-) measurements at the various wavelengths, so that these distibutions are independent and can be multiplied to obtain the joint probability. By numerical integration, one gets $\langle X \rangle$, $\langle Y \rangle$, $\langle X^2 \rangle$ and $\langle Y^2 \rangle$. It can be shown that the 'best' Fourier coefficients have modulus $(\langle X \rangle^2 + \langle Y \rangle^2)^{1/2}$ and phase $\varphi_{ph} = \varphi_h + \mathrm{atan2}\,(\langle Y \rangle,\langle X \rangle)$. In the case of reflections which have, for the particular space group of the crystal of interest, a phase constrained to special values, the treatment is similar except that $\langle Y \rangle = 0$ and $\langle Y^2 \rangle = 0$.

For each method, a set of programmes has been written and implemented on our VAX 780 computer. Both have been extensively tested using artificial data with gaussian-distributed errors of adjustable σ produced as described by Phillips and Hodgson (4).

MAD methods with a single sample produce phases which are poorly determined for some reflections, for instance those for which the structure amplitude of the anomalous scatterers is weak. In this respect, a procedure which adds independent bits of phase information, such as filtering in direct and reciprocal space as described by Wang (6), may be a useful complement. We have now included Wang's algorithm as an option in our programmes but more experience will be necessary to quantify the usefulness of the combination of MAD and filtering in various practical situations.

III- How to perform a MAD experiment?

1) Instrumental requirements

A dipolar magnet or a wiggler source of synchrotron radiation with a critical wavelength below 4Å are suitable X-ray sources to perform properly MAD experiments. The beam stability during periods of several hours is of utmost importance. Orientational instabilities change the angle of incidence of the white beam on the first crystal of the monochromator, thus shifting the photon energy; such shifts should not exceed 0.5eV when edge featutures are exploited to enhance anomalous effects. Positional instabilities change the intensity and the polarization of the beam bathing the crystal sample, so that a long beam lifetime and correcting coils driven by position monitors placed along the synchrotron beam line are highly desirable. The DCI storage ring at LURE is a good source for optimized anomalous scattering with a lifetime of about 40 hours at 1.84 GeV, 0.25 A and a beam of adequate stability.

The monochromator must be easily and reproducibly tuneable with provision for precise wavelength measurements. The exit slit must be fixed.We use, in our current experiment (7), a separate function double crystal monochromator with sagittal bending on the second crystal. With symmetrically cut Ge(111) crystals, the intensity at the sample (1-$5\times10^{5}s^{-1}$) matches nicely the count rate capability of the area detector.

Since the X-ray beam is polarized in the horizontal plane, the multi-circle goniometer should preferably have its equatorial plane vertical. This is not the case at LURE, but the trouble is not too serious because protein crystals diffract only to low or medium resolution.Due to the high parallelism and monochromaticity of the beam emerging from the monochromator, the profiles of Bragg reflections are so narrow

-typically 1-2 $x10^{-2}$ deg FWHM for a well ordered and fresh protein crystal- that it is necessary to have elementary angular steps of 10^{-3} deg. or less on goniometer axis, especially φ and ω.

The intensity of the beam after collimation by the slit system of the diffractometer has to be continuously monitored. The monitor is used to put on a common scale diffraction data measured with an X-ray beam of varying intensity and, eventually, to control the exposure time on electronic pictures ('frames') to get a fixed count-per-frame during the whole experiment. We use a small ion chamber with semi-cylindrical electrodes which it is mounted on the collimator. Charges are collected by a high quality capacitor during the exposure time of any frame.

An electronic area detector (EAD) is of crucial importance for MAD experiments. The diffraction pattern is recorded on line and complex measurements schemes with frequent changes of wavelength are easily performed under computer control. In addition to convenience and speed, an AED achieve optimum signal-to-noise ratio because 'frames' can be multiplied at will, each frame corresponding to a small rotation of the crystal comparable to the rocking width of a reflection. Finally, AED have a detection quantum efficiency close to unity . We use a spherical drift multiwire chamber system developed in collaboration with the team of Charpak at CERN (Geneva); this detector is parallax-free. It has 185,500 useful pixels of $1x1mm^2$ and a typical count rate of 500,000 s^{-1} at 12% count losses (7).

Finally , some additional equipment is necessary. First, the sample must preferably be kept at a constant low temperature; we use an FTS systems gas cooler for this purpose. Second, it is essential to determine the variation of f" with energy across the absorption edge ; this is conveniently performed by measuring with a solid state detector the fluorescence of the sample crystal as the X-ray energy is scanned in the domain of interest.

2) Choice of wavelengths

Largest changes in f' and f" are found in the wavelength range across an absorption edge. Sharp threshold peaks are sometimes observed (8); these so-called 'white lines' appear to be associated with transitions from a deep core level to final states with spacially very compact character. Absorption spectra of all open shell atoms will show prominent discrete resonances from core levels with suitable symmetry into the open shell. To be of practical interest, the transition should lie in the range 0.5—2.5Å: the only good candidates for displaying very strong anomalous dispersion in this range are the 2p—5d resonances in elements from cesium to platinum (9). Among them, lanthanides interact with biological materials in specific ways and, in particular, the ability of Ln^{3+} to replace Ca^{2+} is quite interesting (10).

The values of the anomalous scattering factors f' and f" in the region of interest have to be determined. f" is proportional to the product of the photon energy by the photoabsorption cross section. Far away from any edge, this cross section can be fairly well obtained from theoretical calculations but in the threshold region, where damping effects cannot be neglected, such calculations are very difficult; it is in constrast rather simple to measure the cross section by fluorescence technique on the sample crystal, from which unscaled values of f" are derived. In our experiment, we have also measured the intensity difference between the mates of a Bijvoet pair, because this difference is proportional to f". Unscaled f" values are merged with calculated factors at points remote from the edge. Finally, f' are calculated from scaled f" values with the

Kramers-Kronig relation (11).

Once the variations of f" and f' as a function of λ have been determined, the following questions have to be answered: how many wavelengths should one use? which specific wavelengths should be selected ? Narayan and Ramaseshan (12) have derived a formula for the rms phase error in terms of the coordinates f'_λ and f''_λ in the complex plane of "centres" corresponding to different wavelengths. This formula is useful to guide the experimenter's choice. As an example, in the case of measurements at two-wavelengths of Bijvoet pairs with equal rms errors, the quantity which has to be maximized to minimize the average phase error is $(f'_1 - f'_2)^2(f''^2_1+f''^2_2)$. In agreement with Phillips and Hodgson (4), we believe that a two or three wavelength measurements of Bijvoet pairs with emphasis on accuracy of the data appears most optimal.

IV- Results

Parvalbumin of the superfast swimbladder muscle of the fish Optanus Tau is a small protein of molecular weight 10.1 kdaltons which binds two Ca^{2+} which can be substituted by lanthanides. Crystals of the Tb-labelled protein were grown in ammonium sulphate. The crystal data are: space group $P2_12_12$, a = 56.1, b = 59.4, c = 27.8Å, Z = 4 molecules/unit cell.

Diffraction experiments were performed at LURE using the bending magnet X-ray beam line D1 on the storage ring DCI and an equipment- less sophisticated than the one which is now available- consisting of a Ge(220) 'channel-cut' monochromator, a single-axis goniometer with the Mark I spherical drift multiwire proportional chamber (13). Data collection and a preliminary data reduction were performed on-line with a PDP 11/34 minicomputer. The variation of f" in the vicinity of the L_{III} absorption edge of terbium was determined by measuring Bijvoet differences and rescaling with theoretical values; f' was calculated with the Kramers-Kronig relation. Three wavelengths were selected: λ_1 = 1.6490 Å (f" maximum); λ_2 = 1.6469 Å; λ_3 = 1.6501 Å (|f'| maximum) and diffraction data sets at 2.3 Å resolution were collected at each of these wavelengths. The crystal was oriented so that Bijvoet pairs (hkl and $\overline{hkl}$) could be measured simultaneously with the plane of the detector set normal to the X-ray beam. The three sets were put on the absolute scale using intensity statistics; the average intensity λ_1-set was about 8% weaker than in other sets due to increased absorption.

The Tb partial structure was solved by anomalous Patterson techniques (14) using the λ_1-set and refined by least squares against the strongest anomalous differences (15). Two sites were found, one of them with a low occupancy. Phase calculations were done in three steps:

(a) The initial phasing combined the first statistical method described previously for general reflections and the second method for constrained phase reflections. Electron density maps based on 2439 phased reflections (figure of merit 0.73) were calculated for the two enantiomorphs. A chain tracing could be done on one of the two maps, using the known primary sequence of the molecule. In parallel to the work at Orsay, data were measured by our collaborators at Liege with a diffractometer and Cu Kα radiation and the structure was independently solved at 3.2 Å resolution by molecular replacement (16), using the carp parvalbumin structure as a model (17); the carp parvalbumin structure could thus be superimposed onto our MAD map, and provėd to fit nicely the electron density map, especially in regions of identical amino acid sequences (18, 19).

(b) the second statistical method was extended to the case of acentric reflections (5) and new phases were calculated; the electron density was quite similar to the previous one.

(c) Finally, the phases obtained from (b) were submitted to the filtering procedure of Wang (6). Limited changes were observed with respect to (b), with, in general, an improved connectivity of the electron density.

A new data set at 2Å resolution has been recently collected at Liege and the Orsay group has used them for a careful refinement of the Tb-parvalbumin structure which is now close to completion. At the present stage, the crystallographic R-factor is 0.15, with 150 water molecules included in the model. We aim at obtaining 'reference' phases to be compared with MAD phases.

We believe that this pioneering study is quite encouraging . But it is obvious that the case which has been solved here is a favorable one: the protein of interest was a fairly small one and the anomalous signal was strong. We have performed extensive simulations with artificial data which demonstrate the potential of MAD experiments for more complex structures and/or weaker anomalous effects. But the full expression of this potential depends now essentially on the improvement of detectors and data collection software and on the availability of stable synchrotron beam during long periods of time. Other problems have also to be considered, such as refinement of f' and f" to get more accurate and possibly Bragg angle-dependent values, X-ray dichroism near to an absorption edge (20) which introduces anisotropy in the anomalous scattering and accurate absorption corrections.

Acknowledgements

The Mark I and II detectors are the result of a patient collaboration with R. Bosshard at LURE and G. Charpak at CERN . The monochromator was designed in collaboration with M. Bessières, M. Jouvin, M. Bessenay , S. Lefebvre, J. Frouin and M. Sauvage-Simkin. Otto Dideberg and J. P. Wery at Liège, S. Brunie, J. Janin, J. L. Risler and A. Rimsky contributed to the work on parvalbumin. Discussions with J. C. Phillips at Brookhaven on MAD methods were quite stimulating. Finally, we thank the 'Lord of the Ring' P. Marin and its collaborators.

References

1- D. M. Blow & F. H. Crick, Acta Cryst. 12, 794 (1959)
2- J. Karle , Intl. J. Quant. Chem. 7, 357 (1980)
3- W. A. Hendrickson, J. L. Smith, R. P. Phizackerley, E. A. Merritt & W. E. Love SSRL 1984 Progress Report, SSRL, Stanford University
4- J. C. Phillips & K. O. Hodgson, Acta Cryst. A36, 856 (1980)
5- R. Fourme, M. Chiadmi & R. Kahn, 1983-1985 LURE Progress Report, LURE, Université

de Paris-Sud, Orsay.
6- B. C. Wang in 'Methods in Enzymology' 115, 90 (1985), H. W. Wyckoff, C. H. W. Hirs & S. N. Timasheff eds., Academic Press
7- R. Kahn, R. Fourme, R. Bosshard & V. Saintagne, Nucl. Instrum and Meth., A246, 596, (1986)
8- P. S. Wei & F. W. Lytle, Phys. Rev. B19, 679 (1979)
9- G. Wendin, Physica Scripta 21, 535 (1980)
10- C. H. Evans, TIBS 8(12), 445 (1983)
11- R. de L. Kronig & H. A. Kramers, Z. Phys. 48, 174 (1928)
12- R. Narayan & S. Ramaseshan, Acta Cryst. A37, 636 (1981)
13 R. Kahn, R. Fourme, R. Bosshard, B. Caudron, J. C. Santiard & G. Charpak, Nucl. Instrum. and Meth. 201, 203 (1982)
14- M. G. Rossmann, Acta Cryst. 14, 383 (1961)
15- W. Hendrickson & M. M. Teeter, Nature 290, 107 (1981)
16- M. G. Rossmann & D. M. Blow, Acta Cryst. 15, 24 (1962)
17- P. C. Moews & R. H. Kretsinger J. Mol. Biol. 91, 201 (1975)
18- R. Kahn, R. Fourme, R. Bosshard, M. Chiadmi, J. L. Risler, S. Brunie, O. Dideberg, J. P. Wery & J. Janin, Symposium on "New Methods in X-ray Absorption, Scattering and Diffraction for Application in Structural Biology", Bristol (1984) (proceedings Ed. by H.D. Bartunik and B. Chance, Academic Press (1986) p. 297.
19- R. Kahn, R. Fourme, R. Bosshard, M. Chiadmi, J. L. Risler, O. Dideberg & J. P. Wery FEBS 179, 133 (1985)
20- D. H. Templeton & L. K. Templeton, Acta Cryst. A41, 365 (1985)

X-Ray Diffuse Scattering from Protein Crystals

J. DOUCET[*] [**], J.P. BENOIT[*]

[*] LURE, Laboratoire CNRS-CEA-MEN, Université Paris-Sud, Bât. 209 D, F-91405 Orsay (France).

[**] Laboratoire de Physique des Solides, Bâtiment 510, Université Paris-Sud, F-91405 Orsay (France).

I - INTRODUCTION

It is now well established that the biological activity of proteins is related not only to their mean molecular structure, but also to their intramolecular mobility (1). Nearly all techniques sensitive to dynamics have given evidence for intramolecular mobility in proteins, since these compounds display a large variety of motions within a wide spectrum of vibration frequencies (ranging from 10^{12} Hz down to 10^{-1} Hz). The dynamics of proteins has also been approached by theoretical calculations, either by simulation of molecular dynamics trajectories (2), or by normal mode calculations (3).

The contribution of crystallography to the problem of intramolecular motions is also important. Comparison between structures of a same protein under different conditions (temperature, crystal symmetry, substrate binding) has for instance provided quite valuable information concerning the flexibility of trypsinogen (4), immunoglobulins (5), hexokinases (6) and citrate synthase (7). Crystal structure determination provides a set of thermal factors (B-factors), one for each individual atom, which are proportional to the mean square amplitude $<u^2>$ of the displacements around the rest position when $|u_{max}|$ is small. From extensive studies of the B-factors of two globular proteins : metmyoglobin (8) and lysozyme (9) some interesting results have been obtained, but more precise B-factor analyses, as for example the description of the molecular motion as rigid-body displacements including translations and rotations (10), remain limited by various factors. Firstly, the usual $<u^2>$ determinations hold only for harmonic displacements and small $<u^2>$ values (typically less than 30.10^{-4} nm^2). Another limitation lies in the fact that for macromolecules the B-factors, which are considered as isotropic, exclude the possibility of determining privileged directions of motion. The third limitation is due to the fact that the values obtained for the B-factors are very sensitive to all the errors accumulated during the data collection and analysis. Finally the thermal factors are individual factors, summing up all the types of displacements (dynamic ones as well as static ones), and thus no correlation between the atomic displacements can be derived from their values.

Up to now, we have only been concerned with "classical" crystallography, i.e. crystallography which is based on the analysis of Bragg reflexions. But extra information can be derived from the scattering of X-rays outside Bragg reflexions. Let us emphasize this point by describing an imperfect crystal by the superposition of an ideal periodic crystal, corresponding to its average structure, with small perturbations (atomic displacements).

The scattering pattern of such a crystal displays two components : the diffraction component which is the set of the Bragg reflexions from the ideal mean-crystal, and the diffuse scattering component which is spread out of the Bragg reflexions. The diffuse scattering, which is generally much weaker than the diffraction component, is characteristic of the small perturbations; its study gives access to the deviations from the ideal mean structure. Diffuse scattering has been widely and successfully used for the study of ferroelectric crystals (11), incommensurate crystals (12), liquid crystals (13), one dimensional-crystals (14), etc.. As regards proteins, the importance of diffuse scattering has been mentioned (9, 15, 16), but this technique has only been used once by G.N. Phillips et al who derived a model of filament motion for tropomyosin (17).

Our aim is to extend the use of diffuse scattering analysis for the case of biological macromolecular compounds.

II - DIFFUSE SCATTERING CONCEPTS

The concepts of X-ray scattering by imperfect crystals have been clearly explained by A. Guinier (18). A fundamental point lies in the fact that, because the X-ray frequencies (10^{18} Hz) are at least 10^5 times larger than atomic motion frequencies in condensed matter, each X-photon "sees" an instantaneous picture of the crystal, and thus the dynamic displacements are not distinguishable from the static ones.

Assuming that the jth atom of the nth unit cell is displaced from its mean position $\vec{r}^j$ by the perturbation $\vec{u}_n^j$, the contribution $F_n(\vec{S})$ of the nth cell to the total scattered amplitude of scattering vector $\vec{S}$ can be written :

$$F_n(\vec{S})=\sum_j f_j(\vec{S})\ \exp(-2i\pi\vec{S}.(\vec{r}^j+\vec{u}_n^j)) \qquad (1)$$

where the f_j quantities are the atomic scattering factors and $\sum_j$ sums over all the atoms within each unit cell. The total scattered amplitude is given by :

$$A(\vec{S})=\sum_n F_n(\vec{S})\ \exp(-2i\pi\vec{S}.\vec{r}_n) \qquad (2)$$

where $\sum_n$ sums over the N unit cells (located ar $\vec{r}_n$).The intensity at the extremity of the reciprocal vector $\vec{S}$ is equal to :

$$I(\vec{S})=A(\vec{S}).A^*(\vec{S})=|A(\vec{S})|^2 \qquad (3)$$

Substituting (1) and (2) into (3), $I(\vec{S})$ splits into two terms :

$$I(\vec{S})=I_B(\vec{S})+I_D(\vec{S}) \qquad (4)$$

with :

$$\begin{cases} I_B(\vec{S})=N|\langle F_n\rangle|^2 \sum_n \exp(-2i\pi\vec{S}.\vec{r}_n) & (5) \\ I_D(\vec{S})=N\sum_m V(\vec{r}_m)\Phi_m(\vec{S})\ \exp(-2i\pi\vec{S}.\vec{r}_m) & (6) \end{cases}$$

In the expression of $I_B(\vec{S})$, $\sum_n \exp(-2i\pi\vec{S}.\vec{r}_n)$is equal to zero excepted when $(\vec{S}.\vec{r}_n)$ has an integer value, i.e. when the extremity of $\vec{S}$ is located on a reciprocal node; in that case, we have : $I_B(\vec{S}) = N^2|\langle F_n\rangle|^2$. This expression corresponds to the well-known set of Bragg reflexions with F_n averaged over the N cells. Debye (19) has shown that for small amplitude and harmonic displacements :

$$|\langle F_n\rangle|^2=\left|\sum_j f_j\ \exp(-2i\pi\vec{S}.\vec{r}^j)\ \exp(-8\pi^2\langle u_{nS}^{j2}\rangle\sin^2\theta/\lambda^2)\right|^2 \qquad (7)$$

where u_{nS}^j is the projection of u_n^j along $\vec{S}$ and $\frac{\sin\theta}{\lambda}=\frac{|\vec{S}|}{2}$. The quantity $B=8\pi^2\langle u_{nS}^{j2}\rangle$ is called the temperature factor of atom j.

Let us turn to $I_D(\vec{S})$. $\sum_m$ sums over all the inter-cell distances $\vec{r}_m$; $V(\vec{r}_m)$ is the factor which takes into account the proportion of vectors $(\vec{r}_n+\vec{r}_m)$, the extremities of which lies outside the crystal volume; $\Phi_m(\vec{S}) = \langle \varphi_n\varphi^*_{n+m}\rangle$ with $\varphi_n = F_n(\vec{S}) - \langle F_n(\vec{S})\rangle$. Since Φ_m is a complex quantity , I_D is spread all over the reciprocal nodes. The term I_D is the diffuse scattering intensity; it is roughly equal to square modulus of the Fourier transform of the correlation function Φ_m of the deviations φ of the factor structures F for cells n and n+m. A great variety of disorders can be found in crystals, but all crystals exhibit at least one disorder which is due to the lattice mode vibrations. The associated diffuse scattering area, which is called thermal diffuse scattering (TDS), is located around each reciprocal node; in the approximation of isotropic crystalline elastic constants the TDS contribution to I_D is equal to : $K|\langle F\rangle|^2\ \frac{S^2}{q^2}$ (8)
where q is the distance of the extremity of $\vec{S}$ to the closest reciprocal node and K is a factor depending on the elastic constants. In fact, the TDS contribution is not directly related to intramolecular disorders. The possible other contributions are of interest for our problem.

III - X-RAY DIFFUSE SCATTERING DETECTION

What are the conditions required to detect X-ray diffuse scattering ? Even when it proceeds from highly correlated displacements, its intensity is generally 10^2 to 10^5 times weaker than the Bragg intensities. Thus, in order to both reduce the exposure times and radiation damages, high intensity beams are necessary. In addition, a very low background is necessary, which implies to work with a highly monochromatic beam; in some cases it is also better to avoid the air-scattering working under vacuum or within an helium atmosphere. Finally, a beam with a high spatial resolution is required, since the width of the diffuse scattering features must be sometimes accurately determined. All those conditions can be perfectly fulfilled by the synchrotron radiation source, which is in fact the ideal X-ray source for the diffuse scattering detection.

Two points of interest must be emphasized. Firstly, the diffuse scattering features are not due to a tuning effect of the molecular frequencies with the pulsed structure of the synchrotron beam. We have checked on this point comparing the patterns of several crystal proteins obtained either with a synchrotron beam or with a conventional source : the diffuse features are identical. Secondly, the diffuse scattering from protein crystals is not due to radiation damage effects since fresh and old crystals display the same features; on the contrary, the diffuse scattering intensity seems to decrease in the same way as the Bragg spot intensities.

IV - DIFFUSE SCATTERING FROM LYSOZYME CRYSTALS

We have tested the feasibility of diffuse scattering studies of proteins on hen-egg white lysozyme crystals. We have studied the orthorhombic form of lysozyme which is stable at physiological temperatures. The structure of this crystalline form has been determined at 0.2 nm resolution by J. Berthou et al (20). The data collection of diffuse scattering has been performed at LURE using synchrotron radiation. Because of the much longer exposure times (10^2 to 10^5 times) than for recording Bragg reflexions, only a set of "still" photographs has been obtained. This procedure is rather crude but can be justified by the fact that the number of parameters involved in collective displacements is much smaller than the number of parameters involved in structure determinations and also considering our experiment as a test.

One example of experimental film is shown on fig. 1-a. It has been obtained using the synchrotron radiation of LURE on station D16 (Paris-Sud University). A single bent germanium crystal allows one to work with a low background beam (λ = 0.161 nm).The photographs have been obtained using a STOE Explorer Camera in fixed geometry (distance crystal-film : 75mm) with exposure times as long as 30 minutes whereas the recording of the Bragg reflexions only needed a few seconds; the diameter of the collimator was 0.5mm. For this figure, the "vertical" rows are the traces of reciprocal planes perpendicular to $\vec{c}^*$ and the "horizontal" ones of planes perpendicular to $\vec{a}^*$. Four intensities components can be observed : Bragg reflexions, thermal diffuse scattering around Bragg spots, a large diffuse ring (maximum around $3nm^{-1}$)which proceeds from the sample holder (capillary glass) and finally the diffuse scattering which gives evidence of a disorder specific to the compound. On the films the diffuse scattering can be observed along rows, with a modulated intensity, and is in fact, in the reciprocal space, mainly located along two families of planes : (100)* and (001)*. We will not go here into the details of the analysis of the diffuse features which has been given elsewhere (21). Let us just precise that the qualitative characteristics have been derived from the localization of the diffuse scattering whereas the quantitative parameters have been deduced by comparing the experimental films with simulated films (fig. 1-a and 1-b). In short, molecular displacements, correlated along about 20nm, have been detected. These rigid-body displacements are polarized along $\vec{a}$ for the correlations in the $\vec{a}$-direction and along $\vec{c}$ for the correlations in the $\vec{c}$-direction; their mean square amplitude is of the order of 5.10^{-4} nm^2. Looking at the structure (fig. 2) the molecules which are correlated in the $\vec{a}$ direction form two types of rows : (3 - 1 - 3' - 1' -...) and (4 - 2 - 4' - 2' ...),

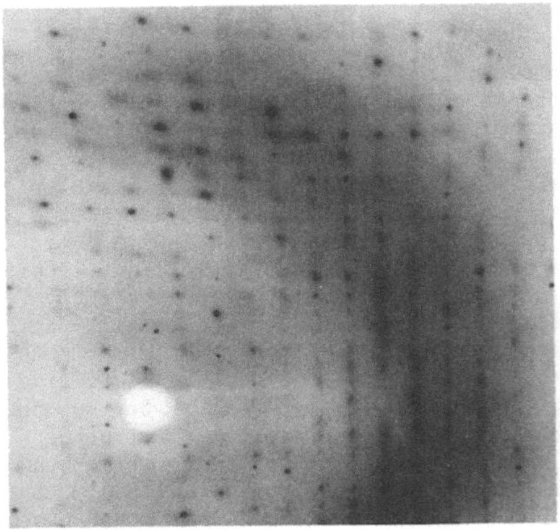

1-a

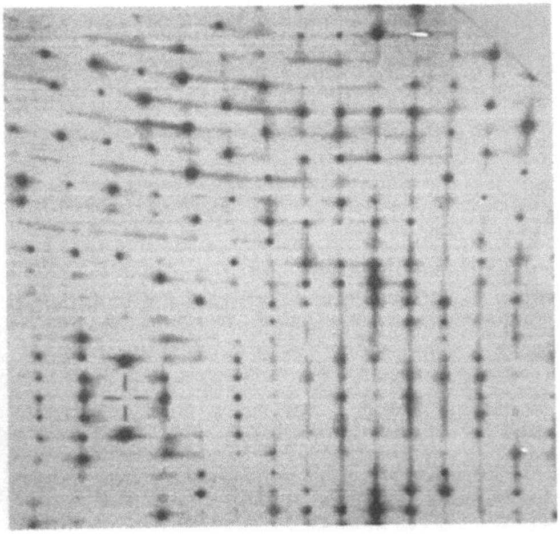

1-b

Figure 1-a Part of experimental film. Diffuse scattering is visible along "horizontal" and "vertical" rows.

Figure 1-b Corresponding simulated pattern.

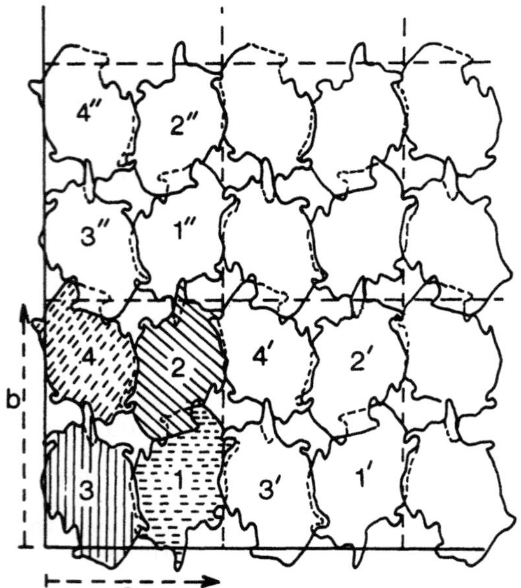

Figure 2 Projection onto the $(\vec{a},\vec{b})$ plane of the molecular envelopes; the unit cell contains four molecules: 1,2,3 and 4.

the displacements between the various rows being independent. In the $\vec{c}$ direction, two types of independent rows are also found : (1 - 2 - 1 - 2 - ...) and (4 - 3" - 4 - 3" - ...); the displacements of molecules 1 and 2 (or 4 and 3") are coupled.

What are the relations between the correlated molecular displacements, the intermolecular contacts and the intramolecular mobility ? A satisfactory aspect of the results just described is that the rows have a physical meaning, i.e. that the molecules forming a row are sterically coupled.

For instance, the contact between molecules 1 and 3 is ensured by about a dozen of intermolecular contacts shorter than 0.4 nm; fifteen to twenty well localized water molecules in the intermolecular zone reinforce this coupling.

On the contrary, no contact between molecules 3 and 4 exists, which explains the absence of correlations in the $\vec{b}$ direction. Regarding superunits c, the coupling between molecules 1 and 2 (or 4 and 3") is also plausible, since there is an overlap of these molecules looking along $\vec{c}$, which implies coupled displacements. What is the cause of the correlated displacements ? They could arise from a hinge-bending mode about an axis passing through carbon atoms 38 and 87 (36), which are roughly located in the ($\vec{a}$, $\vec{c}$) plane. The corresponding motion would display a large component in the b direction which is not observed, so this hypothesis is not probable.

Another possible cause is a local disorder, involving one or more residues in contact with a neighbouring molecule, which would produce a displacement of the latter, all displacements being correlated within one row . Such phenomena occurs perhaps at the intermolecular contact between residues Asp 101 and Asn 103 of molecule 1 and Thr 47 and Asp 48 of molecule 2, which is characterized by distances between the mean atomic positions which are smaller than the Van der Waals distances. The associated disorder is probably localized on residues 101 and 103 which have high B values (around 50.10^{-2} nm^2) contrary to residues 47 and 48. The interesting point is that residues 101 and 103 belong to a loop situated on the top of the active site which is characterized by high thermal factors, not only in the orthorhombic form, but also in the quadratic form of lysozyme (22).

V - CONCLUSION

We have shown that X-ray diffuse scattering from protein crystals can be detected and analysed using a synchrotron X-ray source. In the case of lysozyme, the diffuse scattering features are due to rigid-body molecular displacements correlated along about 20 nm. It must be emphasized that this disorder only represents one kind of disorder among all the kinds which can be analysed by this technique. For instance, it is highly probable that for other compounds the diffuse features will be directly connected to intramolecular motions.

ACKNOWLEDGEMENTS

The authors are very grateful to J. Berthou and A. Lifchitz for preparing very good lysozyme crystals, for access to their lysozyme data and for helpful comments. We also wish to thank S. Megtert for his collaboration during the experiments at LURE.

REFERENCES

(1) Huber R. and Bennett W.S. Biopolymers 22, 261-279 (1983)

(2) Karplus M. and Mc Cammon J.A. C.R.C. Cri. Rev. Biochem. 9, 293-349 (1981)

(3) Tasumi M., Takenchi H., Ataka S., Dwivedi A.M. and Krimm S., Biopolymers 21, 711-714 (1982)

(4) Walter J., Steigemann W., Singh T.J., Bartunik H., Bode W. and Huber R. Acta Cryst. B 38, 1462-1472 (1982)

(5) Poljak R.J., Amzel L.M. Avey H.P., Becka L.N. and Nisonoff A. Nature new Biol. 235, 137-140 (1972)

(6) Anderson C., Stenkamp R.E., Mc Donald R.C. and Steitz T.A., J. Mol. Biol. 123, 207-219 (1978)

(7) Remington S.J., J. Mol. Biol. 158, 111-152 (1982)

(8) Frauenfelder H., Petsko G.A. and Tsernoglou D. Nature 280, 558-563 (1979)

(9) Sternberg M.J.E., Grace D.E.P. and Phillips D.C., J. Mol. Biol. 130, 231-253 (1979)

(10) Schomaker V. and Truebold K.N. Act. Cryst. B 24, 63-76 (1968)

(11) Comès R., Lambert M. and Guinier A. Sol. St. Comm. 6, 715-719 (1968)

(12) Durand D.; Dénoyer F., and Lambert M., J. Phys. 43, 149-154 (1982)

(13) Doucet J., Levelut A.M. and Lambert M., Phys. Rev. Lett. 32(6) 301-303 (1974)

(14) Pouget J.P., Megtert S., and Comès R., Recent Developments in condensed matter Physics (1) 295-309 (1981)

(15) Schevitz R.W., Podjarny A.D., Krishnamachari N., Hughes J.J. and Sigler P.B. Nature 188-190 (1979)

(16) Artymiuk P.J., Blake C.C.F., Grace D.E.P., Oatley S.J., Phillips D.C. and Sternberg M.J.E., Nature 280, 563-568 (1979)

(17) Phillips G.N., Fillers J.P., Cohen C. Biophys. J. 32 (1), 485-502 (1980)

(18) Guinier A. "X-ray diffraction", Freeman and Company, Sao Francisco and London (1963)

(19) Debye P. Ann. Phys. Paris 43, 49-95 (1914)

(20) Berthou J., Lifchitz A., Artymiuk P. and Jollès P. Proc. R. Soc. Lond. B. 217, 471-489 (1983)

(21) Doucet J., Benoit J.P., to be published.

(22) Phillips D.C., Sc. American (5) 215, 78-90 (1966)

Laue Diffraction from Protein Crystals: Theoretical Aspects

Keith Moffat[*o], Durward Cruickshank[+o§] and John Helliwell[o§]
*Section of Biochemistry, Molecular and Cell Biology, Cornell University, Ithaca, NY 14853 U.S.A.
+Department of Chemistry, UMIST, Manchester M60 1QD, England
oDepartment of Physics, University of York, York YO1 5DD, England
§SERC Daresbury Laboratory, Warrington WA4 4AD, England

Introduction

If a stationary crystal is illuminated by a polychromatic beam of x-rays or neutrons, then a Laue diffraction pattern results. Each Laue reflection arises from a central line (a ray) in reciprocal space, which may contain one or more reciprocal lattice points (RLPs) with associated structure factors. It has been clear since the first protein x-ray Laue patterns were obtained (Moffat *et al.*, 1984, 1986; Bilderback *et al.*, 1984; Helliwell, 1984, 1985; Hedman *et al.*, 1985) that the Laue technique possesses a number of advantages. It makes optimum use of a naturally polychromatic spectrum such as that emitted by a bending magnet, wiggler or undulator on a synchrotron or storage ring source. Very brief exposure times in the millisecond range have been obtained on strongly scattering protein samples (Bilderback *et al.*, 1984; Hajdu *et al.*, 1986; Moffat *et al.*, 1986). A stationary crystal yields integrated intensities directly, which are much less sensitive to transient changes in unit cell dimensions or crystal orientation than the (partial) intensities obtained by diffraction of monochromatic radiation. A typical Laue diffraction pattern contains many more reflections than a typical monochromatic, oscillation pattern. These advantages are particularly appropriate for dynamic experiments, in which the diffraction intensities change rapidly with time in response to a structural perturbation (Moffat *et al.*, 1984, 1986; Helliwell, 1985).

Apparently offsetting these advantages are two potential disadvantages. First, if a Laue reflection arises from more than one RLP and structure factor, then the recorded Laue intensity is the sum of the constituent intensities, and the individual structure factors that are required for subsequent structural analysis cannot be extracted directly. Second, even if a Laue reflection arises from only one RLP and structure factor, a series of wavelength-dependent correction factors must be applied to obtain the desired structure factor. This is necessarily a more complex process than that for monochromatic data, and may yield structure factors of lower precision.

We consider briefly here the results of a theoretical investigation of these advantages and disadvantages. The full analyses, together with quantitative comparison with experiment, will be published separately.

Laue and monochromatic oscillation exposure times

By equating the integrated intensities of a reflection $\underline{k}$ obtained by Laue and monochromatic oscillation techniques (Kalman, 1979), the ratio of the Laue exposure time Δt to the monochromatic exposure time ΔT may be shown (K. Moffat and J.R. Helliwell, unpublished) to be given by

$$\Delta t/\Delta T = \text{reflection factor} \times \text{focussing factor} \times \frac{\text{monochromator rocking width}}{\text{angular range of oscillation}}$$

The reflection factor is of order unity, varying with $\underline{k}$ over the range from about 0.3 to 3.0; the focussing factor is unity for equal (or no) focussing of the Laue and monochromatic beams. Under that circumstance, the ratio of exposure times is governed by the last term. A typical single crystal monochromator rocking width is 2.5×10^{-4} radians. For a single reflection, the angular range of oscillation is the sample rocking width, say 8×10^{-4} radians; for one film, the range is typically 3.5×10^{-2} radians; and for equal numbers of stimulated RLPs, the range is $0.6\ d^*max\ \Delta\lambda$ radians, where d^*max is the limiting sample resolution and $\Delta\lambda$ is the wavelength range incident on the sample. For typical values of $d^*max = 0.5$ Å^{-1} and $\Delta\lambda = 2$ Å, the range is 0.6 radians. Thus, the Laue exposure gain, given by the inverse of the above ratio, $\Delta T/\Delta t$ ranges from 3 (one reflection) through 150 (one film) to 2000 (equal number of RLPs). The exposure gain for Laue _versus_ monochromatic oscillation techniques is indeed very substantial, up to two to three orders of magnitude for comparable focussing. This arises primarily from the fact that a reflection is stimulated for the entire Laue exposure, but only for a brief fraction of a monochromatic oscillation exposure as it is passing through the Ewald sphere.

In practice, most monochromatic oscillation data are obtained on synchrotron beam lines with both horizontal and vertical focussing, whereas Laue data have been obtained with no focussing (CHESS, the Cornell High Energy Synchrotron Source) or vertical focussing only (SERC Daresbury). Thus the observed exposure gain has so far been less than that calculated above.

It can be shown that the integrated intensity for a Laue reflection at a Bragg angle of $\Theta_6 = 5°$, a lysozyme crystal volume of $(200\ \mu m)^3$, a unit cell volume of $5.6 \times 10^{-50}\ m^6$, stimulated at 1 Å, for typical CHESS operating conditions, is 4.5 photons s^{-1} (electron)$^{-2}$. For a structure amplitude of 500 electrons, an average reflection, this yields 1.1×10^6 photons s^{-1}. In the absence of detector noise, a precision of 1% in intensity would be obtained in an exposure time of only 9 ms. Such exposure times have been obtained (Moffat _et al_., 1986), though with much lower precision. Even shorter times will be achievable when focussing of the white wiggler beam at CHESS is employed, and when diffraction is recorded using sensitive image plates with very low noise (Miyahara _et al_., 1986), rather than with Polaroid or conventional x-ray film.

Number of reflections stimulated

In a Laue experiment, the total number of stimulated RLPs N_L is (Moffat _et al_., 1986)

$$N_L = \frac{\pi d^{*4}max\ \Delta\lambda}{4V^*}$$

where V^* is the reciprocal unit cell volume. Note that N_L is directly proportional to the incident wavelength range, $\Delta\lambda$. N_L is a substantial fraction of the total observable reflections, given by $0.375\ d^*max\Delta\lambda$. Thus, even for a triclinic crystal, a complete Laue data set will consist of a relatively small number of films (or image plates), each containing numerous reflections. The Laue technique becomes progressively more efficient in stimulating reflections at higher resolution; the fraction stimulated between d^* and $d^* + dd^*$ is $\frac{1}{2}d^*\Delta\lambda$. If the crystal has higher symmetry, then by careful orientation of the crystal it is possible to record _all_ the unique diffraction data beyond a certain resolution in _one_ photograph (K. Moffat, unpublished results). For example, if the diffraction pattern belongs to the Laue group m3, then by aligning the crystal such that the x-ray beam is normal to $(01\bar{1})$ and inclined at 15° to (100), 90% of all unique data at 5.7 Å resolution, 97% at 3.9 Å resolution, and 100% at 2.9 Å resolution or less, are recorded on one photograph. This calculation assumes λ-max = 2 Å and λ-min = 0 Å.

Of the total stimulated RLPs N_L, a certain number N_S give rise to Laue reflections which contain only a single RLP and structure factor, and are thus most readily quantifiable. We (manuscript in preparation) have shown that to a close approximation N_S is also proportional to $\Delta\lambda$. Hence, the maximum number of readily quantifiable reflections is obtained with a wide wavelength range $\Delta\lambda$.

The overlapping orders problem

The overlapping orders problem arises from multiple solutions to Bragg's Law: the pairs d,λ and d/2,λ/2 and d/3,λ/3..... all satisfy Bragg's Law simultaneously. Diffraction from all orders of a fundamental spacing d overlaps exactly, at the same scattering angle. We then ask: What fraction of RLPs gives rise to single Laue reflections containing only single structure factors? If it is small, then the Laue technique is ill-suited to yield precise structure factors.

Although this is a central problem, it does not appear to have been attacked previously. We have now deduced the following results (D.W.J. Cruickshank, J.R. Helliwell and K. Moffat, Acta Cryst.: manuscript in preparation).

1) If the angular acceptance of the detector is unrestricted, then a remarkably large fraction (72.8%) of all RLPs gives rise to single Laue reflections, even for Δλ infinite.

2) This fraction increases to greater than 83.2% when a more experimentally realistic value of Δλ is used, say 2.0 Å.

3) This fraction depends only on the ratio of λ-max to λ-min where Δλ = λ-max - λ-min, and not on the space group, unit cell dimensions, crystal orientation or limiting resolution of the crystal d^*_{max} (provided $d^*_{max} < 2/\lambda$-max). See Fig. 1.

4) The fraction of reflections at a resolution d^* that gives rise to single Laue reflections depends markedly on d^*, and on λ-max/λ-min; it is generally lower at low values of d^* than at high.

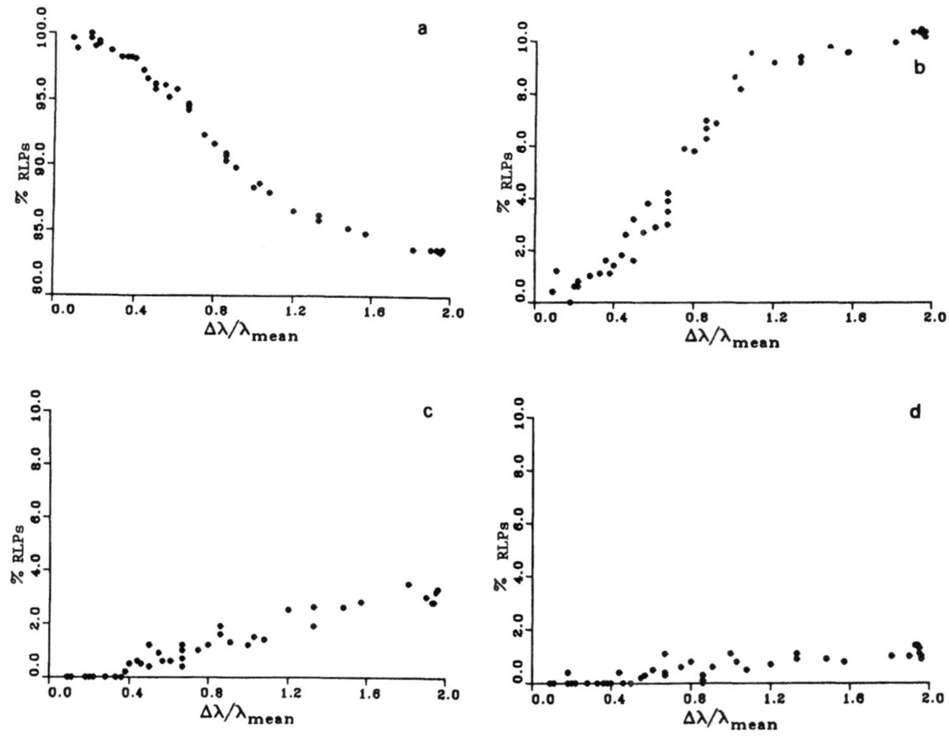

Figure 1 The percentage of RLPs that lie on single (panel a), double (panel b), triple (panel c) and quadruple (panel d) rays, as a function of the ratio of the wavelength range Δλ to the mean wavelength, (λ-max + λ-min)/2. Note the different scale in panel a compared with panels b, c and d. Data points are from extensive computer simulations. From D.W.J. Cruickshank, J.R. Helliwell and K. Moffat, manuscript in preparation.

5) Restricted angular acceptance of the detector can reduce significantly both the fraction and the number of single Laue reflections, N_S.

Agreement between these theoretical predictions and experimental values derived both from analysis of x-ray Laue photographs of protein crystals, and from extensive computer simulations, is good. It is evident that, under a wide variety of realistic experimental conditions, the effect of multiple orders is not a serious limitation on the use of the Laue method for either static structure determination or dynamic, time-resolved studies.

Quantitation of Laue diffraction patterns

The recorded intensity $I(\underline{k})$ of a Laue reflection depends on a variety of factors associated with the incident x-ray spectrum, the sample and the detector. Under certain assumptions it may be written as (Campbell et al., 1986):

$$I(\underline{k}) = H(\lambda)J(\Theta)|F(\underline{k})|^2 \Delta t$$

where Δt is the exposure time, $J(\Theta)$ expresses the Θ-dependence, and $H(\lambda)$ expresses the λ-dependence. The first two are generally known to good precision, but $H(\lambda)$, the wavelength normalization curve or λ-curve, has to be determined in order to obtain $|F(\underline{k})|^2$ from $I(\underline{k})$. It may be obtained in a subsidiary experiment; by comparison with a monochromatic data set; or internally, by comparison of the recorded intensities of symmetry-related reflections (Campbell et al., 1986). Computer programs to accomplish the complete quantitation of Laue patterns, including determination of the λ-curve and the unscrambling of Laue intensities arising from more than one structure factor, are under development at both Daresbury (Campbell et al., 1986 and references therein) and CHESS (B. Smith, D.M.E. Szebenyi, and K. Moffat, unpublished results). Preliminary analyses yield overall R-factors in the range from 8-12% on intensity, with further reductions anticipated when profile fitting is incorporated. It appears that overall R-factors comparable with those obtained by monochromatic oscillation techniques are likely to be readily attainable.

Conclusions

Theoretical and experimental investigation of the two major disadvantages of the Laue technique noted above has suggested that they are more apparent than real; and the prominent advantages are now on a sounder footing. While further work is needed on optimization of the signal-to-noise, and of the wavelength range for a given experiment, it appears that the Laue technique may become the approach of choice for structure determination using a synchrotron source.

Acknowledgements

K.M. gratefully acknowledges support while on leave at SERC Daresbury and the University of York, from the John Simon Guggenheim Foundation and the Royal Society. Supported by grants from the SERC (J.R.H.) and NIH (K.M.).

References

Bilderback, D.H., Moffat, K. and Szebenyi, D.M.E. (1984) Nucl. Instrum. and Methods 222, 245-251.
Campbell, J.W., Habash, J., Helliwell, J.R. and Moffat, K. (1986) CCP4 Newsletter Number 18, SERC Daresbury Laboratory.
Hajdu, J., Machin, P., Campbell, J.W., Clifton, I.J., Zurek, S., Gover, S. and Johnson, L.N. (1986) CCP4 Newsletter Number 17, SERC Daresbury Laboratory.
Hedman, B., Hodgson, K.O., Helliwell, J.R., Liddington, R. and Papiz, M.Z. (1985) Proc. Natl. Acad. Sci. USA 82, 7604-7607
Helliwell, J.R. (1984) Repts. Progr. Physics. 47, 1403-1497

Helliwell, J.R. (1985) J. Mol. Struct. 130, 63-91.
Kalman, Z.H. (1979) Acta Cryst. A35, 634-641.
Miyahara, J., Takahashi, K., Amemiya, Y., Kamiya, N. and Satow, Y. (1986) Nucl. Instrum. and Methods A246, 572-578.
Moffat, K., Schildkamp, W., Bilderback, D.H. and Volz, K. (1986) Nucl. Instrum. and Methods A246, 617-623.
Moffat, K., Szebenyi, D.M.E. and Bilderback, D.H. (1984) Science 223, 1423-1425.

White Beam Laue Diffraction: Data Processing Developments and Results for Single Crystal Data

J W Campbell, I J Clifton, M Elder, P A Machin, S Zurek
(SERC Daresbury Laboratory, Warrington UK)
J R Helliwell (York University and Daresbury Laboratory, UK)
J Habash (York University, UK), J Hajdu (Oxford University, UK),
M M Harding (Liverpool University, UK)

1. INTRODUCTION

Polychromatic synchrotron x-radiation is being used at Daresbury to record single crystal Laue diffraction data for several purposes, time resolved crystallographic studies of proteins being a most exciting example.

The large number of reflections recorded per film by the Laue technique, and the extremely high intensity of the white x-radiation generated via the Wiggler magnet at the SRS (λ_c = 0.9A^o), allow the recording of a substantial fraction of the unique data from a stationary protein crystal, with a single Laue exposure, in a 100 millisecond timescale. This technique therefore opens up completely new territories in the study of structural changes, catalysis and phase transitions in protein crystals.

Computer programs have been developed for the quantitative analysis of these Laue patterns recorded on X-ray film. An overview of this software and the processing strategy is presented in the following sections. A small molecule crystal and the protein Pea Lectin have been used as test data and results are presented for these. Preliminary results of kinetic Laue photography in the study of activity in Phosphorylase b crystals are also reported.

2. THE LAUE METHOD

The Laue technique (stationary crystal and white radiation) is described in detail in The Laue Method (1975). Its use with **Synchrotron radiation** is detailed in several papers and is summarised here. It was applied in a small molecule single crystal refinement study by Wood et al (1983). Moffat et al (1984) at CHESS advocated the use of the Laue method as appropriate for kinetic protein crystallography and emphasised the need to restrict the wavelength band to avoid energy multiple spots. Helliwell (1984), Hails et al (1984) and Helliwell (1985) advocated the use of as broad a wavelength band as possible to stimulate as many reflections as possible since the number of energy multiple spots was not actually a large fraction of the recorded data. However the hypothesis of Greenhough and Helliwell (1983) that the geometric overlap (including spatial and energy overlaps - see Helliwell (1985) for the distinction) would be a limiting problem though true, can in fact be avoided by using a large enough detector for the protein crystal case. Helliwell (1984) and Hedman et al (1985) showed that radiation damage need not be a significant problem and that complete data sets were collected via the Laue method for Pea Lectin and gramicidin. Moffat et al (1986) have obtained similar results with a

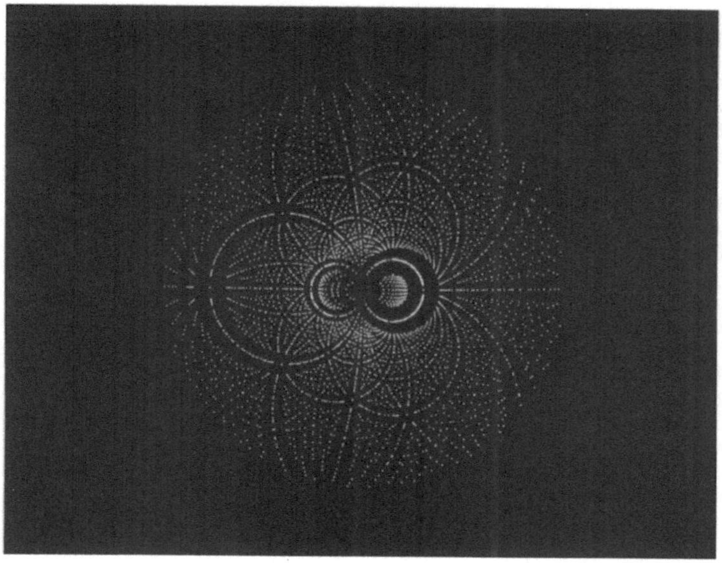

FIGURE 1. Laue diffraction from Pea Lectin. The predicted pattern shows the wavelength variation for 3462 spots. In the original coloured illustration the colour is coded in 8 bins ranging from blue ($\lambda <$ 0.72) to red ($\lambda \geq$ 2.34). Resolution limit = 3.5A° (this is less than that used by Helliwell (1984) (1985) to describe the Pea Lectin Laue patterns purely for illustrative purposes; similarly for figure 2).

FIGURE 2. Predicted Pea Lectin Laue pattern showing singlet spots (3201), doublets (201) and triplets and higher (60).

broad band pass with protein crystals at CHESS. A few important aspects of the Laue method are summarised below.

(a) The different reflections in the Laue pattern result from different wavelengths present in the incident radiation. At the SRS, Daresbury the white beam radiation is effectively the range 0.3 or 0.5 to 2.6 A^{O} for most experiments, and the variation of the wavelength of diffraction data is illustrated in Figure 1 where it can be seen that the short wavelength (blue) reflections on the whole dominate the centre of the pattern and that there is a gradual increase of wavelength, through the spectrum for reflections at greater distance from the centre.

Reciprocal lattice points which are stimulated by different wavelengths but at the same angle of diffraction (for example h k l at λ and nh nk, nl at λ /n) are superimposed in the Laue pattern. This feature is illustrated in Figure 2 where the multiplet spots are highlighted by colour and it is interesting to note that these spots are a small proportion of the total and in general are well separated spatially from other spots. A theoretical analysis of the proportion of multiplet spots under different resolution and wavelength conditions is given by D W J Cruickshank et al (1986).

(b) Laue experiments at Daresbury have usually employed a film pack of 6 films to cover the dynamic range of the intensities at different wavelengths and to make wavelength discrimination possible for multiple spots. A typical pack is:-

2 papers, F, F, Al, F, Al, F, Al, F, Cu, F

Where F stands for film (CEA reflex or Kodak DEF), Al for aluminium foil (150 μ thick), Cu for copper foil (38 μ thick). The properties of this pack are shown in Figure 3 and are described in I J Clifton et al (1985) but note in particular the rapid increase of absorption with wavelength and the discontinuities arising from the silver and bromine edges, due to the elements in the film.

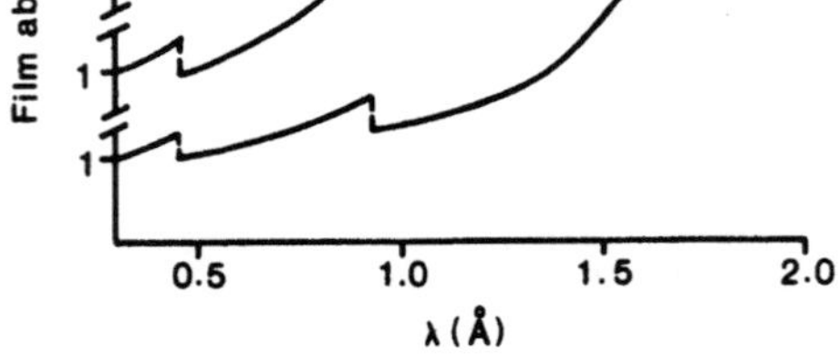

FIGURE 3. Film absorption as a function of wavelength.
Curve I is for two successive films with no separator
Curve II is for successive films separated by one Al foil
Curve III is for successive films separated by one Cu foil
The two discontinuities occur in the film absorption factor at the silver and bromine absorption edges.

(c) The different wavelengths necessarily used affect the intensities of recorded reflections through a variety of processes and the determination of the resulting wavelength normalisation curve possibly poses the greatest challenge in the quantitative analysis of these data. The wavelength dependent processes include:-

(i) the incident intensity in the SRS spectrum
(ii) sample absorption
(iii) film factor, obliquity, absorption by cassette paper
(iv) source polarisation
(v) Lorentz factor $\lambda^2/\sin^2\theta$
(vi) anomalous scatterers if present

The methods currently used for calculating this wavelength normalisation curve are described in Section 5c.

3. OVERVIEW OF THE STAGES IN INTENSITY MEASUREMENT

After scanning all the films in each pack (typically at 50μ on a drum scanner) the stages in processing are:

(a) Prediction of the Laue pattern to match the experimental film of diffraction data.

(b) Refinement of missetting angles, cell parameters etc to obtain accurate positional parameters for spots on the given film.

(c) Measurement of the integrated intensity of each reflection at the predicted position, for all films in the pack.

(d) Derivation of interfilm scale factors, each of which is a function of wavelength, and scaling together of all films in the pack. The result is h, k, l, I, σ, λ for each reflection (after LP correction).

(e) 'Unscrambling' of reflection intensities which are superimposed on the film. (This is described in greater detail in Section 5b).

(f) Normalisation to take account of the variation with wavelength of factors such as incident intensity, listed in Section 2c. (This is described in more detail in Section 5c).

(g) Scaling and merging of film packs (as for monochromatic data).

Computer programs have been set up for each of these stages and they are described in more detail in the Information Quarterly for Protein Crystallography, No. 15 (1985) and in J R Helliwell et al (1986). Details of the broad band pass Laue data processing strategy were given in Helliwell (1985).

Essentially, the software has been derived from techniques and programs for oscillation film data processing, existing routines have been used when possible (MOSFLM, 1980) but many new programs have been written. The software is written in FORTRAN 77 and runs on a VAX 11/750 apart from the prediction routines which utilise the raster screen of an ICL PERQ computer.

TABLE 1 Crystals and data sets used for testing

	Pea Lectin	BPD*
	M.WT 50,000 Orthorhombic $P2_1\ 2_1\ 2_1$ a = 50.37 b = 60.58 c = 135.50	$C_{10}\ H_{15}\ N_2\ P\ S_3$, M.WT 290 Orthorhombic $Pna2_1$, Z = 4 a = 15.037 b = 11.118 c = 7.818
	5 Film packs at 30° intervals around spindle.	6 Film packs at 15° intervals around spindle.
d min	2.6A	1.1A
λ min	0.33A	0.3A
λ max	2.6A	2.7A
Total no. of spots generated	9389	724
Spots with overlap (harmonics)	2177	184
Nodal spots	54	65

* trimethly 1H-2,1,3-benzophosphodiazine-4(3H)-thione 2,2 disulphide from Colin Reynolds, York University

TABLE 2 Summary of R factors for the 2 test data sets

	Pea Lectin	BPD
1. R_{sym} (on I) within the top film (for identical λ pairs)	4%	4 - 9%
2. Film to film scaling R_{scale} (on I) within one pack	8%	8 - 10%
3. Normalisation R (on I) I_{Laue} to I_{mono} within one pack	-	12% (10 - 16%)
4. Merging of packs R merge (on I)	12%	13%
5. Agreement with monochromatic data R mono (on F)	11%	11%

where

$$R_{sym} = \Sigma_{hkl}\ \Sigma_n\ (I_{hkl} - \bar{I}_{hkl})/\Sigma_{hkl}\ n\ \bar{I}_{hkl}$$

and

$$R_{mono} = \Sigma\ (F_{mono} - F_{Laue})/\Sigma F_{mono}$$

4. TEST DATA AND PROCESSING RESULTS

The results obtained so far with two test data sets, the protein Pea Lectin and a small organic molecule BPD, illustrate the quality of the data and some of the problems. This work is described in more detail in CCP4 newsletter No. 15, is summarised in Tables 1 and 2, and essential points are highlighted below.

The problems of obtaining background corrected intensity measurements from Laue data are very similar to those experienced using other camera geometries. However it is worth mentioning that:-

(a) Initial pattern prediction and refinement is often more difficult because there is a need for very accurate prediction because of the high density of recorded spots. Monochromatic still photographs may not be available for determining crystal orientation (but see Section 5a) and certain parameters are inexact by nature (minimum and maximum wavelength and resolution).

(b) Intensity measurement may be difficult because of spatially close spots (only measurable if profile fitting is available), and wavelength overlap spots (which may be treated as described in 5b).

The results which are summarised in Table 2 demonstrate that reasonable statistics may be obtained for measurements from Laue data. Profile fitting was carried out on a fraction of the data and improved the agreement factors (Rsym) from 6% to 4% as well as potentially increasing the quantity of measurable data. The final agreement factors of 11% between F_{Laue} and F_{Mono} for Pea Lectin may usefully be compared with 8% obtained in a comparison between the monochromatic data set collected with film and Synchrotron Radiation to that collected with a four circle diffractometer on a conventional Cu Kα source. Improvements are still being made to the later stages of data analysis and we are confident that White Beam Laue photography on the Synchrotron promises to be a useful quantitative crystallographic technique.

5. NOVEL COMPUTATIONAL DEVELOPMENTS

Three novel software developments are of particular interest and are described here.

(a) Automatic Indexing of Laue Patterns

A technique for determining the crystal orientation from a single Laue photograph has been devised and implemented as an option in the program LGEN (M Elder) which is used for Laue pattern simulation and which runs on a PERQ computer. The angles between low order reciprocal lattice points (nodal spots) are calculated for the given crystal cell and compared with those derived from nodal spot positions on a photographic film. The indices of these nodal spots are thus determined and the orientation of the crystal follows. The method has been used successfully on a series of cubic Insulin Laue films and some organic and organometallic compounds.

(b) Unscrambling of Harmonic Reflection Intensities

A complication of the Laue method is the superposition, in a fraction of the data, of multiple order reflections (harmonics) which result in composite spots on the recorded photograph. The intensity of one of these spots is given by the sum of the intensity contributions from the constituent reflections. Thus a particular spot's intensity may be a function of several wavelengths, which cannot be deconvoluted with the information on one film alone, but may be if measurements are available from several films in a pack.

These multiple spots can account for some 10% - 20% of the Laue pattern data and contain much of the low resolution data, so a method for unscrambling them is certainly useful. An algorithm has been programmed (called UNSCRAM) and tested on Laue data from the protein Pea Lectin. An analysis of the unscrambled intensities produced an Rsymm of 9.5% on doublet reflections (and 18.3% on triplets), which although higher than that for singlets does show that the resulting intensities may usefully be added to complete the dataset.

Reference: S Zurek et al, (1985).

(c) Determination of the Wavelength Normalisation Curve

The recorded Laue intensity for a given reflection (k) may be expressed as a function of the wavelength (λ), the Bragg angle (Θ) and the position (x) as follows:-

$$I(k) = f(\lambda)\ g(\Theta)\ h(x) |F(k)|^2$$

where g (Θ) includes Lorentz, polarisation and obliquity factors. Both g and h are known. Quantifying Laue diffraction patterns therefore depends upon the determination of $f(\lambda)$, known as the wavelength normalisation curve or λ-curve.

The λ-curve may be derived by various strategies, two of which are mentioned here. For the Pea Lectin example the resulting λ curves were very similar.

1) By reference to the monochromatic dataset (program LAUESCALE). Using this method on Pea Lectin data has yielded an overall merging R factor on intensities Rmerge = 12.1%.

2) Derivation from the sample Laue data alone using measurements from symmetry related reflections recorded at different wavelengths (program LAUENORM). For Pea Lectin this resulted in Rmerge = 12.6%.

This work is reported in more detail in J W Campbell et al (1986) by some of the authors with Professor Keith Moffat (Cornell, USA) who has made major contributions to this topic.

6. KINETIC LAUE PHOTOGRAPHY IN THE STUDY OF ACTIVITY OF PHOSPHORYLASE b CRYSTALS

Experiments have been carried out to study catalysis and structural transitions in crystals of the enzyme Phosphorylase b using Laue photography. The crystals have high symmetry and a large proportion (up to 70%) of the complete data to a certain resolution can be recorded as a single exposure. The exposure time for a single photograph (of some 34,000 reflections) is 250 msec.

Figure 4 shows the course of the experiment where a sequence of photographs (17, 18, 19 etc) is taken from a single crystal of Phosphorylase b. The sequence starts with the native crystal and follows the soaking into the crystal of the substrates (hatched areas in the figure) and the subsequent attainment of chemical equilibrium. One interesting feature is the observation of an order-disorder-order transition which occurs rapidly on addition of phosphate, temporarily affecting the crystal lattice.

The data have been processed to 3A° resolution and the first electron density maps have been calculated. The difference maps obtained are fairly noisy, but do show the bound oligosaccharide at the glycogen storage site. The noisiness of the map may be attributed to a) genuine errors in the Laue data and b) the fact that only about 30% of a complete data set to 3A° resolution was available for the calculation (wavelength and spatially overlapping spots were omitted because profile fitting was not available). Clearly there is room for improvement and work is underway to increase the number of reflections which can be integrated, to minimise absorption effects and to optimise data processing strategies. Following reprocessing we hope to re-assess the quality of the data and to calculate better maps which will further demonstrate the exciting possibilities of kinetic experiments using Laue techniques. However, already we have shown that for a set of protein crystal intensity data collected in 0.25 second, recognisable structural features have been seen in a conventional difference Fourier map.

(Reference J Hajdu et al 1986)

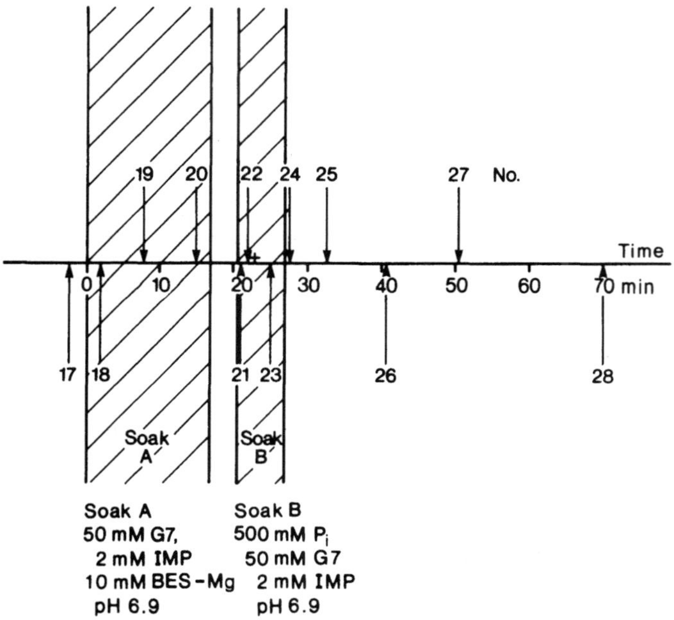

FIGURE 4. Kinetic Laue Experiment on Phosphorylase b (horizontal scale is calibrated in minutes) showing the soaking of substrates and sequence of photographs. (Exposure time = 250 msec).

REFERENCES

1. The Laue Method, J L Amorós, M J Buerger, M C de Amorós, Academic Press (1975).

2. I G Wood, P Thompson, J C Matthewman, Acta Cryst B39, 543 (1983).

3. K Moffat, D Szebenji, D Bilderback, Science 223, 1423, (1984).

4. J R Helliwell, Reports on Progress in Physics 47, 1403 (1984).

5. J Hails, M M Harding, J R Helliwell, Dl/SCI/P428E, Daresbury Laboratory preprint (1984).

6. J R Helliwell, J Mol Structure 130 63 (1985).

7. T Greenhough, J R Helliwell, Progress in Biophysics and Molecular Biology 41, 67 (1983).

8. J R Helliwell, M Z Papiz, I D Glover, J Habash, A W Thompson, P R Moore, N Harris, D Croft, E Pantos, Nuclear Instruments and Method A246, 617 (1986).

9. B Hedman, K O Hodgson, J R Helliwell, R Liddington, M Papiz, PNAS (USA) 82 7604 (1985).

10. K Moffat, D Bilderback, W Schildkamp, K Volz, Nuclear Instruments and Methods A246, 627 (1986).

11. D W J Cruickshank, J R Helliwell, K Moffat, (in preparation).

12. I J Clifton, D W J Cruickshank, G Diakun, M Elder, J Habash, J R Helliwell, R C Liddington, P A Machin, M Papiz, J Appl. Cryst. (1985) 18, 296.

13. Information Quarterly for Protein Crystallography, CCP4 Newsletter No 15, (April 1985). Daresbury Laboratory.

14. J R Helliwell et al (in preparation).

15. MOSFLM - a suite of programs for online evaluation of integrated intensities on oscillation photographs (1980).

16. S Zurek, M Papiz, P A Machin, J R Helliwell. Information Quarterly for Protein Crystallography, CCP4 Newsletter No. 16 (September 1985), Daresbury Laboratory.

17. J W Campbell, J R Helliwell, K Moffat. Information Quarterly for Protein Crystallography, CCP4 Newsletter No. 18 (July 1986), Daresbury Laboratory.

18. Janos Hajdu, Pella Machin, John W Campbell, Ian Clifton, Sebastian Zurek, Sheila Gover, Louise N Johnson. Information Quarterly for Protein Crystallography, CCP4 Newsletter No. 17 (March 1986), Daresbury Laboratory.

A New Area Detector (Imaging Plate) and Its Application to Diffraction Studies at the Photon Factory

Y. AMEMIYA, N. KAMIYA+, Y. SATOW, T. MATSUSHITA, J. CHIKAWA
K. WAKABAYASHI*, H.TANAKA** and J.MIYAHARA***

Photon Factory, National Laboratory for High Energy Physics
Oho-machi, Tsukuba-gun, Ibaraki 305, Japan
* Department of Biophysical Engineering, Faculty of Engineering Science, Osaka University, Toyonaka, Osaka 562, Japan
** Department of Physiology, School of Medicine, Teikyo University Itabashi-ku, Tokyo 173, Japan
*** Miyanodai Development Center, Fuji Photo Film Co.,Ltd. Kaisei-machi, Ashigarakami-gun, Kanagawa 258, Japan

ABSTRACT

A new integrating area detector, originally developed for diagnostic radiography, has been applied to X-ray diffraction experiments using synchrotron radiation. The system is based on a photostimulable phosphor ($BaFBr:Eu^{2+}$) screen which can temporarily store an X-ray image. The stored image is read out by measuring the intensity of luminescence($\lambda \sim 390$ nm) stimulated by a He-Ne laser beam scanning the screen surface. The area detector (250×200 mm^2) has a 100 % detective quantum efficiency for $8 \sim 17$ keV X-rays, a dynamic range of $1:10^5$, a spatial resolution better than 0.2 mm (FWHM) in two orthogonal directions and no counting rate limitation. A high-quality diffraction pattern from a contracting muscle was obtained within a 10-s exposure. Laue diffraction patterns of cytochrome c' and Weissenberg photographs of ω-amino acid:pyruvate aminotransferase were successfully recorded with an X-ray dose of approximately one thirtieth as much as that required by Kodak DEF-5 X-ray film. Small-angle scattering patterns from a polyethylene sheet during stretching were recorded in the time-resolved mode.

1.Introduction

In order to make the best use of high-flux synchrotron X-rays in various kinds of diffraction and scattering experiments, it is necessary to develop an area detector which has a high quantum efficiency, a wide dynamic range and no counting rate limitation. Such an area detector is especially desirable in experiments involving biological specimens where either the exposure time or the amount of the X-ray dose to specimens is the most critical limiting factor. A conceptually new area detector, called an imaging plate, has been developed which meets these requirements. In this system, a photostimulable phosphor screen is used to temporarily store a two-dimensional X-ray image as a distribution of color centers. The image is then read out by an image reader which measures the intensities of luminescence, which is released by the stimulation of a laser beam scanning the surface of the phosphor screen (Fig. 1). In the following the principle and performance of the imaging plate is described; some of the results obtained with it are then presented and a new image reader system is proposed.

+ Present address: The Institute of Physical and Chemical Research
Wako, Saitama 351-01, Japan

2. System description

(a) Imaging Plate (Photostimulable phosphor screen)

The photostimulable phosphor($BaFBr:Eu^{2+}$)[3,4] shows a phenomenon called "photostimulated luminescence"(PSL). When exposed to X-rays, the phosphor temporarily stores a fraction of the absorbed X-ray energy. It emits luminescence of an intensity proportional to the original X-ray intensity when later stimulated by visible light. The mechanism of the PSL can be explained by the energy level diagram shown in Fig. 2. By X-ray irradiation, a part of the Eu^{2+} ions are ionized to Eu^{3+} and electrons which are liberated to the conduction band are trapped at quasi-stable levels of F-centers (a kind of color centers). These levels originate from vacancies of Br^- ions in the phosphor crystal. By visible light stimulation, the trapped electrons at F-centers are again liberated to the conduction band and return to Eu^{3+} ions, converting them to excited Eu^{2+} ions. Then, luminescence is emitted as a consequence of electron transition from 5d to 4f in the Eu^{2+} ions. Figure 3 shows the spectra of the light which stimulates the phosphor and the spectra of the PSL. The response time of the PSL is 0.8 μs. The crystal structure of the phosphor is shown in Fig.4; some of the Ba^{2+} ions are replaced by doped Eu^{2+} ions.

The phosphor has the following characteristics which are suitable for obtaining high-quality images with a low X-ray dose : 1) the absorption coefficient for X-rays is high, 2) PSL radiation is efficiently released under such visible light stimulation as a He-Ne laser beam, 3) the PSL-radiation spectrum falls within the range 300∿500 nm, where the quantum efficiency of the photomultiplier is high, 4) the wavelength of the PSL radiation (λ∿390 nm) is sufficiently separated from that of stimulation light for separate detection and 5) there is no substantial fading of the stored image for several hours.

The imaging plate (IP), approx. 0.5 mm thick, is a flexible plastic plate coated with polycrystals (crystal size: 4∿5 μm in diameter) of the phosphor, typically 150 μm thick, combined with an organic binder. The surface of the phosphor is protected by a 10 um thick polyethylene terephthalate sheet. A typical IP size is about 250 mm × 200 mm.

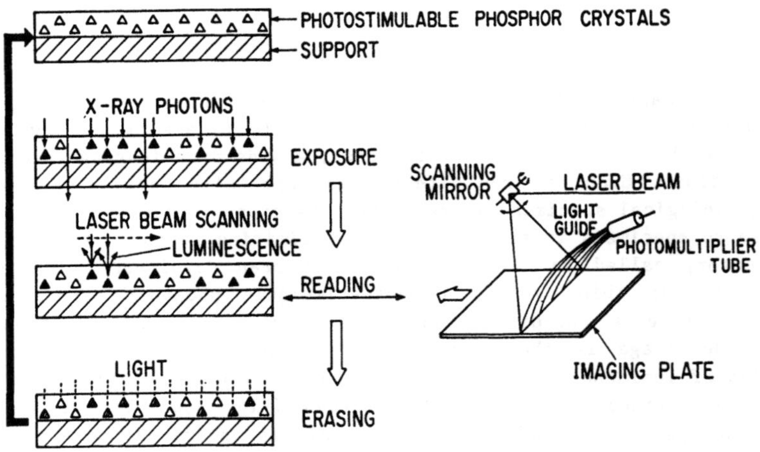

Fig.1 Principles of image recording, reading and erasing with an imaging plate[2)]

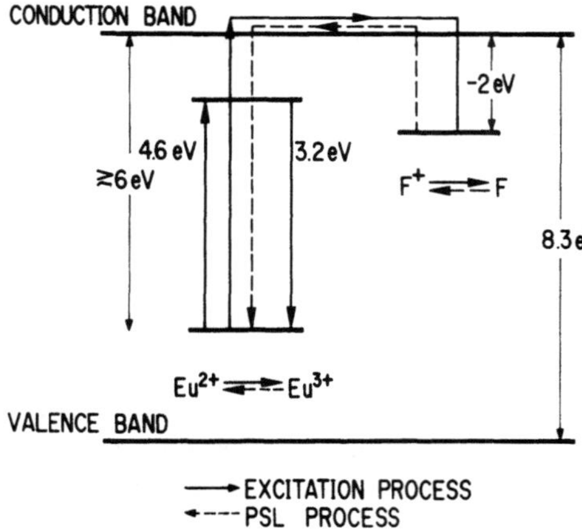

Fig. 2
The mechanism of photostimulated luminescence (PSL) in $BaFBr:Eu^{2+}$ (Ref. 3))

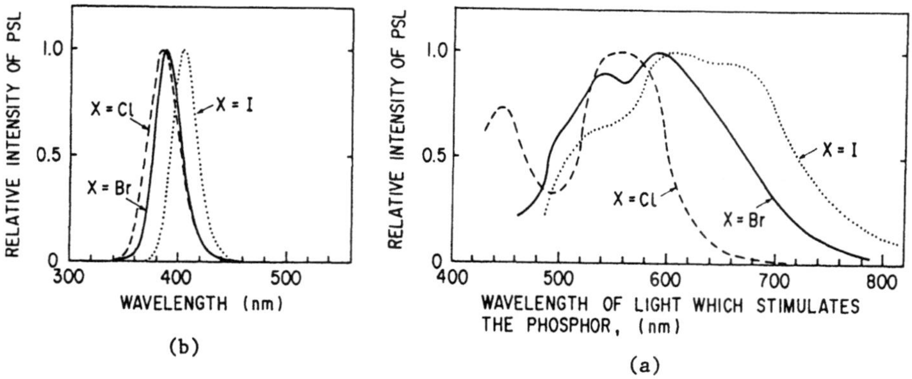

Fig. 3 (a) Stimulation spectra of $BaFX:Eu^{2+}$ (X= Cl, Br, I)
(b) Photostimulated luminescence spectra of $BaFX:Eu^{2+}$ (X= Cl, Br, I)
(Ref. 2))

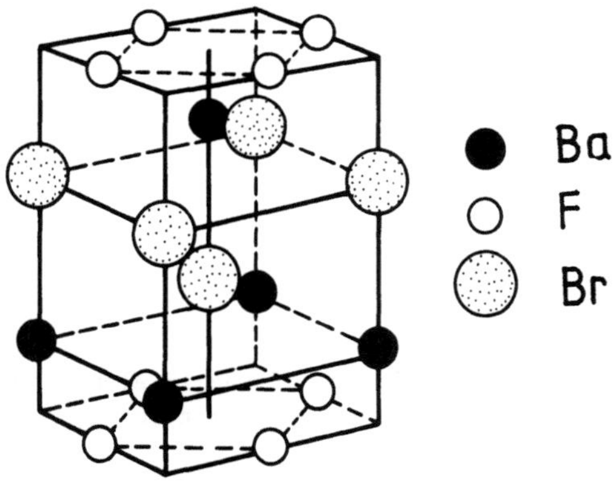

Fig. 4
Unit cell of BaFBr
(a=0.451 nm, c=0.744 nm)

(b) Image reader

The X-ray image stored on the IP is read by a scanning He-Ne laser beam (20 mW) which releases PSL radiation. Scanning of the laser beam is performed with a oscillating mirror, while the IP is traversed so as to form an orthogonal scan (illustrated in Fig. 1). The PSL radiation released by laser scanning is collected through a light guide into a photomultiplier tube (PMT), which converts the PSL radiation into electrical signals. The output signals from the PMT, which are analog signals as a function of time, are logarithmically amplified and converted with an A/D converter into a time-series of digital signals (8 bits/pixel) that form histogram slices of the incident X-ray intensity on the IP. The PMT sensitivity and the amplifier gain can be preset for optimum reading. The scanning speed is about 14 μs per pixel. The minimum pixel size is designed to be 100 μm × 100 μm in the present system. The digital signals are sent to an image processor (a computer) for digital manipulation and are finally stored in magnetic tapes. The X-ray image is displayed on a photographic film by an image writer. In the image writer, another He-Ne laser beam modulated by the image signals is used to scan the photographic film to imprint the image. The residual image on the IP can be erased by irradiation with visible light of a normal light box to allow repeated use (Fig.1). The IP can be used in almost the same way as an X-ray film, except that it can be treated in a light place before exposure to X-rays.

3. Detector Performance [5,6]

(a) Active area size and spatial resolution

The active area size of the IP and the corresponding pixel size are listed in Table 1. All of the following description is concerned with an IP of 251 mm × 200 mm. Figures 5 (a),(b) show intensity profiles on the IP when it is exposed to a 20 μm wide line-shaped monochromatic X-ray beam. X is along the laser scanning direction and Y is perpendicular to it. The full width of half maximum (FWHM) is less than 2 pixels (that is, 200 μm) in either direction. The observed intensity profiles are identical in several arbitrarily chosen regions within the active area. The spread in the intensity profiles is mainly due to the scattering of incident X-rays in a 150-μm thick phosphor layer.

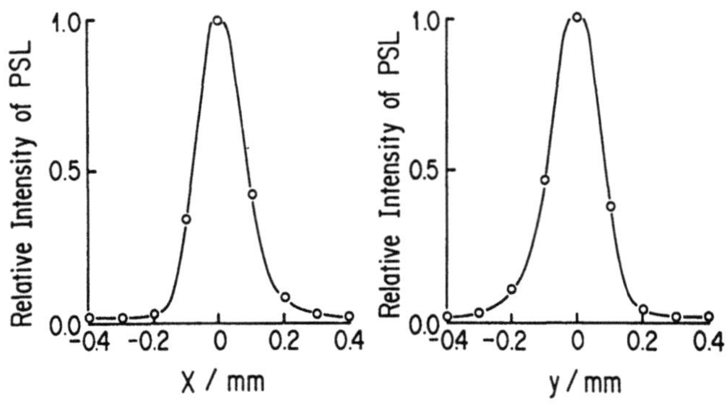

Fig. 5 Intensity profiles which show the spatial resolution of the IP along the laser-scanning direction, x, and along the IP-scanning direction, y.

Table 1. The area size of the imaging plate (IP)

active area size (mm × mm)	251 × 200	251 × 302	352 × 352	352 × 428
pixel size (mm × mm)	0.1 × 0.1	0.15 × 0.15	0.2 × 0.2	0.2 × 0.2
total pixel number	2510 × 2000	1670 × 2010	1760 × 1760	1760 × 2140

(b) Linearity and dynamic range

The linearity and dynamic range of the PSL was measured with X-rays which ranged over six orders of magnitude. The IPs were read under three different sensitivities (E) of the image reader to cover the full intensity range. Figure 6 shows the relationship between the relative intensity of the PSL (y) and the incident X-ray photons/pixel (x) for both CuKβ and MoKα X-rays. According to a least-squares fit of the data to a straight line of the form, $y= a(x+ b)$, the relative intensity of the PSL is linear to the incident X-ray photons per pixel in the range from 8 to 4×10^4 X-ray photons/pixel within a relative error of 0.05. The non-linearity between y and x is not negligible at higher exposure levels. A parabolic curve of the form $y = a\{(x+b) + c(x+b)^2\}$ fits with the data over five orders of magnitude within a relative error of 0.05. The c value, i.e. the non-linear coefficient is of the order of $1{\sim}3 \times 10^{-6}$.

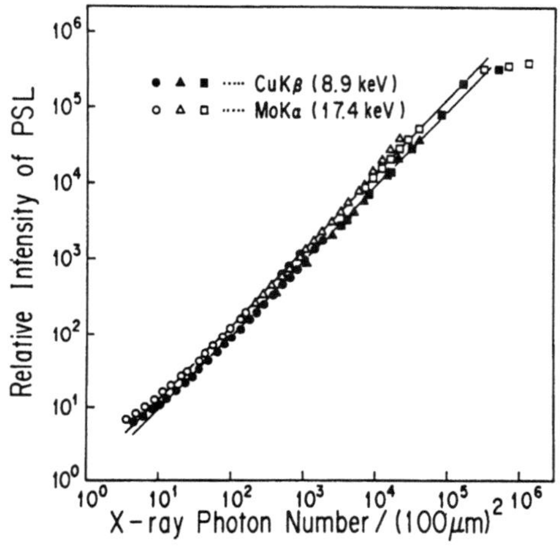

Fig.6
The response of photostimulated luminescence to X-ray exposure level. ○, △, □ for MoKα, ●, ▲, ■ for CuKβ. IPs were read under three different sensitivities (E) (○ ●: E=4000, △▲ :E=400, □■ :E=40) and 3.6 preset dynamic range.

(c) Efficiency

Figure 7 shows the absorption efficiency of the BaFBr as a function of the X-ray photon energy. The absorption efficiency is 100 % for up to about 17 keV X-rays when the phosphor is 150 μm thick. The absorption edge at 37.4 keV is due to barium. The quantum efficiency of integrating detectors, however, can not be determined by the absorption efficiency alone, because the noise level of the system usually deteriorates the quantum efficiency of integrating detectors. In Fig.8, the relative uncertainty for an intensity measurement is plotted as a function of the exposure level for a comparison between the quantum efficiency of detectors. The solid straight line indicates the relative uncertainty in the incident X-ray intensity; it follows Poisson statistics, where the quantum fluctuation is $\sqrt{n}$ for n photons. That is, if there exists an ideal detector, the results should fall on the solid straight line. From this graph, one can easily compare the incident X-ray photons which are necessary to obtain a certain desired accuracy in an intensity measurement; thus, the quantum efficiency can be compared. For example, the high-sensitivity X-ray film Kodak DEF-5 requires about 15 times and 25 times more exposure for CuKβ and MoKα, respectively, than would be needed by an ideal detector to obtain a 10 % relative accuracy. The IP is nearly an ideal detector in the middle exposure range where the exposure level is between 10^1 and 10^3 photons/pixel. The deviation from an ideal detector at lower exposure levels is due to such system background noise superimposed onto the output signal as PMT dark current and stray light from the laser beam which enters the PMT. The background noise of the IP amounts to less than 3 photons/pixel for both CuKβ and MoKα. This value should be compared with an intrinsic chemical fog level equivalent to about 1000 photons/pixel for the X-ray film. The IP system also deviates from an ideal detector at higher exposure levels and the accuracy becomes saturated at around 2 %. This deviation is partially due to errors of the apparatus used in the evaluation experiments and partially due to a non-linearity in the response of the PSL.

The efficiency of detectors can be discussed using a more general term called the detective quantum efficiency (DQE)[7]:

$$DQE = \left(\frac{S_o}{N_o}\right)^2 / \left(\frac{S_i}{N_i}\right)^2,$$

where S = signal and N = noise (standard deviation of signal) and subscripts o and i refer to output and input, respectively. The DQE of the IP system is obtained by comparing the relative uncertainty of the IP system with that of an ideal detector at the same exposure level. The IP system has almost 100 % DQE in the exposure range between 10^1 and 10^3 photons/pixel.

(d) Uniformity of response

The uniformity of the response was measured by exposing a uniform X-ray beam (3.2×10^3 photons/pixel) of 10×10 mm^2 to 20 arbitrarily chosen areas. The integrated intensities of the PSL over the uniformly exposed areas were compared and the macroscopic non-uniformity defined by the relative uncertainty among the integrated intensities was 1.6 %. This macroscopic non-uniformity may be due to non-uniform stimulation by the scanning laser beam deflected with a oscillating mirror in the image reader. The microscopic non-uniformity defined by the relative uncertainty among individual pixel intensities within the uniformly exposed area was 2.2 %. If the incident photon fluctuation is taken into account, the intrinsic microscopic non-uniformity is estimated to be 1.3 %.

(e) Image distortion

The IP was exposed to line-shaped monochromatic X-ray beams with exact separations. The ratio of resulting separations to the true values was examined for several regions over the whole active area to evaluate the image distortion. It was 1.010 ± 0.001 and 0.996 ± 0.001 along the laser-scanning direction and the orthogonal direction, respectively. The image distortion is attributed to an error in the scanning mechanism of the image reader, and can be calibrated for precise spatial measurements.

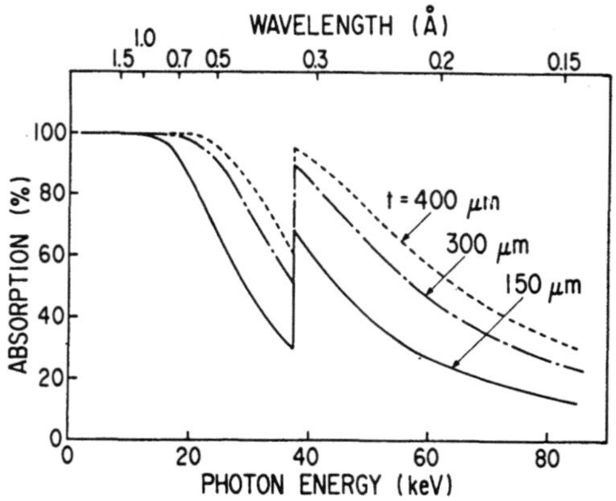

Fig. 7
X-ray absorption efficiency of BaFBr phosphor calculated for different thickness of 400, 300 and 150 μm as a function of X-ray photon energy. The absorption edge at 37.4 keV is due to barium.

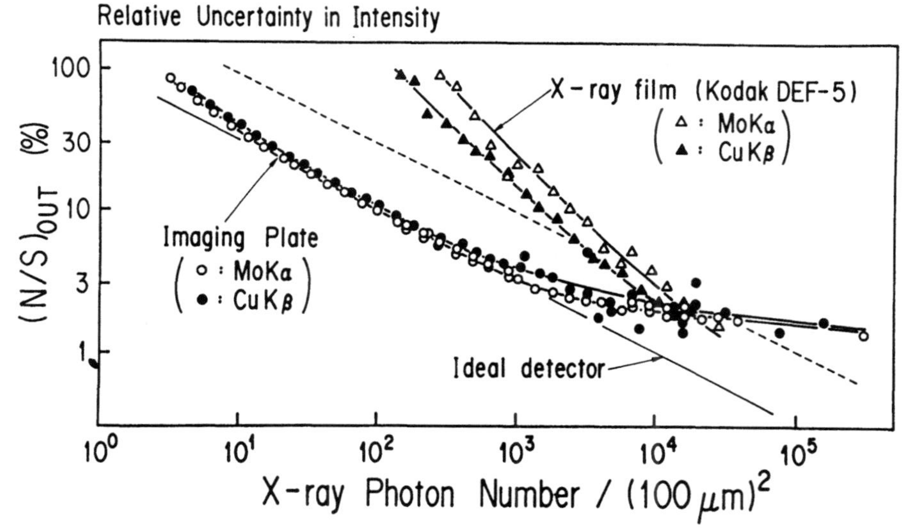

Fig. 8 The relative uncertainty in intensity of detectors as a function of the exposure level. A solid straight line indicates an ideal detector. A dashed line indicates a pulse-counting detector of 10 % efficiency. ○ and ● indicate the IP system for MoKα and CuKβ, respectively. △ and ▲ indicate Kodak DEF-5 X-ray film for MoKα and CuKβ, respectively. X-ray photons required to obtain certain accuracy in intensity measurements are compared.

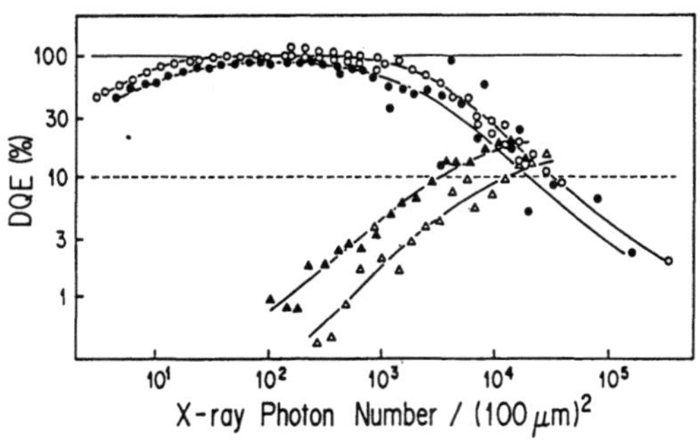

Fig. 9
Detective quantum efficiency (DQE) as a function of the X-ray exposure level. ○,● indicate the IP system for MoKα and CuKβ, respectively. △,▲ indicate Kodak DEF-5 X-ray film for MoKα and CuKβ, respectively.

So far as we experienced, there is no appreciable deterioration in the performance of the IP due to radiation damage or any physical fatigue of the PSL phenomenon. Therefore, the IP can be used many times unless it becomes mechanically damaged. The half time of the stored X-ray image on the IP is about ten hours. There appears small random dots in the image due to a small number of radioactive barium ions included in the phosphor. These dots are negligible if it is read within a couple of hours after the IP is kept in the dark or if the image is read out under the relatively low sensitivity of the image reader.

4.Application

(a) Small angle diffraction from contracting muscle [8,9]

In order to investigate the molecular structure during muscle contraction, it has been attempted to record a two-dimensional high-resolution diffraction pattern from a contracting muscle. Nevertheless, such a diffraction pattern has not been successfully recorded from a single specimen, because it takes about a 10-min exposure time, even with synchrotron radiation. This is far beyond the lifetime of the contracting state of a muscle. Using the IP, a diffraction pattern of up to 2.0 nm axial spacing was obtained from a contracting frog skeletal muscle in as little as 10 s exposure time with a doubly focusing camera.[10] The muscle was isometrically tetanized at the sarcomere length of 2.4 µm for 1.3 s and this was repeated ten times every 15 s. A diffraction pattern(Fig. 10(b)) was repeatedly recorded ten times with 1.0 s exposure time during a steady tension. The tension of the 10th contraction was 83 % of the initial tension. A pattern during a resting period(Fig. 10(a)) was recorded from the same muscle before the contraction. The incident flux of X-rays (λ=0.155 nm) to the specimen was about 8 $\times 10^{10}$ photons/s when the PF storage ring was operated at an energy of 2.5 GeV with 145 mA. The diffraction intensities of all the observed actin layer lines increased without noticeable changes in their peak positions along the layer lines and axial spacings. For instance, the integrated intensities increased by 20 $\sim$ 50 % and $\sim$ 100 % in the prominent 5.9 and 5.1 nm reflections, respectively; these correspond to the pitches of the two genetic helices of the actin monomers in the actin filament. Also found were the intensity increases of the first (-1/37 nm^{-1}) and the second (-1/18 nm^{-1}) actin layer lines which has been attributed to the structural change of the tropomyosin in the actin filaments. These intensifications of the actin layer lines are mainly due to an interaction of actin with the myosin heads, since the amount of the increases was much less in the muscle stretched to the non-overlap length of the sarcomere.[11,12] However, the structure of the actin filaments which interact with the myosin heads during a contracting state differs from that in a rigor state, since the shapes and the peak positions along the actin layer lines were quite different between the two states. As for the changes of the myosin layer lines during contraction, off-meridional parts and the so-called forbidden meridional reflections decreased greatly in intensity or disappeared. However, the 42.9 nm first off-meridional reflection remained with about 20 % of the resting intensity, whereas it completely disappeared in a rigor state. Most meridional reflections indexed to 3n orders of 42.9 nm repeat increased in intensity, indicating the enhancement of the 3-fold screw symmetry. These results suggest that during contraction the myosin heads retain the axial periodicity of the myosin filament and become aligned more perpendicularly along the actin filament, being in agreement with the observation by the rapid freezing.[13]

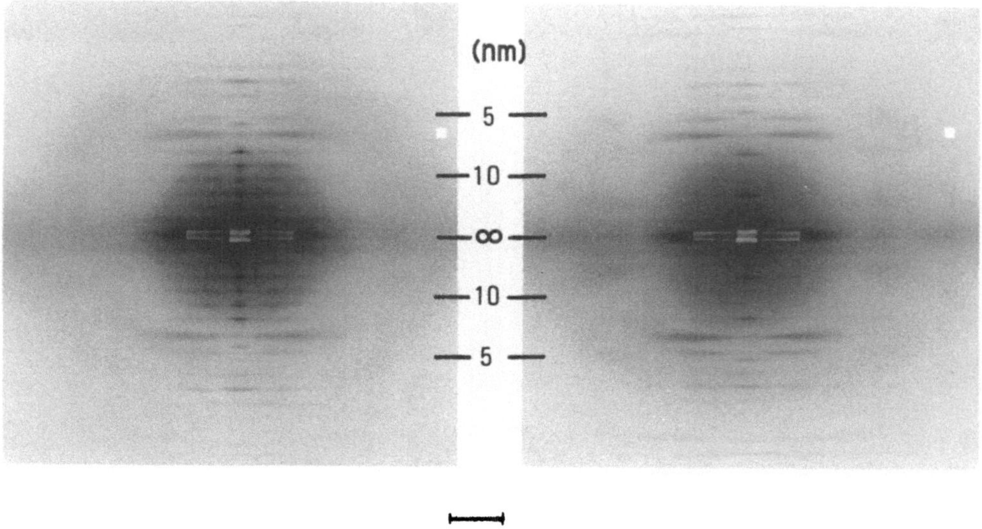

(a) 20 mm (b)

Fig. 10 Small-angle diffraction patterns from a frog sartorius muscle in the resting state (a) and in the tetanically contracting state (b)
Exposure time : 10 s X-ray wavelength : 0.155 nm
Camera length : 146 cm

(b) Protein Laue diffraction of cytochrome c'

Utilizing a white synchrotron radiation beam, preliminary Laue diffraction patterns from cytochrome c' were recorded (Fig.11). The crystal ($0.05 \times 0.07 \times 0.2\ mm^3$) was exposed to the white beam ($0.12 \sim 0.2$ nm) which had been transmitted through an Al filter and reflected by a plane mirror of fused quartz. The patterns were obtained with an X-ray dose of approximately one thirtieth as much as required by a high-sensitivity X-ray film. More than 50 exposures could be obtained from one crystal specimen without any noticeable radiation damage. The other advantage of the IP is that i) the response to X-rays of different wavelengths is relatively uniform, because the absorption efficiency of the IP is almost 100 % for X-rays of wavelength longer than 0.07 nm and that ii) due to the wide dynamic range of the IP, the use of multiple film packs is not required to cover the full range of intensities; thus, a correction for intensity analysis becomes much simpler.

(c) Weissenberg photograph of ω-amino acid:pyruvate aminotransferase

Figure 12 shows a Weissenberg photograph of ω-amino acid : pyruvate amino transferase (space group:I222, cell dimensions:a=12.46, b=13.79, c=6.145 nm).[14] The photograph was obtained with a Weissenberg camera designed for macromolecular crystallography.[15] A full data set was collected from eight such patterns with the computer program WEIS[16] and the R_{sym} value for the intensities was 5.0 %. The exposure time for each pattern was 4.7 min. By using the IP system, the phase determination of this protein could be performed by the anomalous dispersion method with the multi-wavelength data sets collected from only one crystal.

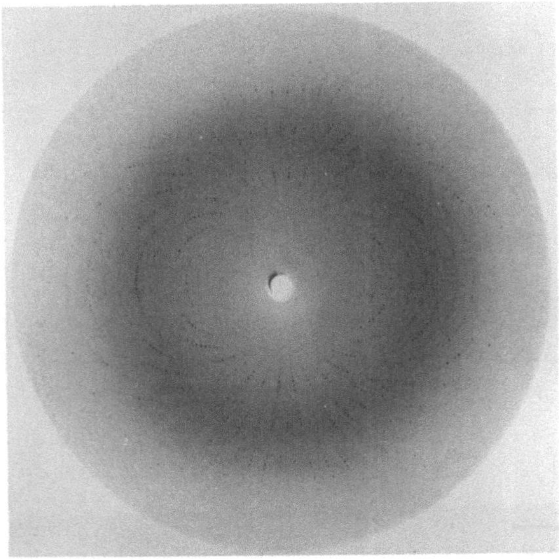

20 mm

Fig. 11
Laue diffraction from cytochrome c' using quasi-white X-ray radiation (λ:0.12 $\sim$ 0.2 nm)

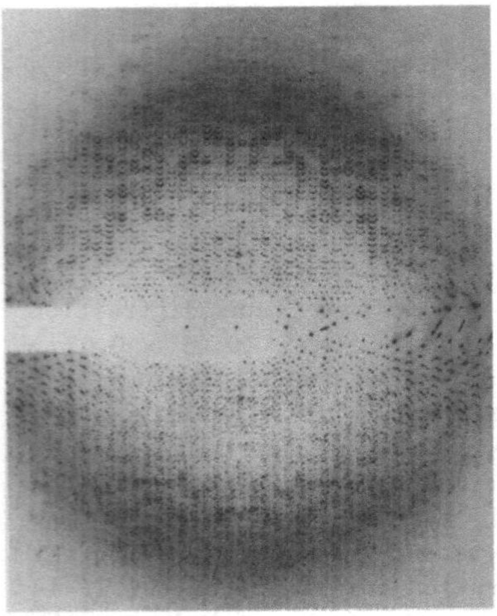

20 mm

Fig. 12
A Weissenberg photograph of ω-amino acid: pyruvate aminotransferase (mercury derivative). A single bent triangular Si (111) monochromator and a Weissenberg camera for macromolecular crystallography were used.
the oscillation angle : 30°
the radius of the camera : 287 mm
exposure time : 4.7 min

(d) Time-resolved SAXS from polyethylene sheet during stretching

By rapidly exchanging many IPs successively, time-resolved diffraction or scattering experiments can be carried out, although the achievable time-resolution is poorer than electrical position sensitive detectors. Such a method is especially useful in the case where diffraction patterns are asymmetric and have wide intensity ranges. Figure 13 shows time-resolved small angle X-ray scattering (SAXS) patterns from a polyethylene sheet under stretching. The polyethylene sheet ($\rho \sim 0.918$ g/cm^3, Melt flow rate(MFR) $\sim$ 0.9 g/min) of 2 mm thickness was placed in the doubly focusing camera[10] and stretched by a speed of 12 mm/min (32 % stretch/min) in the horizontal direction. The exposure time for each IP was controlled by a mechanical shutter placed in front of the specimen and eight IPs were held on a turntable. Each SAXS pattern was recorded with 1.0 s exposure time at 70 s intervals The interval time was determined by the speed of the stretching device. In order to achieve a sub-second time-resolution using IPs, we are now designing a rapid plate-changer and a rapid mechanical shutter.

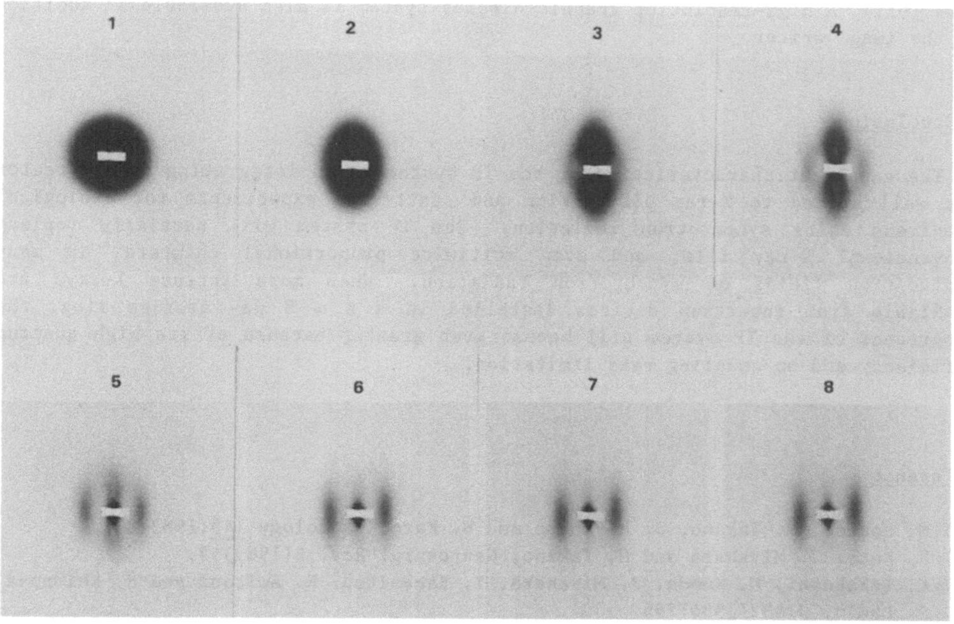

Fig. 13 Small-angle scattering patterns of polyethylene recorded in succession during stretching in an horizontal direction. Exposure time : 1 s.

5. Development of a new image reader

As the original IP system was designed for diagnostic radiography, there is still some room for improvement for the purpose of diffraction studies. A new image reader is now being built at the Photon Factory by modifying a conventional drum-type film densitometer. In the new image reader, the following points are to be improved: (i) The pixel sizes of 25×25 μm^2 and 50×50 μm^2 in addition to 100×100 μm^2 are available by making both a focus size and a scanning pitch of the laser beam changeable. The thickness of the phosphor layer of the IP should be decreased so as to match the spatial resolution expected with smaller pixel sizes. (ii) A 12-bit A/D converter is used in place of a 8-bit A/D converter to improve the precision in X-ray intensity measurement. The uncertainty in the intensity measurement due to the digitization process is reduced from 3.3 to 0.2 %, assuming that the full range of the A/D converter is adjusted to 3.6 orders of magnitude (the dynamic range of a photomultiplier tube). (iii) In order to fully utilize the dynamic range of the IP, the entire intensity range of the PSL radiation is detected simultaneously by two photomultiplier tubes (PMTs) at different sensitivities. The second PMT, receiving roughly 1 % of the total amount of the PSL, is used to detect the PSL at a higher intensity range where the first one is saturated. (iv) Arbitrary sizes of the IP can be read. (v) In order to decrease the image distortion, a drum type film densitometer is utilized which provides more precise scanning pitches than the flat one. (vi) The non-uniformity of response is improved. This is because in the drum-type image reader the incident angle of the laser beam against the IP surface is always constant; thus, the laser beam can stimulate the phosphor more uniformly everywhere on the surface of the

IP. (vii) A high-resolution graphic display system is also available in addition to the image writer.

6.Conclusion

The excellent characteristics of the IP system as an integrating area detector are well suited to X-ray diffraction and scattering experiments for biological specimens using synchrotron radiation. The IP system will partially replace conventional X-ray film, and even multiwire proportional chambers, in many application fields of synchrotron radiation. When more intense X-rays are available from insertion devices installed in a 6 ∿ 8 GeV storage ring, the importance of the IP system will become even greater because of its high quantum efficiency and no counting rate limitation.

Reference

1) M. Sonoda, M. Takano, J. Miyahara and H. Kato, Radiology 148(1983)833.
2) H. Kato, J. Miyahara and M. Takano, Neurosurg. Rev. 8(1985)53.
3) K. Takahashi, K. Kohda, J. Miyahara, Y. Kanemitsu, K. Amitani and S. Shionoya, J. Lumin. 31&32(1984)266.
4) K. Takahashi, J. Miyahara and Y. Shibahara, J. Electrochem. Soc. 132(1985) 1492
5) J. Miyahara, K.Takahashi, Y.Amemiya, N.Kamiya and Y. Satow, Nucl. Instrum. Methods A246(1986)572.
6) Y. Amemiya, N. Kamiya, Y. Satow, J. Miyahara, K. Takahashi and N. Mori, Photon Factory Activity Report (1984/85) 265.
7) J. C. Dainty and R. Shaw, Image Science(Academic Press, New York, 1974).
8) Y. Amemiya, K. Wakabayashi, H. Tanaka, A. Fujishima, T. Kobayashi, N. Mori and J. Miyahara, Photon Factory Activity Report (1984/85) 235.
9) K. Wakabayashi, Y. Amemiya and H. Tanaka, Proc. 11th Taniguchi Int. Symp. Biophys. (ed. T. Yanagida) (Taniguchi Foundation, 1986) in press.
10) Y. Amemiya, K. Wakabayashi, T. Hamanaka, T. Wakabayashi, T. Matsushita and H. Hashizume, Nucl. Instrum. Methods 208(1983)471.
11) K. Wakabayashi, H. Tanaka, Y. Amemiya, A. Fujishima, T. Kobayashi, T. Hamanaka, H. Sugi and T. Mitsui, Biophys. J. 47(1985)847.
12) M. Kress, H. E. Huxley, A. R. Faruqi and J. Hendrix, J. Mol. Biol. 188(1986) 325
13) S. Tsukita and M. Yano, Nature 317(1985)182.
14) Y. Morita, S. Aibara, K. Yonaha, S. Toyama and K. Soda, J. Mol. Biol. 130(1979)211.
15) N. Sakabe, J. Appl. Cryst. 16(1983)542.
16) T. Higashi, N. Kamiya, M. Matsushima, K. Sakabe and N. Sakabe, Photon Factory Activity Report (1983/84) p.VI-16.

High Voltage Electron Diffraction and Computer Graphics of Heavy Meromyosin Crystals

M. TANIGUCHI

Department of Physics, Faculty of Science, Nagoya University, Chikusa-Ku Nagoya 464, Japan

Y. KAMIYA

Department of General Education, Faculty of Engineering, Toyota Technological Institute, Tenpaku-Ku, Nagoya 468 Japan

INTRODUCTION

It is well known that myosin is one of main proteins which has a central role in muscle contraction and cell motility. Myosin molecule is composed of heavy and light meromyosin (HMM and LMM). Heavy meromyosin has two heads and each head contains an ATP hydrolyzing site and an actin binding site (1, 2). Heavy meromyosin (HMM) can be divided by proteolytic reagents into subfragment-1 and subfragment-2 (S1 and S2). It is important to know on the mechanism of muscle contraction whether any conformational change of myosin heads and/or actin filaments occur or not. Therefore, three dimensional reconstructions from electron micrographs of the reconstituted actin containing filaments with myosin, heavy meromyosin or S1 have been investigated (3-7). However, it is difficult to separate actin and myosin with a high spacial resolution in a complexed pattern. Thus, it is necessary to assign intra- and intermolecular structures of myosin heads in detail. Many approaches were undertaken to obtain structural information of the myosin heads using electron microscopy (8-11), X-ray scattering (12) and Hydrodynamic techniques (13). The size and shape of myosin heads are, however, slightly different with different analysis.

The crystals of myosin subfragment-1 (14) have been obtained, and the three dimensional structural analysis of the crystals has been studied by electron microscopy (15).

We have crystallized heavy meromyosin prepared from rabbit skeletal muscle. We have undertaken high voltage electron microscopy and computer reconstruction of these crystals. We analyzed intra molecular structure and intermolecular molecular arrangement of heavy meromyosin crystals using two dimentional fast Fourier transformation and computer graphics.

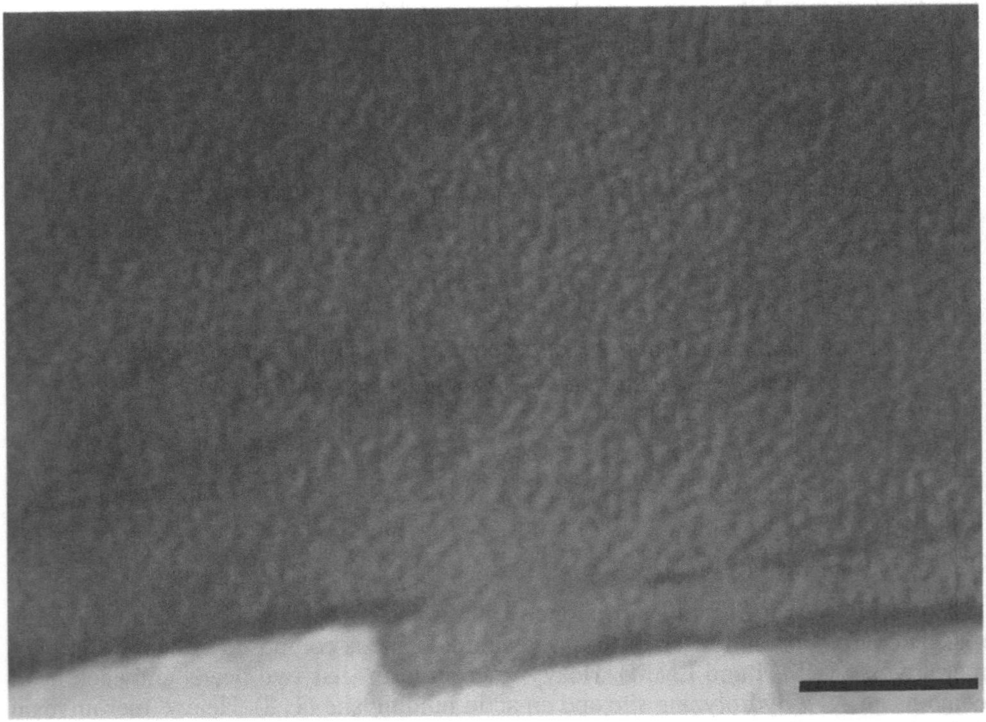

Fig. 1 - A high voltage electron micrograph of IAEDNS labelled heavy meromyosin (HMM) crystals. Bar=1,000 Å.

MATERIALS AND METHODS

a) Preparation and crystallization of HMM

Myosin was prepared from rabbit skeletal muscle, as described previously (16). Heavy meromyosin (HMM) was prepared by the method of lowey and Cohen (17). The HMM powder was dissolved in 0.2 mM $CaCl_2$, 2mM Tris-HCl (pH 7.0), 0.1 mM ATP and 2 mM NaN_3. The concentrations of heavy meromyosin were 50-100 mg/ml. Crystals of heavy meromyosin have been grown by fluorescent dye and acetone at room temperature. The crystals grow as long needle like rods.

b) High Voltage Electron Microscopy

The specimens of heavy meromyosin crystals stained negotively with 1% uranyl acetate were mounted on carbon coated microgrid. High voltage electron microscopes (HU-1000 or HU-500) were operated at accelerating voltage of 400 kV. In many cases a cooling specimen stage at 120 K was equipped in the electron microscopes to protect radiation damage (18).

The electron photographs were recorded at magnification of 30,000 or 60,000 and then the microscope was switched from selected imaging area to diffraction mode. Exposures at low dose were made over periods ranging from 1 to 15 s for diffraction pattern and 10 s to 2 min for image. Photographic recordings were made on Fugi X-ray films. The calibration of magnification was done by measuring the diffraction pattern in Aurum.

c) Image Proceeding

After selecting the electron micrographs, each chosen image was scanned digitally using a computer driven video densitometer (VDM-MT, nac.K.K.) and each digitized image was displayed on a 480 x 640 pixel array, and area were selected for its Fourier transformation. The computation was performed using Fujitu M356 Computer at the Computer Center of Nagoya University.

RESULTS

HMM crystals were obtained by adding acetone as a precipitant to HMM solutions at room temperature. The powder of HMM (mg/ml) was dissolved in solutions containing 0.2 mM $CaCl_2$, 2 mM Tris-HCl (pH 7.0), 0.1 mM ATP and 2mM NaN_3, with or without 300 μM IAEDANS (N-iodoacetylamino-1-naphthylamine-5-sulfonic acid) which is bound to the HMM enabling the process of crystallization to be easily followed under a fluorescence microscope. The crystals grow as long birefringent needle-like rods. Details of the crystallization will be described elsewhere.

Fig.1 shows an electron micrograph of a HMM crystal. The micrograph was at first transformed to digitized images with a 480 x 640 pixel array. Fig.2 (a) shows the digitized image of HMM crystals after the subtraction of background noise. Fig.2 (b) displays the diffraction pattern obtained by two dimensional Fast Fourier Transformation (FFT) from a square area with 256 x 256 pixels. The reconstituted images of HMM crystals were made from these diffraction
patterns using inverse FFT, as shown in Fig.3 (a) and (b). These are made of a rectangular lattice with a typical rectangular lattice parameters of 629 ± 3 Å x 629 ±3 Å and $\gamma = 90°$, the angle between the axis of the lattice and the long axis of the rod is about 60° to 73° (Fig.3). The reconstituted images revealed that HMM contains arraies of antiparallel substructure. The computer diffraction pattern from electron micrographs of HMM crystals was confirmed with electron diffraction patterns taken directly.

DISCUSSION

HMM crystals have been examined by high voltage electron microscopy to dermine how the molecules arrange in the crystals and to obtain the size and shape of molecule and the computer reconstructions of lattice images from crystalline HMM.
In this experiment we have obtained a length for the part of S1 of about 170 Å to 190

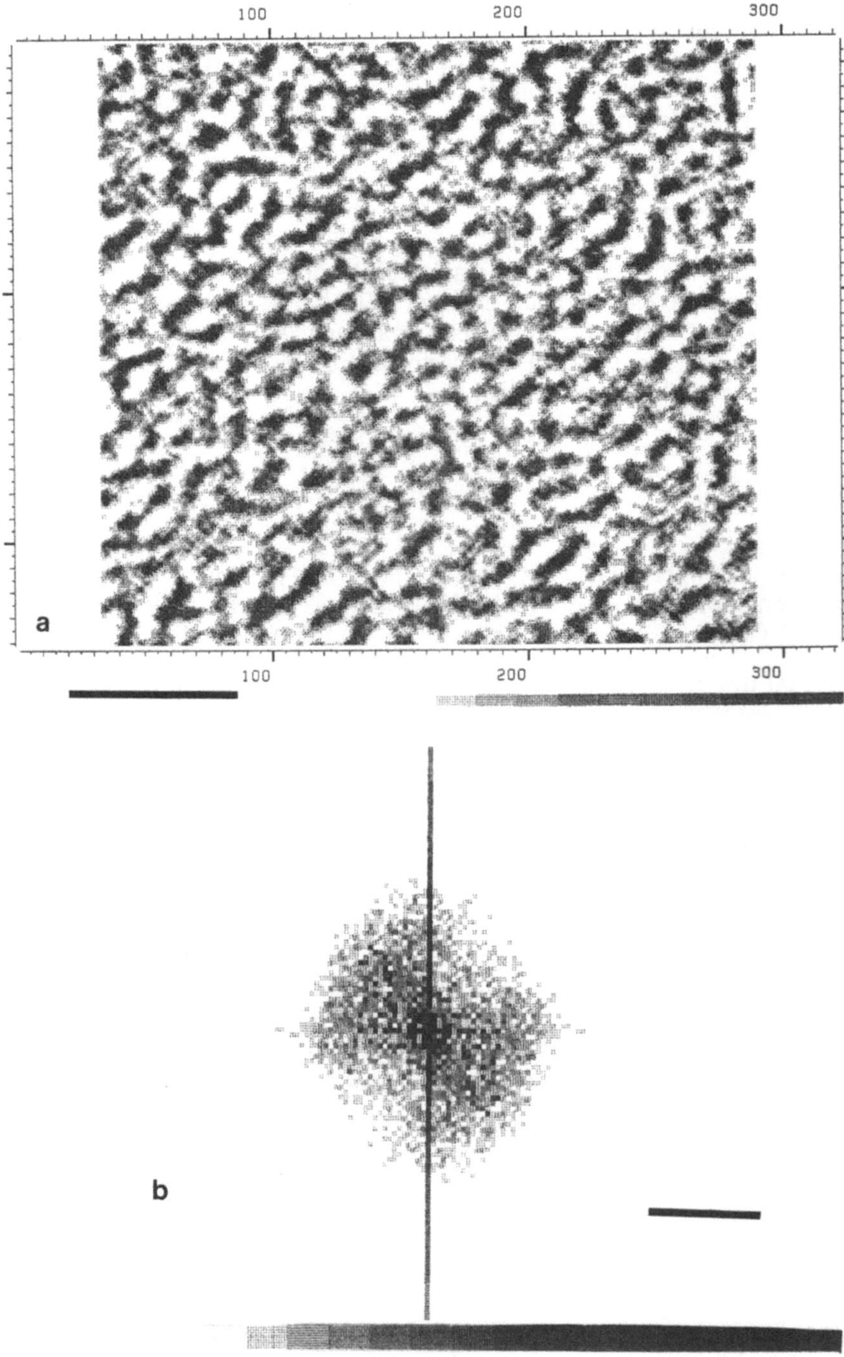

Fig. 2. (a) A digitized image after the treatment of background noise from Fig. 1. Bar = 500 Å.
(b) A computer diffraction pattern obtained from the image in Fig. 2 , Bar = 20 $Å^{-1}$.

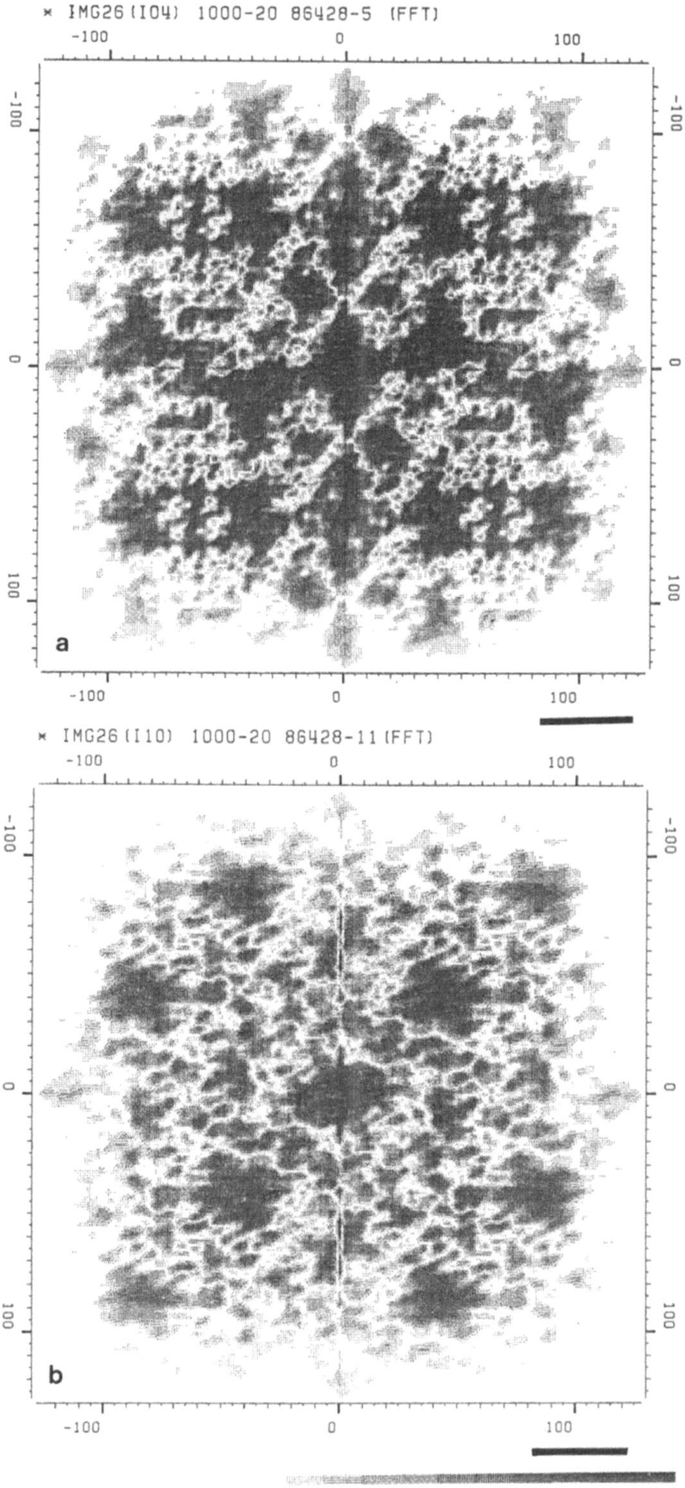

Fig. 3 a -3 b Reconstituted images of heavy meromyosin crystals obtained by inverse Fast Fourier Transformation (FFT) Bar = 200 Å.

Å and a width of about 90 Å to 120 Å . The shape and size of heads in HMM molecule is nearly same order as reported by Elliott and Offer (8).

The width of S1 part in our experiments is slightly thicker than other models (8-10). The value reported here is close to the one obtained by Takahashi (11). He estimated a width of 95 Å from the analysis of negatively stained electron micrographs of myosin molecule.

It is concluded that the computer graphics from electron micrographs is a powerful tool to draw the molecular shape and arrangement of HMM crystals.

ACKNOWLEDGEMENTS

We thank Mr. T. Goto of Nagoya University for his technical assistance.

REFERENCES

(1) S.S. Margossian and S. Lowey (1973) *J. Mol. Biol.,* **73,** 313
(2) A.G. Weeds and R.S. Taylor (1975) *Nature,* **257,** 54
(3) P.B. Moore, H. E. Huxley and D.J. De Roosier (1970) *J. Mol. Biol.,* **50,** 279
(4) K.A. Taylor and L.A. Amos (1981) *J. Mol. Biol.,* **147,** 297
(5) E. Katayama and T. Wakabayashi (1981) *J. Biochem.,* **90,** 683
(6) T. Wakabayashi and C. Toyoshima (1981) *J. Biochem.,* **90,** 683
(7) P. Vibert and R. Craig (1982) *J. Mol. Biol.,* **157,** 299
(8) A. Elliott and G. Offer (1976) *Proc. R. Soc. Lond. B.,* **193,** 45
(9) A. Elliott and G. Offer (1978) *J. Mol. Biol.,* **123,** 505
(10) G. Offer and A. Elliott (1978) *Nature,* **271,** 325
(11) K. Takahashi (1978) *J. Biochem.,* **83,** 905
(12) R.A. Mendelson and K. M. Kretzschamr (1980) *Biochemistry,* **19,** 4103.
(13) J.T. Yang and C.C. Wu (1977) *Biochemistry,* **16,** 5785.
(14) I. Rayment and D.A. Winkelmann (1984) *Proc. Natl. Acad. Sci.,* **81,** 4378.
(15) D.A. Winkelmann, H. Mekeel and I. Rayment (1985) *J. Mol. Biol.,* **181,** 487.
(16) M. Taniguchi (1976) *Biochim. Biophys. Acta,* **427,** 126.
(17) S. Lowey and C. Cohen (1962) *J. Mol. Biol.,* **4,** 293.
(18) M.Taniguchi, and Y. Kamiya (1983) *Nucl. Meth. Instr.,* **208,** 541.

X-RAY SPECTROSCOPY

Higher Order Correlations in Proteins by X-Ray Spectroscopy

A. Bianconi

Dipartimento di Fisica, Università degli Studi di Roma "La Sapienza", 00185 Roma, Italy

INTRODUCTION

EXAFS (Extended X-ray absorption Fine Structure) has been used to measure the interatomic distances of active sites of proteins because it probes the pair distribution function, i.e. the first order correlation function of local atomic distribution, since the beginning of x-ray spectroscopy of biological molecules(1-4) and it is currently used to investigate metalloproteins(5-9). It was pointed out that the x-ray absorption spectra of biological molecules can give information on site geometry by switching the focus of the data analysis from the EXAFS region to the XANES (X-ray Absorption Near Edge Structure) region(10-13).

Theoretical analysis of XANES by multiple scattering approach has confirmed that XANES is a probe of structural information beyond the pair distribution function(14-16) and therefore the bond angles in proteins in solution can be measured.

Hemoglobin (Hb) and myoglobin (Mb)(17) and their interaction with diatomic molecules have been object of many fundamental studies on the relation between the microscopic atomic structure and biological function. The capability of x-ray absorption spectroscopy to probe the immediate environment of Fe in solution has stimulated the research to overcome the limitation to protein crystals of x-ray diffracion methods. Moreover the errors in the measure of interatomic distances by EXAFS are smaller than in protein crystallography.

The measure of the iron displacement out of the pophyrin plane in deoxy-Hb has been put to a test by EXAFS(3,18) but a controversy rose because of EXAFS intrinsic limitations to probe geometry effects beyond the pair distribution function(19,20). On the contrary XANES of deoxy-Hb has been shown to be sensitive to the iron displacement(21) and different values of the displacement in deoxy-Mb (21) and in foetal deoxy-Hb (22) have been found. However no variation of the iron displacement within the sensitivity of the measure was found going from low to high affinity forms of carp deoxy-Hb(23) as well as the average interatomic distance which remains constant within 0.01 Å in agreement with early EXAFS studies on high and low affinity hemoglobins (3,18).

While the protein conformation induces subtle structural changes at the active site the ligand binding determines a largely different atomic arrangement of the iron site.

The contraction of the average interatomic distance between iron and its first neighbours upon oxygen binding was found by EXAFS to be 0.07 Å [3,18] and it has been confirmed by XANES data analysis [24] which has pointed out the fact the iron effective charge is close to that of Fe^{3+} sytems both in the oxided and oxygenated forms[24-26].

The bonding geometry of ligand in myoglobin has been studied by EXAFS[20] and by XANES [16,27,28] and the Fe-C-O bond angle has been found to be 150°. Time dependent XANES measurement of CO recombination in photolized MbCO has been reported[29]. The time dependent dispersive XANES study of the irreversible process of denaturation of HbCO by reducing pH is reported in this book.[30]

Here a review of the study of the Fe-C-O bond angle in hemoglobin by XANES is presented. The XANES of carbonmonoxyhemoglobin (HbCO) in solution is compared with the spectra of synthetic porphyrins and with the angular resolved XANES of a single crystal of carbonmonoxymyoglobin (MbCO). The MbCO crystal shows a large dicroism which is interpreted by the multiple scattering theory of XANES. The multiple scattering resonances of the photoelectron in the direction of CO have been identified and the analysis of these resonances in high signal-to-noise-ratio spectra of hemoproteins in solution has allowed to extract the Fe-ligand bond angles.

The detection, by this method, of subtle changes of the Fe-ligand bond geometry modulated by protein structure opens the way to better establish a correlation between localized structural changes in solution, close to the native state, and the mechanism of protein control of ligand binding.

X-RAY ABSORPTION AND MULTIPLE SCATTERING

According with the Fermi golden rule for optical transition the transition rate from a core level to an unoccupied state at energy E the absorption coefficient in <u>k-space</u> is given by $\mu(E) \sim P(E) \cdot D_{l'}(E)$, where P(E) is the matrix element and $D_{l'}(E)$ the density of states of selected angular momentum l'. Because of the dipole transition rules ($\Delta l = \pm 1$) XANES spectra probe the unoccupied density of states at the selected absorbing site (local) and of selected angular momentum l' (partial). The total absorption cross section can be factorized into an atomic part $\mu_a(E)$ determined by the the single absorbing atom and a part due to the crystal structure $\mu(E) = \mu_a(E) \cdot \mu_s(E)$. Where the atomic cross section $\mu_a(E)$ has strong resonances it becomes the dominant term and can give strong white lines at the $L_{2,3}$ and $M_{4,5}$ edge of transition metals and rare earths. Generally the at the K-edge of transition metals $\mu_a(E)$ does not exhibit resonances and $\mu_s(E)$ generates the main structures observed in K-edge XANES.

The total energy bandwidth $\Delta_t(E)$ for the final states, with an electron of kinetic energy E, is given by the sum of the core level width Δ_h and the width of the excited electron state Δ_e. These are determined by the core hole life time τ_h and the lifetime of the electron τ_e of kinetic energy E using the uncertainty relations.

$$\Delta_t(E)=\Delta_h+\Delta_e=(h/2\pi)(1/\tau_h+1/\tau_e)=\Delta_h+(h/2\pi)(2E/m)^{1/2}/\lambda(E) \quad (1)$$

Where the electron mean free path for inelastic scattering $\lambda(E)=(2E/m)^{1/2}/\tau_e$

It has been demonstrated that the absorption cross section for core transitions can be solved in the real space using the Green function approach.(14) In this approach the wavefunction of the excited electron is constructed by multiple scattering by muffin tin potentials of neighbour atoms within a cluster of atoms with the central absorbing atom.The important aspect of the solution of the absorption cross section in the real space is that the spatial arrangement of neighbour atoms and the interatomic distances, determining the unoccupied density of states of condensed matter, appear explicitly.

If the maximum modulus ρ of the multiple scattering matrix satisfies the condition $\rho<1$, the absorption coefficient can be expanded in a series over the EXAFS and XANES energy range(31,32)

$$\mu(E)=\mu_a(E)\cdot[1+X_2(E)+\Sigma_{n\geq3}X_n(E)] \quad (2)$$

where $\mu_a(E)$ is the atomic absorption, $X_2(E)$ is the "EXAFS signal" in spherical wave due to single scattering, and $X_{MS}(E)=\Sigma_{n\geq3}X_n(E)$ is defined as the "XANES signal" due to multiple scattering, where $X_n(E)$ are the modulation given by the multiple scattering pathways of order n>3 involving n-1 neighbour atoms.

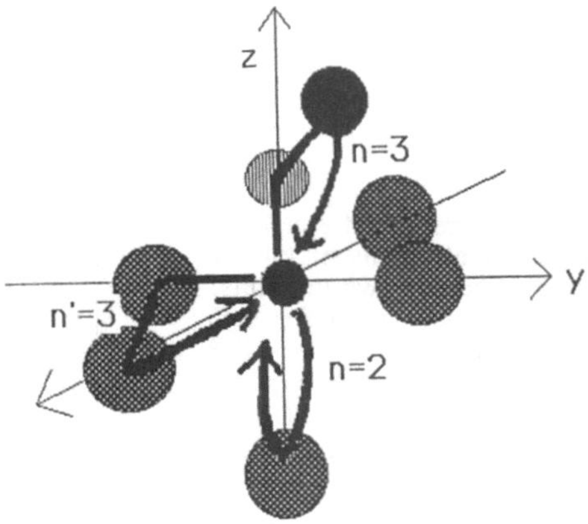

Fig.1 Pictorial view of the electron single scattering (n=2) and multiple (double) scattering (n=3) pathways which begin and end at the central absorbing atom.

Fig. 1 shows a pictorial view of photoelectron pathways which begin and end at the absorbing site. The pathway indicated by n=2 is one of the pathways which contributes to the EXAFS modulation term $X_2(k)$.Because n=2 atoms are involved in each elementary scattering process $X_2(k)$ probes the pair distribution function. Because only n-1=1 neighbour atom is involved in each elementary scattering process this term is called single scattering term. The total pathlength is 2R, therefore its Fourier transform can give the interatomic distances R.

In fig.1 two pathways n=3 and n'=3 which could contribute to the double scattering term $X_3(k)$ are shown. $X_3(k)$ is the first multiple scattering term and it probes the triplet correlation function because three atoms are involved. The pathlength of double scattering pathway is $R_3=(R_1+|\mathbf{R}_2-\mathbf{R}_1|+R_2)$. The half of total pathlength R_3 is defined as the effective radius $R_{eff}=R_3/2$ for $\chi_3(k)$. There are selection rules on the multiple scattering pathways because the amplitude of multiple scattering term is proportional to $(\cos\beta)^2$ where β is the angle between the outgoing and the incoming directions of the photoelectron at the central site. Where $\beta=90°$, like for the pathway n'=3 in fig. 1, the amplitude of the corresponding $\chi_3(k)$ is zero.

$\chi_2(E)$ is present all over the absorption spectrum both at high and at low kinetic energy E. The high energy EXAFS regime (usually for $k> 4$ Å^{-1}) the absorption coefficient can be approximated by $\mu(E)=\mu_a\cdot(1+\chi_2(k))$ and the plane wave approximation is valid.

The XANES energy range can be defined quantitative as that where the amplitude of the "XANES signal" $\chi_{MS}(E)$ can be detected.The extension of the XANES energy range is determined by attenuation of the amplitude of multiple scattering terms $\chi_3(E),\chi_4(E)$, ... increasing the photoelectron kinetic energy E. In open local structures, with no atoms in collinear configuration, by decreasing the kinetic energy of the photoelectron to lower energies, first the $\chi_3(E)$ component at about 100 eV and successively the higher order terms are switched on up to reach the "full multiple scattering region" where multiple scattering terms up to many orders are relevant. In local structures with atoms in collinear configurations the amplitude of multiple scattering terms $\chi_{n\geq3}(E)$ can extend up to very high energies but negative interference effects between $\chi_{n\geq3}(E)$ can suppress their signal at high energies.

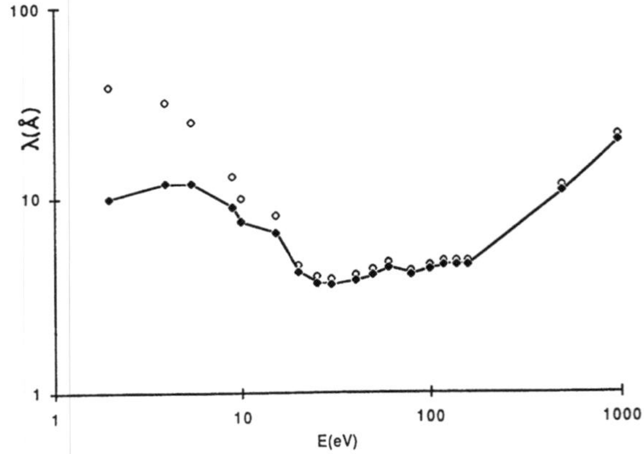

Fig.2 Mean free path for inelastic scattering of electrons of kinetic energy E in silicon (white points) and the effective mean free path for the cluster relevant for XANES spectra including the core hole life time of the Si 1s core level 0.4 eV (black points) following eq. 2.

In order to determine both the size of the cluster for the XANES and the higher multiple scattering term of the series in equation 2, an effective total path length for the excited electron $\lambda_{eff}(E)$ can be defined:

$$\lambda_{eff}(E) = (h/2\pi)\,(2E/m)^{1/2}/\,\Delta_t(E) \qquad (3)$$

The finite size of the cluster of neighbour atoms and the finite number of multiple scattering terms contributing to $\Sigma_{n\geq3}\chi_n(E)$ are determined by $\lambda_{eff}(E)$.

In figure 2 the mean free path $\lambda(E)$ in silicon crystal and the effective $\lambda_{eff}(E)$ for the Si K-edge XANES, including the core hole 1s width Δ_h=0.4 eV ,are shown.

It is clear from this figure that at low energies, in the XANES range, the core hole life time is the main term determining the finite size of the cluster. Moreover the $X_3(E)$ is the dominant multiple scattering term for the silicon k-edge XANES because the effective radius of higher terms is larger than $\lambda_{eff}(E)$.

THE MULTIPLE SCATTERING RESONANCE WITHIN Fe-C-O IN HbCO and MbCO

The polarized absorption spectra can be interpreted as due to multiple scattering in the direction of the electric field ε of the photon beam.In the myoglobin single crystals (space group $P2_1$) it is possible to find special orientations $\varepsilon //$ **a***-axis where the angle ϕ between ε and the heme normal **n** is $\phi = 28°$ and $\varepsilon //$ **c** -axis where the electric field is nearly parallel to the porphyrin plane $\phi = 86°$.

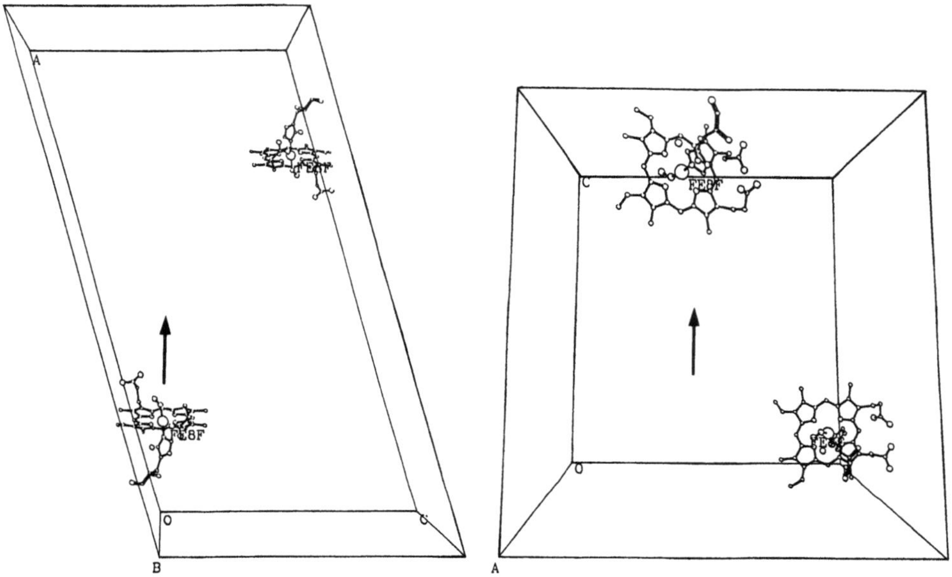

Fig. 3 The two hemes in the unit cell of the Mb crystal , space group $P2_1$, and the two orientations of the electric field ε are indicated by the arrows in the right $\varepsilon //$ **a*** and left $\varepsilon //$ **c** panel.

Figure 3 shows the orientation of the two hemes in the unit cell in the MbO_2 crystal(33) and the arrows the two orientations of the electric vector ε for which have been used to measure the XANES spectra in fig. 4(16).

The polarized XANES spectra in fig. 4 show a large dicroism. Two characteristic peaks C_1 and C_2 are observed for the multiple scattering in the direction of the heme normal **n** (dashed line). The spectrum due to multiple scattering in the heme plane (solid line, $\varepsilon //$ **c**) exhibits the peaks A,B,C,D,E . The upper part of the figure shows the results of full multiple scattering calculations of polarized XANES. The porphyrin plane has been taken to be planar and the coordinates from diffraction data heve been used (34). A satisfactory agreement with the experiment has been obtained taking into account that the broadening, $\Delta_t(E)$, has not been included. The size of the cluster has been found by increasing the number of neighbour shell up to the point that all experimental features were reproduced. The peak C_1 is a given by a full multiple scattering resonance where multiple scattering terms $X_n(E)$ up to high orders contribute with many scattering processes within the CO molecule.

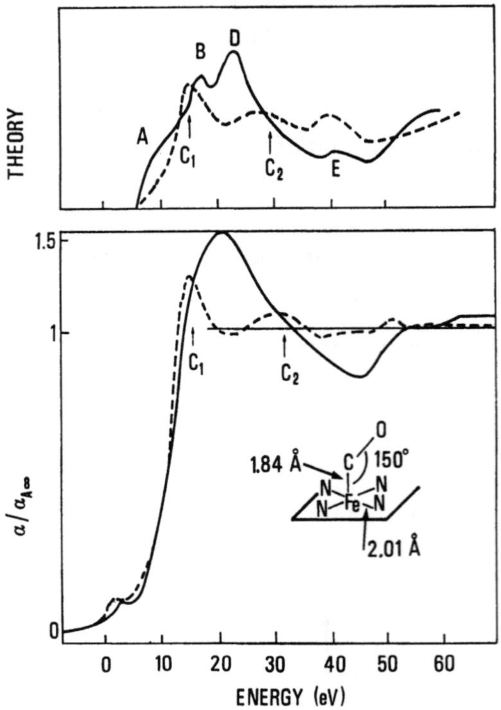

Fig.4 Experimental spectra (lower panel) and multiple scattering calculations (upper panel) of the polarized Fe XANES for $\varepsilon // \mathbf{a}^*$ (dashed line), $\varepsilon // \mathbf{c}$ (solid line). The peaks C_1 and C_2 are mainly due to multiple scattering within the Fe-C-group. The best agreement between theory and experiments for the scattering in the direction along the normal has been found with a bent Fe-C-O configuration as shown in the insert.

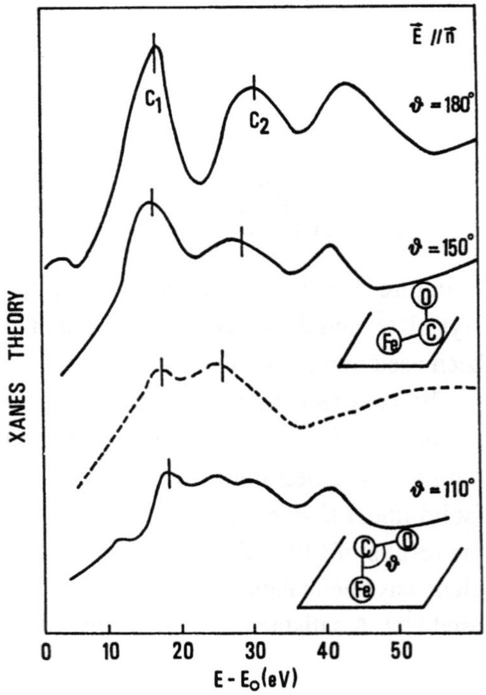

Fig.5 Multiple scattering calculations of the polarized Fe XANES for $\varepsilon // \mathbf{n}$ for different angles of the Fe-C-O configurations as shown in the insert. The dashed curve has been obtained by using the coordinates of neutron diffraction experiment[34] The peaks C_1 and C_2 mainly due to multiple scattering within the Fe-C-O group exhibit a large dependence on the Fe-C-O bond angle determined by multiple scattering pathways along the z axis within the CO molecule (see fig. 1).

The peak C_1 is therefore a god probe of the Fe-C-O orientation. The coordinates of the ligand CO have been varied, as shown in fig. 5, up to obtain the theoretical spectrumin fig 4 in good agreement with the polarized ($\varepsilon //$ **a***, along the heme normal) spectrum (dashed curve). Finally the spectrum of the HbCO in solution is shown in fig. 6. It is compared with the carboxy basket handle porphyrin and the Fe(TPIVPP)THF where the Fe-C-O configuration is linear. The peak C_1 due to multiple scattering along the normal **n** is clearly stronger than the peak D due to multiple scattering in the heme plane in agreement with multiple scattering calculations. In HbCO the intensity of the peak C_1 is as large as that of the peak D showing that the the Fe-C-O configuration is slightly bent. Comparing these results with the spectra of MbCO it is clear that in MbCO the Fe-C-O angle is smaller than in HbCO.

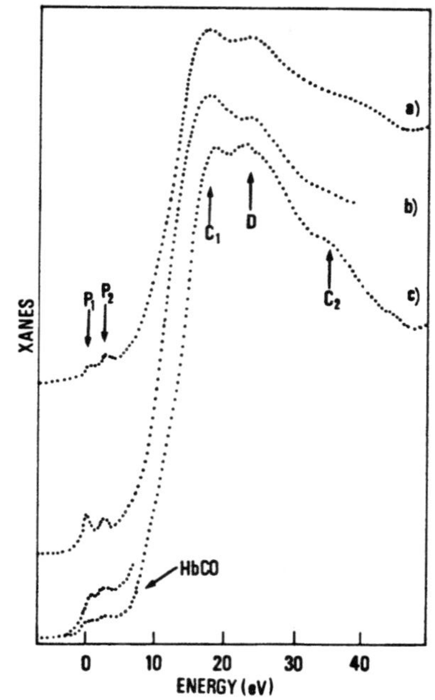

Fig. 6 From top to bottom: a) Fe XANES of the carboxy basket handle porphyrin (ref.30) of carboxy Fe (TPIVPP) THF, from the thesis of S.P.Cramer, 1977 Stanford University unpublished, and of c) HbCO in solution . The spectrum of a thicker HbCO sample to get the pre-edge peaks P_1 and $P_2$is shown.

In conclusion it has been shown that XANES spectroscopy is a direct structural probe which can provide us with unique information on higher order correlation function of proteins. In particular it can be used to study the spatial arrangement of diatomic ligand molecules at the iron site in hemoproteins in solution close to the native state and its variation via time dependent XANES studies.

REFERENCES

1) Kincaid, B.M., Eisenberger, P., Hodgson, K.O., and Doniach S.(1975) *Proc. Natl. Acad. Sci. USA* **72**, 2340
2) Shulman, R.G.,Eisenberger, P., Blumberg, W.E. and Stombaugh, N.A., (1975) *Proc. Natl. Acad. Sci. USA* **72**, 4003
3) Eisenberger, P., Shulman, R.G., Brown, G.S.,and Ogawa S. (1976) *Proc. Natl. Acad. Sci. USA* **73**, 491
4) Hu, V.W.,Chan, S.I., and Brown, G.S,(1977)*Proc. Natl. Acad. Sci. USA* **74**, 3821
5) Doniach,S.,Eisenberger,P., and Hodgson, K.O.,(1980) in "*Synchrotron Radiation*

Research" H. Winick and S.Doniach eds. Plenum Press , New York, p. 425
6) Powers L.(1982) *Biochem. Biophys. Acta* **683** , 1
7) Cramer S.P.and Hodgson K.O. (1979) *Prog. Inorg. Chem.* **25**, 1
8) J.Bordas (1982) in *"Uses of synchrotron radiation in Biology"* edited by Sthurmann, H.B., Academic Press, London, 107
9) Garner C.D. (1986) J. Phys (Paris) **47**, colloque C-8,1111
10) Bianconi A., Doniach S., and Lublin D. (1978)*Chemical Physics Lett.* **59**, 121
11) Belli M., Scafati A., Bianconi A., Mobilio S., Palladino L., Reale A., and Burattini E. (1980)*Solid State Commun.* **35,** 355
12) Bianconi A., Alema S., Castellani L., Fasella P., Giovannelli A., Mobilio S. and Oesh B., (1983) *J. Mol. Biol.* ,**165** 125
13) Bianconi, A., (1987) in *"X-RAY ABSORPTION: Principle, Applications,Techniques: EXAFS, SEXAFS and XANES"* Eds. R.Prins, D.C.Koningsberger, John Wiley & Sons Inc. Publishers, New York,
14) Durham, P.J., (1987) in *ref. 13*
15) Durham, P.J., Bianconi, A., Congiu-Castellano, A., Giovannelli, A., Hasnain, S.S., Incoccia, L., Morante, S., Pendry, J.B., (1983) *EMBO Journal,* **2,** 1441
16) Bianconi, A., Congiu Castellano, A., Durham, P.J., Hasnain, S.S., Phillips, S., (1985) *Nature* , **318,** 685
17) Antonini, E., Brunori, M.,(1971)*"Hemoglobin and Myoglobin in their reactions with ligands "* Eds. A. Henberger and E.L.Tatum, North-Holland, Amsterdam
18) Eisenberger, P., Shulman, R.G.,Kincaid, B.M., Brown, G.S.,and Ogawa S. (1978) Nature **274**, 30
19) Perutz, M.F., Hasnain, S.S., Duke, P.J., Sessler, J.E.,and Hahn, J.E. (1982) Nature **295** , 535
20) Chance,B. Fischetti, R. and Powers L. (1983) *Biochem.* **22**, 3820
21) Bianconi, A., Congiu Castellano, A., Dell'Ariccia,M., Giovannelli, A., Durham, P.J., Burattihi, E., Barteri, M., (1984) *FEBS Lett.* ,**178**, 165
22) Bianconi, A., Congiu Castellano, A.,Dell'Ariccia,M., Giovannelli, A., Burattini, E., Castagnola, M., Durham, P.J., (1985) *Bioch. Biophys. Acta* ,**831,** 120
23) Bianconi, A., Congiu Castellano, A., Dell'Ariccia, M., Giovannelli, A., Morante, S., Burattini, E., Durham, P.J., (1986)*Proc. Natl. Acad. Sci. USA* , **83**, 7736
24) Bianconi, A., Congiu Castellano, A., Dell'Ariccia, M., Giovannelli, A., Burattini, E., Durham, P.J., (1985) *Biochem. and Biophys. Res. Commun.*, **131**, 98
25) Pin, S., Alpert, B., and Michalowicz, A., (1982) *FEBS lett.*, **147,** 106
26) Congiu Castellano, A.,(1987) in *"Biophysics and Synchrotron Radiation"* Springer , Berlin p.92 This Volume
27) Pin. S.,Valat P., Cortes R., Michalowicz A.,Alpert B.(1985)*Biophys. J.* , **48,** 997
28) Bianconi, A., Congiu Castellano, A., Giovannelli, A., Dell'Ariccia, M., Burattini, E., Durham, P.J.,Giacometti, G., (1986) *Eur. Biophysical Journ.* **14**,7
29) Mills, D.M., Lewis, A., Horootunian, A., Huang, J., and Smith, B., (1984) *Science,* **223**, 811
30) Ascone, I., Fontaine, A., Bianconi, A., Congiu Castellano, A., Giovannelli, A. Momentau,M., (1987) in *"Biophysics and Synchrotron Radiation"* Springer , Berlin p.125 This Volume
31) Bianconi, A.,Garcia, J., Marcelli, A., Benfatto, M., Natoli, C.R., and Davoli, I. (1985) *J. Phys. (Paris)* **46**, C9-101
32) Benfatto, M., Natoli, C.R., Bianconi, A., Garcia, J.,Marcelli, A., Fanfoni, M., and Davoli I. (1986) *Phys. Rev. B* **34** , 5774
33) Phillips, S. (1980) *J. Mol. Biol.* **142,** 531
34) Hanson, J.C., and Shoenborn,B.P.J. (1981) *J. Mol. Biol.* **153**, 117

Oxygen Binding Site Structure in Hemoproteins by XANES

A. Congiu Castellano
Dipartimento di Fisica, Universita` degli Studi di Roma"La Sapienza",
00185 Roma Italy.

INTRODUCTION

The experimental determination of the bonding geometry of the diatomic molecules O_2, CO and CN in hemoproteins in solution is still an open problem in spite of extensive experimental research. Diffraction experiments are limited to crystallized proteins, on the contrary x-ray absorption near edge structure (XANES) spectroscopy is a tool for determination of local structure of active site in metalloproteins in solution.[1-8].

The oxygen binding structure, concerning the determination of Fe-O-O bonding angle and the electronic structure of the Fe-oxygen bonding in hemoproteins, has been object of many theoretical and experimental works for understanding the reversible binding of O_2.

Here we report the results of an extensive investigation of the structure of the oxygen binding site to test how the protein structure can modulate the local chemical bonding. We have studied the oxygenated human foetal and adult hemoglobin, carp hemoglobin, and the separated α and β subunits.

The XANES measurements were performed at the Frascati synchrotron radiation facility using the Wiggler beam line. The synchrotron radiation was monochromatized by a Si(111) channel cut single crystal. The absorption spectra were collected in transmission . The spectra were normalized to α / α_0 where α_0 is defined as the atomic absorption above the absorption jump obtained by linear fitting of EXAFS oscillations in the range 50-150 eV. The ligated forms of the hemoproteins have been prepared by standard methods[9]. The subunits of the HbA were separated according to the method of Geraci et al.[10] .Experimental details have been published elsewhere [3-6].

GEOMETRY OF IRON OXYGEN BOND

In fig. 1 the experimental XANES spectra of oxygenated proteins: human foetal hemoglobin ($HbFO_2$), human adult hemoglobin ($HbAO_2$), carp hemoglobin ($carpHbO_2$) and myoglobin (MbO_2) are reported, in fig. 2 their derivative spectra are shown.

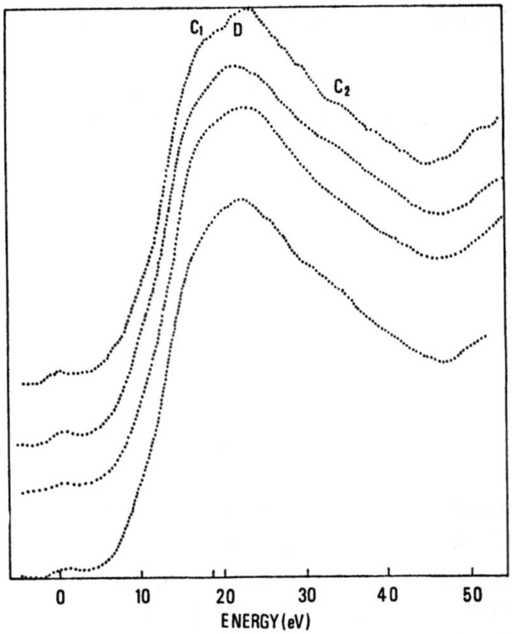

Fig. 1 Iron K-edge XANES spectra of oxygenated hemoproteins : human foetal hemoglobin ($HbFO_2$), human adult hemoglobin ($HbAO_2$), carp hemoglobin (carpHbO_2) and myoglobin (MbO_2).

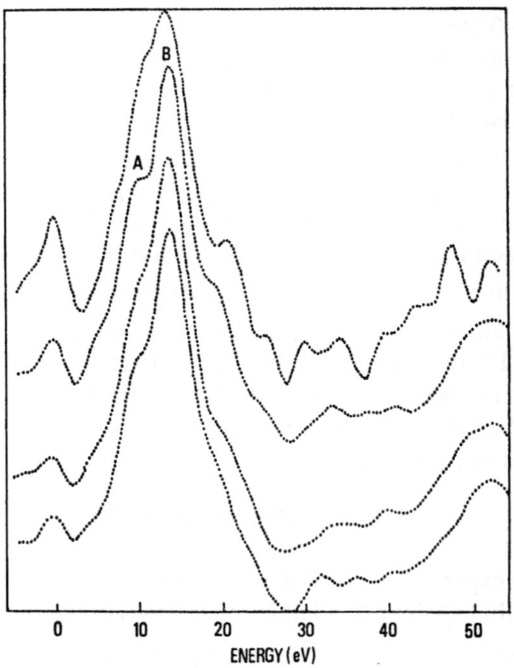

Fig. 2 Derivative spectra of the XANES of $HbFO_2$, $HbAO_2$, carp-HbO_2 and Mb

The zero of the energy scale has been fixed at the absorption threshold of the iron metal K-edge. All proteins show a weak structure at threshold due to transitions to unoccupied molecular Fe-3d derived states. The features above 10 eV are very similar for MbO_2, carp-HbO_2, and adult human $HbAO_2$. The spectrum of foetal hemoglobin shows some differences concerning the lineshape of the spectrum at about 18 eV. In fig. 3 it is shown the spectrum of the oxygenated porphyrin model compound $Fe(O_2)$ TpivPP) (1-NeIm) reported by Cramer in his thesis [11]. The structure of this compound was determined by x-ray diffraction[12].

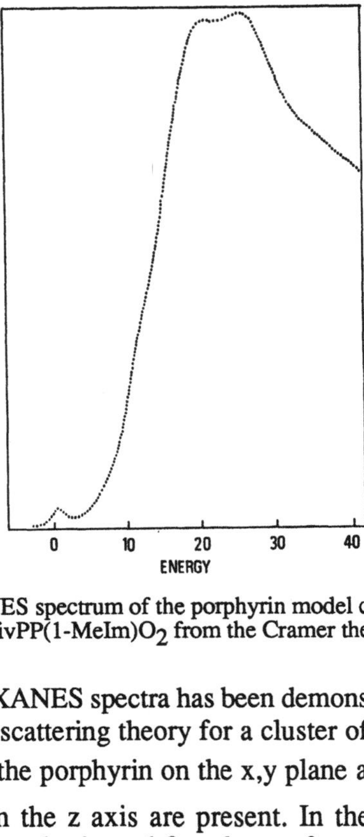

Fig. 3 XANES spectrum of the porphyrin model compound FeTpivPP(1-MeIm)O_2 from the Cramer thesis ref. 11.

The interpretation of XANES spectra has been demonstrated to be feasible on the basis of the full multiple scattering theory for a cluster of about 30 atoms.[2] In the first shell 4 nitrogens of the porphyrin on the x,y plane and one nitrogen N_ε of the proximal histidine F8 on the z axis are present. In the second shell the cluster includes 12 carbons of porphyrin and 2 carbons of proximal histidine F8. In the third shell the cluster includes 8 carbons of the porphyrin. The structure of the heme was taken to be planar and four fold symmetrical. This is an idealized model for the porphyrin where the heme is expected to be distorted with ruffling and doming. These effects play a minor role in XANES as showed by model calculations[13].

In fig.4 the calculated XANES spectra for different bonding angles are reported.The main effect of rotating the bonding angle Fe-O-O is the variation of the intensity of the peak C_1. The peak C_1 and C_2 are due to the multiple scattering of the electron ejected in the z direction and interacting with the oxygen molecule. These peaks are observed in the experimental spectra at the predicted energies.

From the variation of the intensity of the peak C_1 in the experimental spectra of the different proteins and model compounds it is possible to detect variations of the Fe-O-O bonding angle. The main peak D , which is due to multiple scattering in the porphyrin plane, is not affected by the oxygen orientation in the z direction.

The intensity ratio between the C_1 and D peaks in MbO_2 gives a Fe-O-O bond angle 115^0 in agreement with diffraction data for myoglobin (14).

The intensity ratio C_1/D in the porphyrin in fig.3 indicates that the Fe-O-O configuration is bent but with an angle larger than in myoglobin in agreement with crystallographic results (12) giving an angle of 131^0. These results confirm the validity of the theoreticals calculations. The intensity ratio between the C_1 and D in the HbO_2 spectrum is the same as in MbO_2 in fig. 1. This result for HbO_2 in solution is in opposition with the diffraction of hemoglobin in the crystal phase giving 156^0.(15,16)

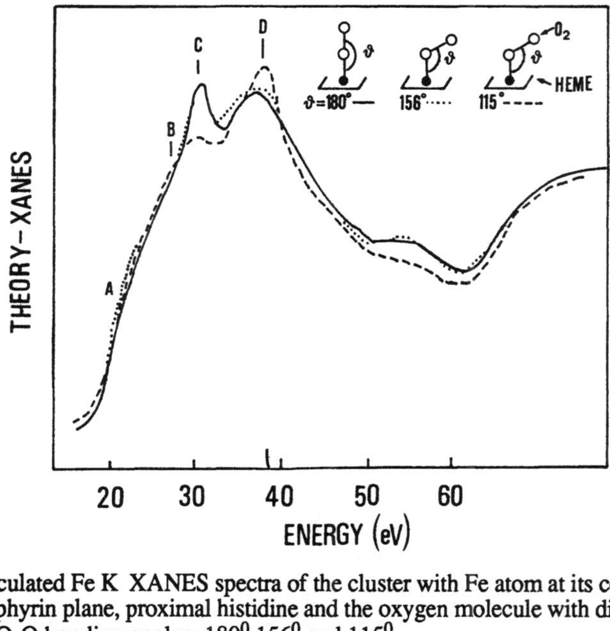

fig. 4 Calculated Fe K XANES spectra of the cluster with Fe atom at its center, the porphyrin plane, proximal histidine and the oxygen molecule with different Fe-O-O bonding angles: 180^0,156^0 and 115^0

The intensity ratio C_1/D in the XANES spectrum of the foetal hemoglobin is smaller than in the adult hemoglobin. Therefore oxygen bonding angle in the foetal hemoglobin is deduced to be smaller than in adult hemoglobin. This effect can be assigned to the substitution of the β chains with the γ chains which affects the local structure of the heme as it has been reported in the deoxy form [4].

A possible variation of the F-O-O bonding angle in different chains has been tested by studying the separated α and β chains. We have measured the XANES spectra of α and β subunits of human hemoglobin in the oxygenated form. The results are shown in fig. 5 with their derivatives. These results are in good agreement with previous studies by Pin et al.[7]

The bonding angle Fe-O-O was found by Shaanan in HbO_2 crystal to be 153° and 159° for the α and β chain respectively[15,16].

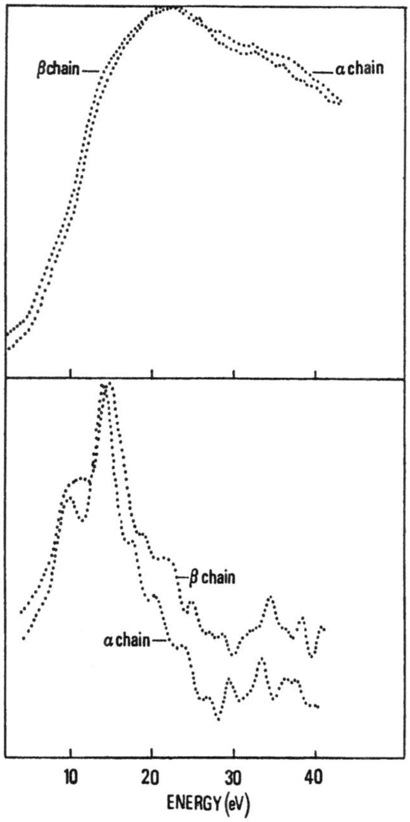

Fig.5 Fe k-XANES of α and β chains in the oxygenated forms and their first derivative.

Brzozowski et al.[17] have found a bonding angle of 160° in the α chains of a crystal of partially oxygenated HbO_2 (α-oxy, β-deoxy) in the T conformation.

Moreover the contraction of either the Fe-$N_ε$ distance (going from 2.07 Å in β chains to 1.94 Å in α-chains) and the Fe-O distance (going from 1.87 Å in β chains to 1.66 Å in α chains) has been found by Shaanan. This contration in the Fe coordination distances in z axis in the α–chain is compensated by the elongation of the Fe-N_P distances in the porphyrin plane,so that the average interatomic distance in the Fe coordination sphere is unchanged. Mossbauer experiments on isolated subunits have shown that the quadrupole splittings of oxy-β chains are smaller than that of oxy-α chains [18,19]. On the contrary there is no difference either in quadrupole splitting and in isomer shift between HbO_2 and isolated oxy-α chains. These results can be correlated by a different Fe-O_2 bonding.

It is also reported that in the absence of organic phosphates there is no preferential oxygen binding to the β-chains or α-chains.[20] On the contrary in the presence of organic phosphates the α hemes have higher affinity for oxygen as compared with the β hemes as it is in the hemoglobin tetramer where the α-chains are the first to bind the oxygen[20-22]. The phosphates in our samples of oxy-β chains, oxy-α chains and HbO_2 were removed by passage through a column of Sephadex G-25 equilibrated with Tris HCl buffer.

Comparing the XANES spectra of the isolated subunits is not possible to observe large difference in the energy region of the peaks C_1 and D in agreement with the fact that the variation of the Fe-O-O bonding angle is expected to be small. We have focused our interest in the energy region of the peaks A and B on the absorption rising edge which can be clearly resolved in derivative spectra.

We observe that in the derivative spectra the two peaks are better separated in the α-chains spectrum which exhibits a clear minimum between A and B. The absorption spectra show that there is a shift of the rising edge toward higher energy of the spectrum of the α-chain. In agreement with the calculation reported in ref.23 this effect is interpreted as due to the movement of the proximal histidine toward the Fe in the α-chain. This structural change of the local structure of the active site should be correlated with the Mossbauer results. Work is in progress to study also the site structure of the oxygenated γ-chain.

The investigation of the oxygen binding site structure by XANES has pointed out: i) a difference between the oxygen bonding configuration in human adult hemoglobin going from the solution to the crystal phase is indicated by comparison of XANES in solution and the crystallography; ii) the oxygen binding geometry in the foetal hemoglobin is slightly different from that of adult hemoglobin indicating that the sequence changes induce variations of the local structure in the deoxygenated form [4] as well in the oxygenated form; iii) the small difference between the active sites in separated oxy α and β-chain can be assigned to the movement of the proximal histidine.

THE IRON EFFECTIVE CHARGE IN THE Fe-OXYGEN BOND

It has been shown[1] that the XANES spectra can give a contribution for understanding the problem of the electronic structure of the Fe-O_2 bond.In fact at threshold, XANES spectra are determined by excitations of core Fe 1s electrons to the lowest unoccupied states at the Fe site.We have demonstrated[5] that the energy shift of the XANES multiple scattering features can be due to two different possible effects: 1) a contraction in the interatomic distance, and /or 2) a different binding energy of the 1s core initial state.

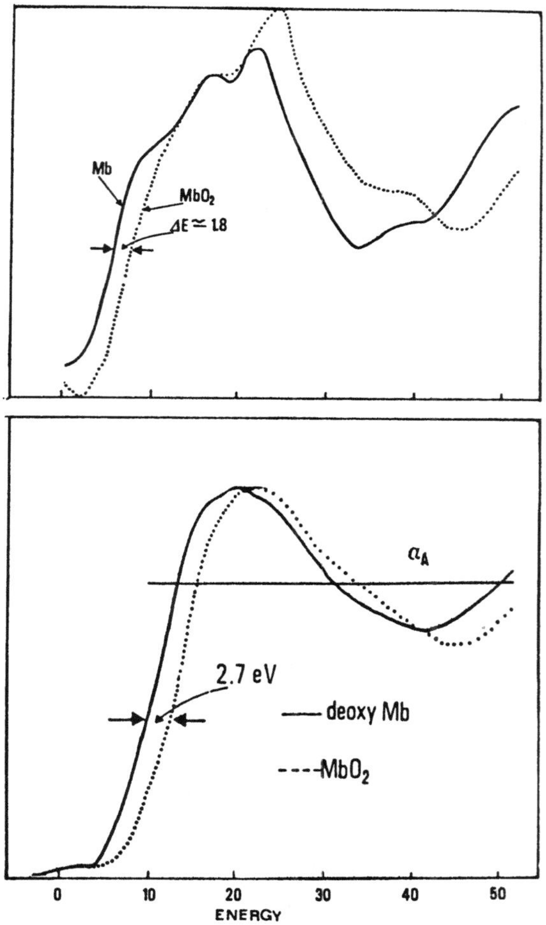

Fig.6 Theoretical (in the upper part) and experimental (in the lower part) XANES spectra of $MbO_2$and deoxy-Mb .

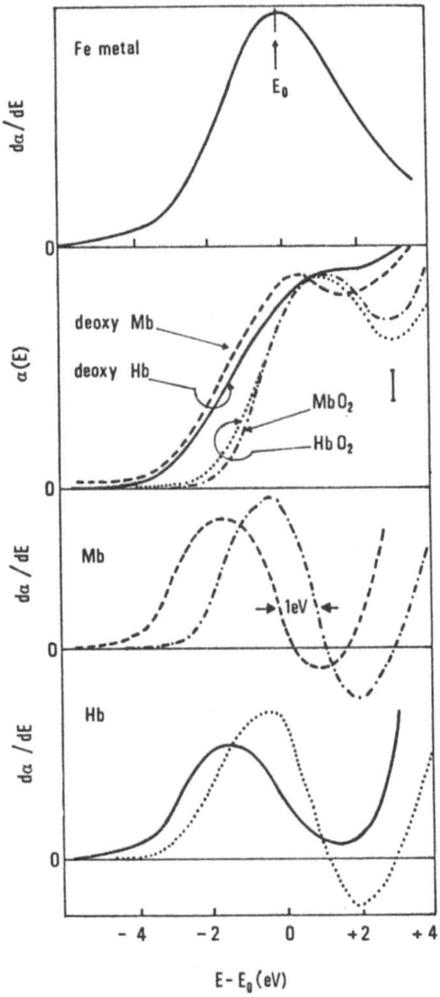

Fig.7 (panel a) the derivative spectrum of absorption threshold of Fe K-edge); panel b) The 1s--3d peaks of experimental XANES of Hb and Mb in the oxy and deoxy forms; panel c and d) the derivative spectra of these peaks

We calculated the XANES spectra for several models both in the deoxygenated and oxygenated forms , the Fe-N_p distance ranging from 2.06 Å for deoxy-Mb to 1.96 Å for MbO_2 (1.98 Å for HbO_2) and the orientation of the O_2 molecule ranging from 180^0 to 115^0, to analyze the combined effects of Fe-N_p contraction and O_2 orientation on the shift of the rising absorption edge and XANES features. In fig.6 the theoretical XANES spectra of MbO_2 and deoxy-Mb in the upper part, and the experimental XANES spectra in the lower part, are reported.

We have found that the contraction of 0.1 Å of the Fe-N_p distance following O_2

bonding, induces an energy shift which is ~1.0 eV smaller than that predicted for variations of interatomic distances and coordination geometry. Therefore there is an additional contribution to the energy shift of the XANES features due to the larger binding energy of the 1s core level both in the HbO_2 and MbO_2.This binding energy is controlled by the electronic charge of the occupied valence bonding orbitals in the atomic spatial region of Fe site.

Than we have addressed our attention to the weak excitonic peak that appears before the rising absorption jump edge. In all the experimental spectra the zero of the energy scale was carefully fixed at the absorption threshold of Fe metal K-edge defined as the first maximum of its derivative spectrum shown in the upper part of fig.7. The half width at half height of this peak is ~ 1.4 eV and is due partly to intrinsic width of Fe (1s) core level and partly to instrumental resolution. In fig. 7b we show a comparison between the 1s--3d peaks in $HbO_2$and MbO_2 and the deoxy-Hb and deoxy-Mb spectra.In fig. 7c and 7d we can observe shifts of ~ 1.0 eV between oxy and deoxy forms of Hb and Mb and its derivative spectra

This shift can be assigned to the variation of the core 1s binding energy. In fact the final states are the lowest unoccupied states which in both Hb and HbO_2 is a molecular orbital due to mixing between Fe(3d) and porphirin π^* orbital and it remains at the same energy going from Hb to HbO_2[(24)]. Because of core hole-electron attraction, the first excited state in molecules is a localized bound state and its energy shift follows the effective charge of the iron.

In conclusion we obtain a chemical shift of ~ 1.0 eV for the 1s core level both in MbO_2 and in HbO_2 due to change of the effective charge on Fe atom in hemoproteins in solution, during oxygenation process. Our results are in agreement with Mossbauer[(18,19,25)] and infrared[(26)] spectroscopies and molecular orbital calculations [(27)]wich show evidence of electronic charge transfer from Fe to O_2 upon oxygenation which can be due to formation of delocalized molecular orbitals with larger electron density on the oxygen,giving a Fe^{3+} O^{2-} diamagnetic configuration.

References

1) Bianconi, A.,*"X-RAY ABSORPTION: Principle, Applications,Techniques: EXAFS, SEXAFS and XANES" Eds. R.Prins, D.C.Koningsberger,John Wiley & Sons Inc. Publishers,* New York, (1987)
2) Durham, P.J., Bianconi, A., Congiu-Castellano, A., Giovannelli, A., Hasnain, S.S., Incoccia, L., Morante, S., Pendry, J.B., (1983) *EMBO Journal,* **2,** 1441
3) Bianconi, A., Congiu Castellano, A., Dell'Ariccia, M., Giovannelli, A., Durham, P.J., Burattini, E., Barteri, M., (1984) *FEBS Lett.* ,**178**, 165
4) Bianconi, A., Congiu-Castellano, A.,Dell'Ariccia, M., Giovannelli, A., Burattini, E., Castagnola, M., Durham, P.J., (1985) *Bioch. Biophys. Acta* ,**831,** 120
5) Bianconi,A., Congiu-Castellano,A., Dell'Ariccia,M., Giovannelli,A., Burattini, E., Durham, P.J., (1985) *Biochem. and Biophys. Res. Commun.,* **131**, 98

6) Bianconi, A., Congiu Castellano, A., Durham, P.J., Hasnain, S.S., Phillips, S., (1985) *Nature* , **318**, 685
7) Pin, S., Alpert,B., and Michalowicz,A., (1982) *FEBS Lett.* **147,** 106
8) Pin.,S., Valat,P., Cortes,R., Michalowicz,A., Alpert,B., (1985) *Biophys. J.* , **48**, 997
9) Antonini, E., Brunori, M., *"Hemoglobin and Myoglobin in their reactions with ligands "* (Eds. A. Henberger and E.L.Tatum) North-Holland Publ., Amsterdam (1971)
10) Geraci, G., Parkhurst, L., J., Gibson, Q.,H., (1969) *J. Biol.Chem.,* **244,** 4664
11) Cramer, S.P., (1978), PhD Thesis, Stanford University, *SSRL Report* **78/07** Stanford Synchrotron Radiation Laboratory, Stanford.
12) Jameson, J.B., Molinaro, F.S. Ibers, J.A., Collman, J.P., Brauman, J.I., Rose, E., and Suslick , K.S.(1987) *Jour. Am. Chem. Soc.* ,**100**, 6769
13) Bianconi ,A., Congiu Castellano, A., Dell'Ariccia, M., Giovannelli, A., Morante, S., Burattini, E., Durham, P.J., (1986) *Proc. Natl. Acad. Sci. USA* , **83**, 7736
14) Phillips, S.E.V., (1980) *J. Mol. Biol.,* **142** , 531
1 5) Shaanan, B., (1982) *Nature* , **296**,683
16) Shaanan, B., (1983) *J. Mol. Biol.* ,**171**,31
17) Brzozowski, A., Derewenda, Z., Dodson, E., Dodson,G., Grabowski, M., Liddington, R., Skarzynski,T., Valley, D., (1984) *Nature* , **307**, 74
18) Tsai,T.E., Groves, J.L., Wu ,C.S., (1981) *J. Chem. Phys.* , **74**, 4306
19) Hoy, G.R., Cook, D.C., Berger ,R.L., and Friedman, F.K., (1986) *Biophys. J.,* **49**, 1009
20) Viggiano, G., and Ho, C.,(1979)*Proc.Natl.Acad.Sci.USA* ,**76**,3673
21) Huestis, W.H., and Raftery, M.A., (1972) *Biochem. Biophys. Res. Commun.* ,**49**, 1358
22) Peller L.,(1982) *Nature* , **300**, 661
23) Morante, S., Congiu Castellano, A., Dell'Ariccia, M., Durham,P.J., Giovannelli, A., Burattini, E., and Bianconi, A., in this book
24) Eaton,W.A., Hanson, L.K., Stephens, P.J., Sutherland, J.C.,and Dunn, J.B.R., (1978)*J.Am.Chem.Soc.* ,**100**, 4991
25) Barlow, C.H., Maxwell, J.C., Wallace, W.J., and Caughey, W.S.,(1973) *Biochem.Biophys.Res.Commun.,* **55,** 91
26) Bade, D. and Parak, F.,(1978) *Z.Natuforsch.*,**33C**, 488
27) Kashiwagi, H., (1985) *"Biomolecules"(Nagota et al. eds.) Japan Sci.Soc.Press Tokyo-Elsevier Amsterdam* pp.31-50

Local Structure of Heme-Iron Studied by High-Resolution XANES: Thermal Spin Equilibrium in Myoglobin

Hiroyuki Oyanagi, Tetsutaro Iizuka*, Tadashi Matsushita**, Satoshi Saigo***, Ryu Makino* and Yuzuru Ishimura*

Electrotechnical Laboratory, Sakuramura, Niiharigun, Ibaraki 305, Japan
*Department of Biochemistry, School of Medicine, Keio University, Shinanomachi, Shinjuku, Tokyo 160, Japan
**National Laboratory for High Energy Physics, Ohomachi Tsukubagun, Ibaraki 305, Japan
***Department of Physics, Jichi Medical School, Yakushiji, Minamikawachi, Tochigi 329-04, Japan

ABSTRACT

The local structure of heme-iron of alkaline met-myoglobin (MbOH) during the thermal spin equilibrium between high and low spin states has been studied by a high-resolution XANES. The spin-state sensitive near edge features have been found in the pre-edge 1s-3d transition peak and absorption threshold region. These characteristic features are dependent on the high spin concentration and are related to the d-electron configuration and local structure of heme-iron, i.e. the distortion of octahedral ligand field caused by a displacement of heme-iron from the heme plane. The spin-state sensitive near-edge structures were also found for other myoglobin derivatives and can be used as a spin state marker. From a systematic high resolution XANES study on MbCO, MbCN and oxy-Mb), near-edge features which are sensitive to the angle between the diatomic ligand molecule and heme normal were found. These characteristic features can provide the elaborate local structure of heme-iron such as the displacement of iron atom from the hemeplane or the bond angle of ligand molecule when combined with EXAFS experiments.

INTRODUCTION

Some ferric hemoproteins such as hydroxide complexes of ferric hemoglobin and myoglobin show the intermediate values of spin susceptibilities between those characteristic of 5- and 1-unpaired d electrons. The intermediate paramagnetic susceptibilities and light absorption spectra of these complexes have been interpreted on the basis of thermal spin equilibrium between two magnetic isomers, one in high spin and the other in low spin states [1,2]. Fluoride and cyanide complexes were found to be in purely high and purely low spin states, respectively. On the other hand, hydroxide, azide, imidazole and cyanate complexes exhibited the intermediate magnetic susceptibilities and optical spectra.

Although the thermal spin equilibrium has been extensively studied for a variety of hemoproteins and their model compounds from the thermodynamical viewpoints, the relation between the local structure of heme-iron and spin states has not been established yet. Heme-iron is expected to be out of hemeplane in high spin states due to extended distribution of d-electrons.

The local structure of heme-iron has been successfully studied for a variety of hemoproteins. The magnitude of this displacement is, however, difficult to determine from EXAFS experiments only, which are insensitive to site symmetry. Therefore we have utilized the XANES to study the iron site which can provide the information on also site symmetry [3].

The temperature dependence of Fe K-XANES has been measured for alkaline metmyoglobin (MbOH) which shows a thermal spin equilibrium between 80 K where heme-irons are purely in low spin states and 300 K where most of them are in high spin states [2] to correlate the local structure of heme-iron with its spin states. Studies were also extended to other myoglobin derivatives with purely low spin and purely high spin states.

The correlation between XANES and the angle between the ligand molecule and heme-normal [3] has been systematically studied for myoglobin derivatives with a diatomic ligand molecule such as CN, CO and dioxygen for which the tilt angle varies from 180°(CN) to 110°(O_2).

EXPERIMENTAL

Efforts have been taken to measure Fe K-XANES spectra with a high resolution (<2 eV) to resolve fine structures. A sagittally bent crystal monochromator [4] and a fluorescence-detection spectrometer [5] were used. A typical energy resolution at the Fe K-edge (7.11 keV) is 1.5 eV using silicon (111) monochromator crystals. All the measurements were carried out when the storage ring was operated at 2.5 GeV, 200-100 mA. Care has been taken to reduce the higher harmonics contribution by detuning the two crystals by 15%. A scintilation counter array consisting of 9 NaI detectors covering 18 % of the total solid angle and a Si(Li) detector were used for the concentrated (>5 mM) and dilute (<2mM) samples, respectively. Mn filters were used for both detector systems to reduce the background. The use of a solid state detector (SSD) is essential to record weak preedge features with a high signal-to-background ratio, counting only the fluorescence signal.

MbOH solution (2-24 mM) was rapidly frozen in an aluminum cell with a Kapton window by liquid nitrogen and immediately transferred to a cryostat which was cooled down to 80 K with a closed cycle helium refrigerator. The temperature dependence of Fe K-XANES and EXAFS was made between 80 K and 300 K. Myoglobin derivatives such as MbCN, MbCO and oxy-Mb were prepared prior to the measurements. Fe K-XANES spectra of these myoglobin derivatives were also measured at 80 K. Optical absorption spectra were taken after the measurement, which showed no indication of radiation damage during the experiment.

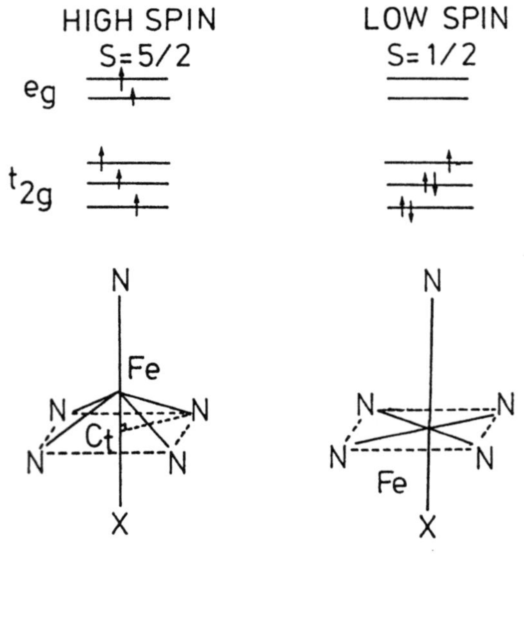

Fig. 1 Local structure of heme-iron in Fe-protoporphyrin IX. Heme-iron is coordinated with four nitrogen atoms of heme-plane and one nitrogen of proximal histidine. X denotes the position of 6th ligand.

RESULTS AND DISCUSSION

Local structure and spin states in MbOH

Heme-iron of metmyoglobin is coordinated with four nitrogen atoms of pyrrole rings and another nitrogen atom of proximal histidine as indicated in Fig. 1. The 6th ligand of MbOH is a hydroxy ion. In myoglobin (Mb), heme-iron is displaced from the heme plane by 0.40 Å and from nitrogen plane by 0.27 Å according to the structure analysis by Takano [6]. Because of a doming of pyrrole ring, the iron-nitrogen distance is slightly shorter than the iron-hemeplane distance. The displacement of heme-iron from the mean hemeplane was reported for various high-spin porphyrin compounds [7]. On the other hand, heme-iron in low-spin myoglobin derivatives such as cyanide is expected to be close to the hemeplane. The magnitude of this displacement in deoxyhemoglobin (Hb) has been of considerable interest in relation with the affinity of oxygen binding [8,9]. A general problem in the determination of the deviation from the heme plane is the lack of information on site symmetry in EXAFS data. In this respect, XANES can be a complementary means. However, since XANES is also sensitive to bondlengths, both EXAFS and XANES experiments are necessary to estimate the magnitude of displacement using either band-structure or full multiple scattering approaches.

Temperature dependence of Fe K-XANES of MbOH

Figure 2 shows the temperature dependence of Fe K-XANES for MbOH. Solid and dotted lines indicate the raw spectra and their first derivatives. At 80 K where MbOH is purely in low spin states, several characteristic features are observed which are denoted after Bianconi et al. [3].

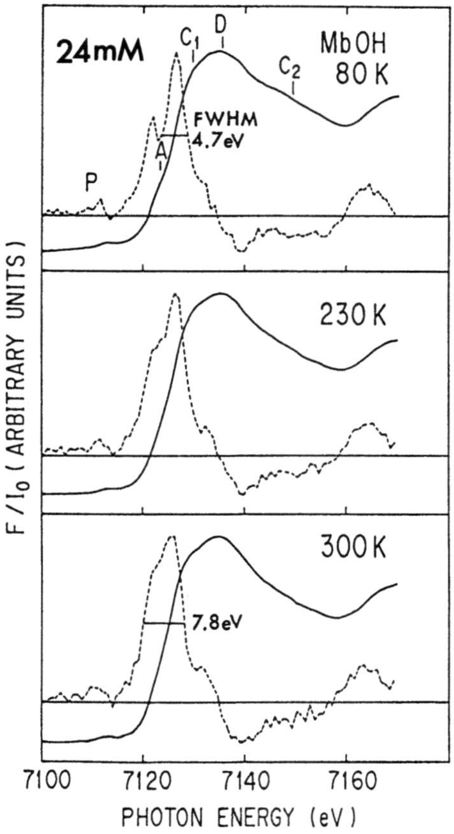

Fig. 2 Temperature dependence of Fe K-XANES of MbOH. At 80 K (low spin states), characteristic features A and C_2 are clearly observed, which gradually disappear with the increase of temperature.

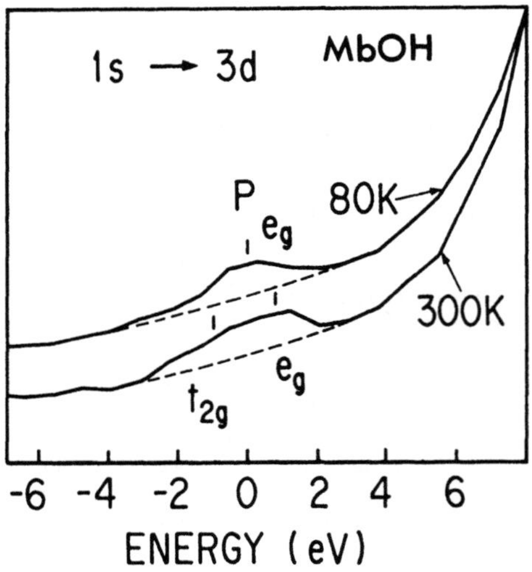

Fig. 3 Blow-up of 1s-3d transition peak of MbOH. The lower energy side broadens for high spin states.

Peak P observed at at 7112 eV is a quadrupole-allowed transition from Fe 1s to empty 3d states. Figure 3 shows blow-up of 1s-3d peak. This weak peak broadens on going to high temperature. As indicated in Fig. 1, 3d orbitals split into two e_g orbitals and three t_{2g} orbitals in an octahedral ligand field. In high spin states, there are empty states in both e_g and t_{2g} states whereas only one out of five empty states is in t_{2g} orbital in low spin states. Therefore the width of 1s-3d transition peak is expected to increase in high spin states and shift toward lower energy, which is in agreement with the observation indicated in Fig. 3. The eg orbital splits into z^2 and x^2-y^2 orbitals. The crystal field splitting between z^2 and x^2-y^2 orbitals is likely to be much smaller than the e_g-t_{2g} splitting in case of a weak ligand like a hydroxy ion.

In Fig. 2, A indicates the position of a shoulder structure observed for low-spin MbOH at 7122 eV, which appears as an inflection point in the first derivative. This shoulder gradually disappears with the increase of high spin content and at 300 K, no shoulder structure is observed. After a careful normalization, a difference spectrum was obtained by subtracting 80 K data from spectra at higher temperature. In a difference spectrum, a sharp positive peak with the FWHM (full width at half maximum) of 5 eV is observed at 7122 eV. The magnitude of this peak is plotted as a function of temperature in Fig. 4 and compared with the calculated high spin content which is normalized to the data at 300 K. The high spin content was calculated using thermodynamical data [2]. Although the magnitude of difference peak deviates slightly from the calculated temperature dependence of high spin content at lower temperatures, the overall behavior is well predicted. Therefore, this difference peak is related to the spin state.

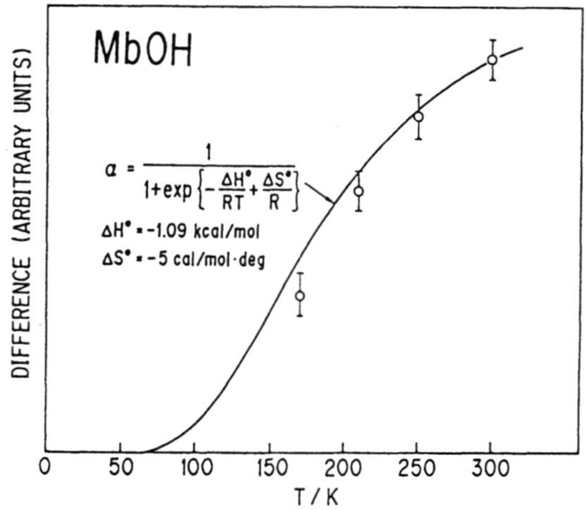

Fig. 4 Temperature dependence of the magnitude of difference peak between T and 80 K. High spin concentration is calculated using thermodynamical parameters and plotted after the normalization at 300 K.

Temperature dependence of Fe K-XANES was performed for both concentrated (24mM) and dilute (<2mM) samples. These two sets of data agreed well with each other and the effect of sample concentration was not observed.

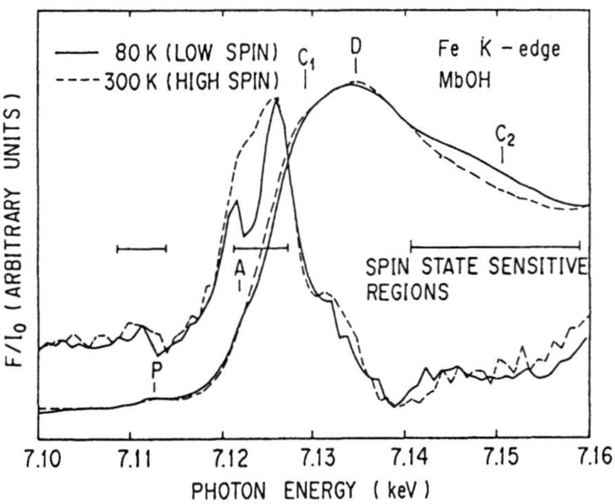

Fig. 5 Comparison of Fe K-XANES spectra between low spin and high spin states. Solid and dashed lines indicate the data for low spin and high spin states, respectively.

At 80 K, a broad bump structure C_2 is observed at 7148 eV. This peak decreases its intensity on going to high spin states and it disappears at 300 K. This bump appears as a broad negative peak in a difference spectrum with a FWHM of 14 eV. The magnitude of C_2 and A in a difference spectrum is inversely correlated with each other while their magnitudes are proportional to the high spin content . The distortion of octahedral ligand field is expected to affect the near edge structures because of the multiple scattering within the heme-plane. In Fig. 5, the Fe K-XANES spectra of MbOH in low spin (solid line) and high spin states (dashed line)are compared. As indicated in this figure, there are three regions which are spin-state sensitive. Among them 1s-3d peak is affected by both 3d electron configuration and p-d hybridization. We note that these spin state sensitive near edge features indicated in this figure are different from the ligand field indicator regions found by Chance et al. [10]. The relative intensity ratio of two bump structures located at 7171 eV and 7190 eV is also spin state sensitive.

Myoglobin derivatives in high spin states

Figure 6 shows Fe K-XANES spectra for $MbOH_2$, Mb(II) taken at 80 K and MbOH taken at 300 K. The absence of A and C_2 in those spectra suggests that these derivatives are in high spin states, indicating that the above criteria can be extended to other high-spin myoglobin derivatives with ferric and ferrous heme-irons. A more clear method to determine the spin state is the FWHM of a sharp peak in the first derivative spectra at 7123 eV. For low spin states, the FWHM is 3.4-4.7 eV while for high spin states, this value is 6.5-8.1 eV. For instance, the FWHM for low-spin MbOH is 4.7 eV which increases to 8.1 eV for predominantly high-spin MbOH. For Mb(II) and $MbOH_2$, the FWHM is 6.5 eV and 6.8 eV, respectively. A little bump structure is observed at 7117 eV for Mb(II) which becomes a shoulder peak in the first derivative. This structure does not correspond to A but corresponds to P' which are observed for MbCN and MbCO.

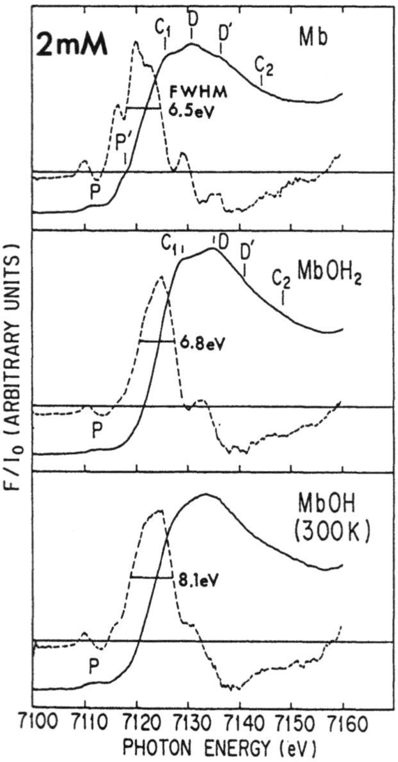

Fig. 6 Fe K-XANES spectra of high-spin myoglobin derivatives taken at 80 K.

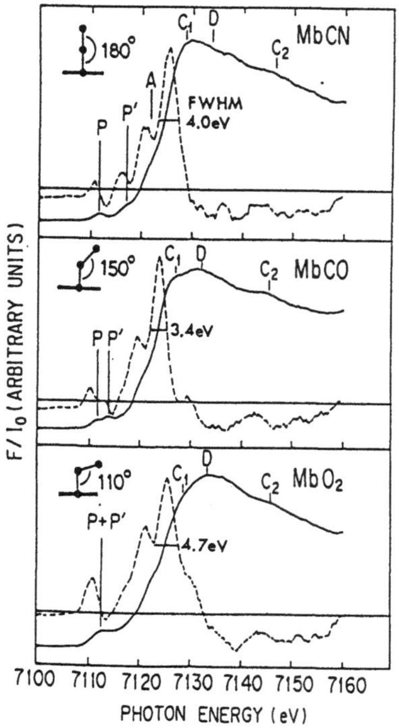

Fig. 7 Fe K-XANES spectra of low-spin myoglobin derivatives. Going from CN to dioxygen, the tilt angle changes from 180°to 110°. C_1/D ratio is correlated with this angle.

Myoglobin derivatives in low spin states

Fe K-XANES spectra for various low-spin myoglobin derivatives are shown in Fig. 7. A and C_2 are present in all spectra indicating that these "spin state markers" can be applied to other low-spin myoglobin derivatives. The absorption maximum for MbCN denoted as C_1 is located at 7129 eV. On going from CN to O_2, C_1 decreases its intensity while C2 and A are not affected. For MbCN, Fe-C-N angle is 180°while Fe-C-O and Fe-O-O angles are 150°and 110°, respectively. As suggested by Bianconi et al. [3], C_1 originates from a strong focusing scattering in the linear chain consisting of heme-iron and a diatomic molecule. Therefor the intensity of C_1 can be used to determine the angle between the ligand molecule and the heme normal. C_1/D is 1.09 for MbCN , 0.96 for MbCO, and 0.83 for oxy-Mb. These values are in good agreement with the calculated ones by Bianconi et al. [3]. Since this ratio is not dependent on whether the heme-iron is ferric or ferrous, the tilt angle of diatomic ligand molecule can be determined from the interpolation.

A broad bump structure denoted as P' is observed 5.4 eV above the 1s-3d transitin peak P for MbCN. On the other hand, a doublet structure separated with each other by 2.3 eV is observed in the prepeak region for MbCO while only a single peak is observed for oxyMb as shown in Fig. 7. Since the position of P'in MbCN coincides with that of Mb(II), the origin of these two structures are the same. The 1s-3d transitions are allowed between 1s states and x^2-y^2, z^2 and xy d-orbitals for low-spin complexes. According to the polarized experiments for MbCO single crystal [3], P and P' are polarized along the heme normal and along the heme plane, respectively, suggesting that P and P' correspond to z^2 and x^2-y^2 orbitals. The split between P and P' increases on going from a weak ligand (dioxygen) to a strong ligand (CN). The reason for a large split of 5.4 eV observed for both high-spin Mb(II) and low-spin MbCN is an interesting problem to be solved. It is also another problem that the transitions to lower tg orbitals in high-spin Mb(II) are not observed as an extra peak or broadening in the lower energy side.

ACKNOWLEDGEMENT
The authors express their greatest thanks to Prof. A. Bianconi and Dr. Fujikawa for valuable discussions and encouragements.

REFERENCES

1 P. George, J. Beetlestone and J.S. Griffith, Rev. Mod. Phys. 36 (1964) 441.
2 T. Iizuka and T. Yonetani, Adv. Biophys. 1 (1970) 157.
3 A. Bianconi, A. Congiu-Castellano, P.J.Durham, S.S. Hasnain and S. Phillips, Nature 318 (1985) 685.
4 T. Matsushita, T. Ishikawa and H. Oyanagi, Proc. of 2nd Int. Conf. on VUV and X-Ray Synchrotron Radiation, Stanford, 1985.
5 H. Oyanagi, T. Matsushita, H. Tanoue, T. Ishiguro and K. Kohra, Jpn. J. Appl. Phys. 24 (1985) 610.
6 T. Takano, J. Mol. Biol. 110 (1977) 537.
7 J.L. Hoard, M.J. Hamor, T.A. Hamor and W.S. Caughey, J. Am Chem. Soc. 87 (1965) 2312.
8 P. Eisenberger, R.G. Shulman, B.M. Kincaid, G.S. Brown and S. Ogawa, Nature 274 (1978) 30.
9 M.F. Perutz, S.S. Hasnain, P.J. Duke, J.L. Sessler, J.E. Hahn, Nature 295 (1982) 535.
10 M.R. Chance, L.J. Parkhurst, L.S. Powers and B. Chance, J. Bio. Chem. 261 (1986) 5689.

The High and Low Spin Iron Sites and the T_l to R_l Transition in Carp Hemoglobin

S. Morante*, A. Congiu Castellano+, M. Dell'Ariccia,
P.J . Durham++, A.Giovannelli+, E. Burattini**, A. Bianconi+

**Dipartimento di Fisica, Universita' degli studi di Roma "Tor Vergata", Roma, Italy*
+Dipartimento di Fisica, Universita' degli Studi di Roma, "La Sapienza", 00185 Roma, Italy
++SERC Daresbury Laboratory, Daresbury Warrington WA44AD (U.K)
***INFN and CNR , Laboratori Nazionali di Frascati, 00044 Frascati, Italy*

INTRODUCTION

The molecular mechanism of the modulation of the protein activity and in particular the control of oxygen binding in hemoglobin (Hb) at the level of atomic structure has been object of large interest. Most of the studies have been addressed to the location of the cooperativity free energy required to switch between the two quaternary structure : the low affinity unligated (deoxy) tense T_d structure and the high affinity ligated relaxed R_l structure. The free energy can be located: a) at the heme site(1) b) at the subunit interface(2) and/or c) distributed over other bonds of the protein (3-5). X-ray absorption spectroscopy provides via the analysis of EXAFS (x-ray absorption fine structure) and XANES (x-ray absorption near edge structure) information on the local structure of active sites in solution. The early studies of high and low affinity forms of deoxygenated adult hemoglobin (HbA) and Hb Kempsey by EXAFS(6,7) and the more recent studies of low and high affinity carp hemoglobin (carp Hb) by XANES (8,9) have shown that the average distance between iron and its first neighbours does not change more than 0.01 Å comparing the two conformations in the unligated (or ligated) forms. These studies show that most of the cooperative energy associated with the change of the quaternary structure is not located at the active site but subtle important structural changes at the active sites, induced by the quaternary conformation , are expected from the indication of changes of magnetic susceptibility.

In carp hemoglobin, both in the unligated and in the ligated (azide met) form, it is possible to induce, by allosteric effectors, the transition from one to the other quaternary conformation indipendently from the ligand presence. Measurements of magnetic susceptibility in the frozen(10,11) and liquid(12) state have confirmed the dependence of the spin state equilibrium of carp azide methemoglobin on the quaternary conformation of the hemoglobin.The transition from the tense unligated form T_d to the relaxed ligated form R_l is inhibited by adding inositol hexakisphosphate (Ins-P_6) at pH below 6.5. As shown in fig.1 in presence of Ins-P_6 and lowering the pH below 6.5 the effective magnetic moment increases from 2,83 μ_B to 4.2 μ_B indicating a transition from low spin to intermediate high spin going from T_l to R_l conformations. The correlation between the magnetic moment and iron coordination geometry has been established in a large series of chemical compounds(13).

Here we present a careful study of the XANES spectra of carp hemoglobin in order to establish the small variations of the local structure at iron sites which are related with the variation of spin state in the ferric form.

SAMPLES AND EXPERIMENTAL METHOD

Carp blood was obtained as described(14). The blood was centrifuged for 15 minutes at 1500 x g to separate the plasma. The plasma was removed and the packed red cells were washed four times by centrifugation at 1500 x g with three parts of a 0.1 M NaCl solution. The packed red cellswere frozen in liquid nitrogen after washing. Then two parts of distilled water was added to obtain the lysis. To remove the cell ghosts after lysis the cells were centrifuged at 25.000 x g for about 20 minutes.

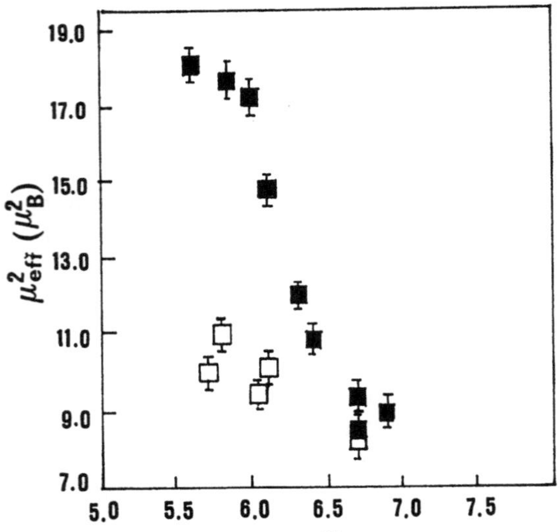

Fig.1 The paramagnetic susceptibility of carp azide met-hemoglobin. The square of the effective magnetic moment μ_{eff} in units of μ_B is plotted as a function of pH under different experimental conditions: (□) in the absence of inositol hexakisphosphate (Ins-P6) at 18 °C and (■) in the presence of Ins-P_6 at 18 °C

The sample was passed through a Sephadex G-25 coarse column equilibrated with 1mM Tris at pH=8 to remove organic phosphate and then through a ionic change Dintzis column to remove salts. First the ferric hemoglobin was prepared: the hemolysate was left at 4 °C with a 5-fold excess of $NaNO_2$ for about 20 minuts followed by a 20 minutes centrifugation at 25.000 x g. The ferric form so obtained was passed through a Sephadex G-25 coarse and a Dintzis column.

To obtain the two forms R_l and T_l we proceeded from the ferric form as follows:

Carp HbN_3 in R_l state : 0.075 ml Bistris buffer (pH=6.6) and 0 .075 ml 1M NaN_3 were added to 0.6 ml Hb^{3+} .

Carp HbN_3 in T_l state : 0.075 ml Bistris buffer (pH=4.9), 0.075 ml 1M NaN_3 and 0.02 ml of 200 mM inositol hexakisphosphate (Ins-P6) were added to 0.6 ml Hb^{3+}.

The sample concentration was about 10 mM.The XANES measurements were performed at the Frascati 'wiggler' facility using synchrotron radiation monochromatized with a Si(111) channel cut crystal. The trasmission method was used to collect spectra. The energy of the threshold k-edge has been defined as the first maximum of the derivative of the metal iron k-edge. The absorption coefficient was normalized in all spectra the atomic absorption coefficient beyond the absorption jump obtained by linear fitting of the spectrum in the EXAFS region. The above normalization procedure allows the comparison between different spectra to extract unambiguos variations of relative intensities of the XANES peaks. The energy resolution was about 1 eV but the accuracy in determination of the energy position was ΔE=0.2 eV.

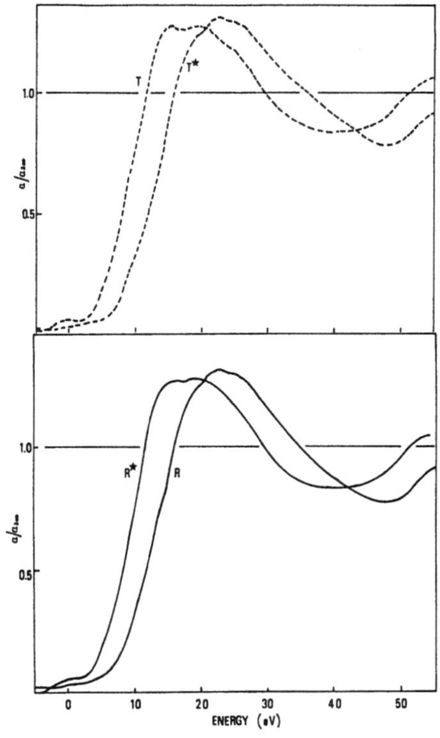

Fig.2 The effect of azide binding on iron XANES spectra of carp hemoglobin: curve T of unligated carp Hb in presence of Ins-P_6 at pH 6.2 in the T_d conformation; curve T* of ligated carp HbN_3 in presence of Ins-P_6 at pH 4.9 in the T_l form; curve R* of unligated carp-Hb at pH 8.2 in the R_d conformation; curve R of carp HbN_3 at pH 6.6.

THE EFFECT OF N_3 BINDING ON THE IRON SITE

In fig. 2 we report the variation of the Fe K-edge XANES spectrum of deoxygenated carp hemoglobin induced by azide binding. In the upper part of the figure the variation of the XANES spectra for carp Hb at low pH in presence of Ins-P_6 is reported. In this condition the transition from the T_d to R_l conformation is inhibited and the protein is in the T conformation both in the ligated and unligated phase. The azide ligated T_l conformation is indicated by T^* in fig. 2. In the lower part the spectrum of the unligated, high affinity carp Hb at pH 8.2 in the R_d conformation (indicated in the figure by R^*) is compared with the ligated carp HbN_3 in the R_l conformation. From fig. 2 it is clear that the large energy shift and the large change of the XANES spectra is determined by the binding of the azide molecule at the iron site. The large energy shift of the aborption rising edge is due to the contraction of the average Fe-N distance by about 0.07 Å. A multiple scattering resonance at energy E_r moves at higher energy with decreasing interatomic distance following the rule $(E_r-E_b)d^2$=constant, where E_b is the energy of a bound state at threshold.(15-16) The bound state in the XANES spectra of hemoglobin can be taken the weak 1s-3d excitation at the threshold. It was founded (17) that in human adult hemoglobin a contraction of 0.1 Å of the average Fe-N distance gives a blue shift of 1.8 eV of the rising absorption edge.

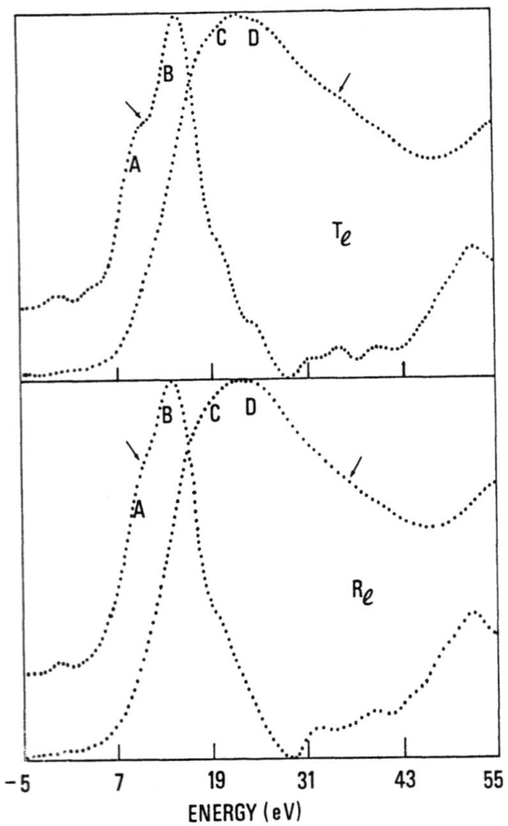

Fig.3 Effect of the transition from the low to high spin transition in azide ferric carp hemoglobin on XANES. In the upper figure is the XANES spectrum of high spin carp HbN_3 in presence of Ins-P_6 at pH 4.9 and its derivative. In the lower figure is the spectrum of the low spin state at pH 6.6 and its derivative.

The change of the iron effective charge from the unligated to the formally Fe^{3+} ferric ligated state contributes with a shift of about 1 eV. These results compared with the spectra of deoxy-Hb in the T_d conformation changing in the R_l conformation(17), in the oxy-Hb, show that the large contraction of the Fe -N distance upon ligand binding is present also if the quaternary structure of the protein remains in the low (or high) affinity conformation.

THE EFFECT OF LOW TO HIGH SPIN TRANSITION ON IRON SITES

In fig.3 careful comparison between derivative spectra of ferric carp HbN_3^+ in the two R_l and T_l forms shows that there are subtle changes (indicated by arrows) in the region of the shoulder B at the rising edge of the x-ray absorption threshold and in the region at about 30 eV . Only in these two energy regions the quaternary structure transition induces changes in the XANES spectra. It is clear that part of the spectrum remains unchanged by the spin transition. This indicates that only few neighbour atoms move with the transition while the others remain fixed. Recent results on the effect of the high to low spin transition on the XANES of met-myoglobin(18) induced by decreasing the temperature indicate that the spin transition induces small variations on the spectra in the same energy regions as that indicated in fig. 3.

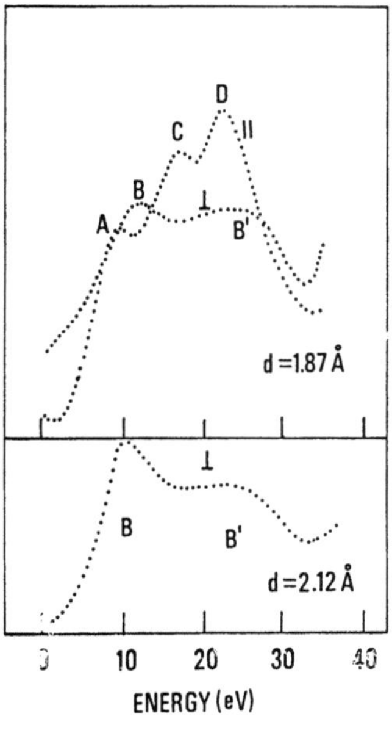

Fig.4 Calculated XANES using the full multiple scattering theory for a pentacoordinated iron in an idealized flat porphyrin in the x,y plane and one nitrogen N_ε in the z direction with the other two nearest atoms of the proxymal histidine.The upper curve shows the polarized XANES spectrum with the electric field parallel to the porphyrin plane (II) and perpendicular (⊥) to the porphyrin plane. The distance between Fe and N_ε is d=1.87 Å in the upper curves. The spectrum due to multiple scattering in the z direction (⊥) is calculated in the lower curve for increased distance between Fe and N_ε d=2.12 Å. The peak B decreases and the peak B\` increases with decreasing d.

We have performed a model XANES multiple scattering calculation in order to understand this variation. We calculated the effect of the variation of the distance between iron and proximal histidine by assuming that the heme remains unchanged. XANES calculations have been done for a cluster with the iron atom at the center including the porphyrin and 3 atoms of the proximal histidine. The features A,C,D and E of the spectrum are determined by multiple scattering in the heme plane (x,y) as shown in the upper part of the figure. The corresponding features at the same energies in the experimental spectrum are unchanged in the low to high spin transition.

The XANES multiple scattering calculations for the photoelectron scattering the z direction has to be considered only a model calculation that does not include the effects of the azide (N_3) molecule (probably tilted like in azide myoglobin[(19)]) and of the other atoms of the proximal histidine. However the z polarized spectrum indicates that the multiple scattering from the most significative atoms of the proximal histidine gives origin to the features B and B'. In fig.4 we can see that on going from higher to lower distances between Fe and the nitrogen atom of proximal histidine N_{ε} the intensity of peak B (at low energies) falls down while there is an intensity growth of peak B' (at higher energies). These are exactly the features that changes in the experimental data

The variation of the experimental data can be assigned to a decrease of the structure B between A and C features and the increase of the structure B' at about 35 eV going from the R_l to the T_l conformation. We can conclude that on going from R_l to the T_l state in the azide carp hemoglobin a contraction of Fe-N_{ε} distance, due to movement of the proximal histidine could interpret the data in the light of theoretical calculations. However we want to point out that also other effects due to the distortions of the full iron site structure including heme doming could also induce similar small effects.

Where the changes in the XANES spectra in the region of peaks B and B' are correlated with variation of the Fe-N_{ε} distance, correlation between the variation of the Fe-N_{ε} distance and the spin state are aspected. The observed similar effect on the XANES of the spin transition in met myoglobin observed by Oyanagi et al. [18)] could be associated with proximal histidine movement. This has to be considered a preliminary result and it is necessary to refine the calculations to give a quantitative measure of the proximal histidine movement correlated with the spin state and the protein quaternary conformation. However the XANES spectroscopy has been able to indicate that the spin state is associated with a subtle structural transitions keeping the average interatomic distance within the iron coordination shell constant within 0.01 Å.

References

1) Perutz M.F. (1970), *Nature*, **28**, 126.
2) Pettigrew D.W., Romeo P.H., Tsapsis A., Thillet J., Smith M.L., Turner B.W., and Ackers G.K. (1982), *Proc. Nat. Acad. Sci. USA*, **79**, 1849.
3) Parkhust L.J., Goss D.J.,and Perutz M.F. (1983), *Biochemistry*, **22**, 5401.

4) Perutz M.F., and Brunori M. (1982), *Nature*, **299**, 421.
5) Rousseau D.L., and Ondrias M.R. (1985), *Biophys. J.*, **47**, 537.
6) Eisenberger P., Shulman R.G., Brown G.S., and Ogawa S. (1976) *Proc. Nat. Acad. Sci.* **73**, 491
7) Eisenberger P., Shulman R.G., Kincaid B.M., Brown G.S., and Ogawa S. (1978) *Nature* **274**, 30

8) Bianconi A., Congiu Castellano A., Dell'Ariccia M., Giovannelli A., and Morante S. (1985) *FEBS letters* **191**, 241.
9) Bianconi A., Congiu Castellano A., Dell'Ariccia M., Giovannelli A., Morante S., Burattini E., and Durham P.J. (1986), *Proc. Nat. Acad. Sci. USA* **83**, 7736
10) Perutz M.F., Sanders J.K.M., Chenery D.H., Noble R.W., Pennelly R.R., Fung L.W.M., Ho C., Giannini I., Porschke D., and Winkler H. (1978) *Biochemistry* , **17**, 3640.
11) Messana C., Cerdonio M., Shenkin P., Noble R.W., Fermi G., Perutz R.N., and Perutz M. (1978) , *Biochemistry*, **17**, 3652.
12) Noble R.W., De Young A., Di Iorio E.E., Winterhalter K.H., Cerdonio M., Morante S., and Vitale S. (1983) , *Eur. J. Biochem.* , **133**, 475.
13) Scheidt W.R.,and Reed C.A., (1981), *Chem. Rev.* **81** , 543
14) Noble R.W., Parkhust L.J., and Gibson Q.H. (1970), *J. Biol. Chem.* **25**, 6628.
15) Bianconi A., Dell'Ariccia M., Gargano A., and Natoli C.R. (1983), in " *EXAFS and Near Edge Structure" Springer Series in Chem. Phys.* , **27**, 57
16) Bianconi A., Fritsch E., Calas G. , and Petiau J., (1985) *Phys. Rev.* B **32,** 4292
17) Bianconi A., Congiu Castellano A., Dell'Ariccia M., Burattini E., and Durham P.J. (1985), *Biochem. Biophys. Res. Commun.* **131**, 98.
18) Oyanagi H., Iizuka T., Matsushita,T., Saigo S., Makino R., and Ishimura Y. (1986) *J. Phys. (Paris)* **47** , C8-1147
19) Stryer L., Kendrew J.C., and Watson H.C. (1964) *J.Mol. Biol.* **8,** 96

EXAFS and XANES Studies of High-Valent Metal-Oxo Porphyrins Relevant to Horseradish Peroxidase

J.E.Penner-Hahn[1], K.Smith Eble[2], J.H.Dawson[2], T.J.McMurry[3], J.T.Groves[3], M.Benfatto[4], S.Doniach[4], B.Hedman[5] and K.O.Hodgson[5]

1) Department of Chemistry, Michigan University, Ann Arbor, MI 48109-1055, USA
2) Department of Chemistry, South Carolina University, Columbia, SC 29208, USA
3) Department of Chemistry, Princeton University, Princeton, NJ 08540, USA
4) Department of Applied Physics, Stanford University, Stanford, CA 94305, USA
5) Department of Chemistry, Stanford University, Stanford, CA 94305, USA

INTRODUCTION

Horseradish peroxidase is a heme-containing enzyme that catalyzes the oxidation of phenolic substrates, utilizing hydrogen peroxide as the ultimate electron acceptor. The resting form of HRP contains a high-spin ferric heme. During the catalytic cycle, HRP is initially oxidized by two electrons to a green species known as Compound I (HRP-I). One electron reduction of HRP-I yields the red species known as Compound II (HRP-II), that retains one oxidizing equivalent above the resting enzyme. There is substantial interest in understanding the chemistry of high-valent transition metal species in general and the peroxidases in particular because of their importance in oxidation and oxygenation reactions. In this report we review our recent EXAFS and XANES studies of high-valent metalloporphyrins which are relevant to horseradish peroxidase compounds I and II.

Extensive physical and chemical characterization has revealed the major features of the HRP reaction cycle [1]. The resting form of HRP contains a five-coordinate heme, with the fifth coordination site occupied by an axial imidazole ligand [2]. HRP-I is two oxidation equivalents above the ferric resting state and therefore formally contains Fe(V). Magnetic susceptibility measurements [3] indicate that there are three unpaired electrons ($S=3/2$) in HRP-I. This has been explained by ferromagnetic coupling of low-spin Fe(IV) ($S=1$) to a porphyrin π-cation radical ($S=1/2$). Consistent with this assignment is the similarity of the optical spectrum of HRP-I and those of synthetic porphyrin π-cations [4]. Investigations of HRP-I by NMR [5] and ENDOR [6] as well as iterative extended Huckel (IEH) calculations [7] all favor the π-cation radical formulation.

HRP-II is formed by one electron reduction of HRP-I and is therefore formally Fe(IV). The Mössbauer spectra of HRP-I and HRP-II are very similar and are different from that of the ferric enzyme [8], supporting the assignment of Fe(IV) for both of the oxidized compounds. Isotopic labeling studies have shown unequivocally that HRP-I and HRP-II contain one atom of oxygen [9], but the protonation state of this oxygen has been a matter of controversy. HRP-I is generally believed to contain an oxo (double bonded oxygen) ligand, an assignment which is supported by ^{17}O ENDOR measurements [10]. Both oxo and hydroxyl ligation have been proposed for HRP-II, however recent NMR [11] and resonance Raman [12] studies appear to be consistent only with the oxo assignment.

Initial XAS studies of HRP [13] showed that both of the high valent HRP species contain highly oxidized iron porphyrins (>Fe(III)) and a coordination environment that could only be modeled by inclusion of a short (~1.6 Å) Fe-O interaction. A key feature of this work was the finding that HRP-I and HRP-II, and related Compound I and Compound II models, had identical structures from the perspective of XAS. A recent report by Chance et al. confirmed these results for HRP-I but suggested that HRP-II has a significantly different structure, characterized by a much longer (1.93 Å) Fe-O interaction [14]. On the basis of this metric information, Chance et al. proposed a peroxidase mechanism involving formation of a double bonded (Fe=O) complex in HRP-I and, on reduction, a single bonded Fe-O complex in HRP-II. This mechanism is in conflict with findings from our laboratories [13,15] and with other evidence [8] that the Fe sites in HRP-I and HRP-II are very similar. In addition, this new mechanism, which would indicate hydroxyl ligation in HRP-II, contradicts a substantial body of evidence [11,12,16] suggesting oxo ligation in this complex. We present here a summary of recent EXAFS evidence confirming the Fe=O structure for both HRP-I and HRP-II [15] together with XANES studies of HRP-I and HRP-II [13] and related highly oxidized metalloporphyrins [17].

EXPERIMENTAL

HRP was prepared and concentrated as described previously [13]. HRP-I was prepared by addition of ethyl hydroperoxide to ferric HRP, giving a characteristic green solution. Addition of one equivalent of reductant (ascorbate) to HRP-I resulted in formation of a red solution, characteristic of HRP-II. Sample composition was verified using UV-visible absorption spectroscopy at 4°C on a diluted aliquot. By the UV-visible criterion, sample homogeneity was greater than 95% for both HRP-I and HRP-II. After preparation, the samples were frozen and maintained at -80° C throughout data collection. Following completion of XAS data collection, the protein samples were immediately reanalyzed by UV-visible absorption spectroscopy; no significant degradation in sample homogeneity was observed. High valent iron porphyrin model compounds with formal oxidation states Fe(IV) and Fe(V) are illustrated schematically (1-3, right). These were prepared [18] according to literature procedures [19]. As a complement to the Fe studies, the crystallographically characterized high-valent Cr porphyrins (TTP)Cr(IV)O (4) and (TTP)Cr(V)N (5) were also studied. Data were calibrated, averaged, and reduced as described previously [20]. The normalized, background subtracted data were converted from energy E to photoelectron wave-vector $k = [(2m/\hbar^2)(E-E_0)]^{1/2}$ with E_0 = 7130 eV (Fe) or 6010 eV (Cr).

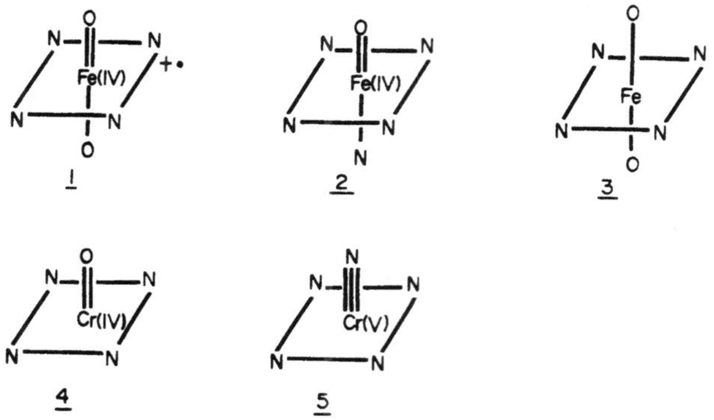

The two different procedures used for data analysis are described in detail elsewhere [15]. In brief, the first method is an empirical procedure utilizing parameterized amplitude and phase functions determined from model compounds. In this method, the number of scatterers and the absorber-scatterer distance are adjusted as variable parameters [21]. The second analysis procedure is similar to Teo's FABM method [22], in which ab initio parameters are used, with the scale factor and ΔE_0 empirically optimized by comparison with model compounds.

RESULTS

EXAFS Spectra

The EXAFS spectra for HRP-I, HRP-II, 1, 2, 3, 4, and 5 are shown in Fig. 1. The spectra for HRP-I and HRP-II and for 1 and 2 are extremely similar, demonstrating that the change in formal oxidation state from Fe(V) to Fe(IV) has a negligible effect on the structure of the Fe site. This is consistent with the principal difference between these species being the presence or absence of a porphyrin π-cation. In addition, all four of these spectra and the spectra for 4 and 5 are quite similar, with a characteristic principal maximum at k ~6.1 Å^{-1}. This suggests that the metal sites in all 6 species must share similar environments. In contrast, the EXAFS spectrum for 3, with a maximum at k ~6.4 Å^{-1}

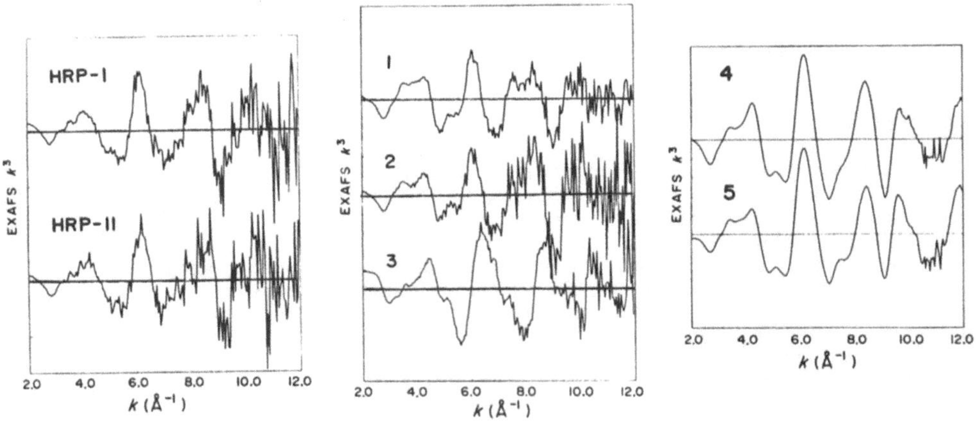

Fig. 1. EXAFS spectra for HRP-I, HRP-II, 1, 2, 3, 4, and 5.

is noticeably different from the other 6 spectra, especially over the range $\underline{k}$ ~5-9 Å^{-1}. This indicates the presence of a distinctly different metal site in 3.

In order to test these qualitative interpretations, we have carried out a complete curve fitting analysis of the EXAFS data. The resulting structural parameters are given in Table I. For all species except 3, the mean-square deviation between the data and the fit decreased 2-fold or more for 2 shell fits (N+O) compared to 1 shell (N only) fits. The short Fe=O distance determined for HRP-I, HRP-II, 1, and 2 is ~1.64 Å. The availability crystallographic structures for 4 and 5 allows us to estimate the uncertainty in this distance as $\leq \pm 0.03$ Å [15].

Table I. Iron-Porphyrin Structural Parameters

	EXAFS[a]		Crystallography	
Sample	M-N	M-O	M-N	M-O
1	2.03	1.62-1.66		
2	2.03	1.64-1.66		
3	1.92	----		
4	2.01	1.52	2.03	1.57
5	2.02	1.49	2.04	1.57
6			2.005	1.604
HRP-I	2.00	1.61-1.64		
HRP-II	2.00	1.60-1.64		

[a]EXAFS results are the extreme values obtained by any of the fitting methods employed. M-N is the average of M-N(pyrrole) and M-axial atom bond lengths. M-O is the M=O (M≡N for 5) bond length. Errors in M-O for 4 and 5 results from errors in E_o [15]. If E_o is allowed to vary, M-O for 4 and 5 changes to the crystallographic value; no such change in M-O occurs for the Fe compounds.

In contrast with the EXAFS data for the other high-valent species, the data for 3 were well modeled using only a single shell of nitrogen, with a relatively short (1.92 Å) Fe-N distance. Attempts to improve this fit by including a shell of oxygen at a short distance resulted in the disappearance ($\sigma \rightarrow$ ~1) of either the N or the O shell. These fitting results, together with XANES spectra (vide infra) indicating a highly symmetric Fe environment, are most consistent with a symmetric

6-coordinate structure for 3, with an average Fe-ligand distance of 1.92 Å. The Fe-N(porphyrin) distance expected for a 6-coordinate porphyrin is ~1.98 Å. In order to obtain the observed average Fe-ligand distance, this would require an Fe-O axial distance of ~1.80 Å in 3. This is consistent with recent NMR and Mössbauer data showing that 3 is in fact a bismethoxy Fe(IV) porphyrin [23].

The most significant result of this work is the finding that a short Fe-O bond with a bond length of 1.64±0.03 Å is present in HRP-I and HRP-II and in their high-valent Fe porphyrin models. The corresponding Cr-O distance in the chromyl analog **4** [24] and in oxo-Cr(V)(salen) complexes [25] is 1.55-1.57 Å. An increase of ca. 0.08 Å in the M-O distance is completely consistent with the addition of two 3d electrons in moving from Cr(IV) to Fe(IV). Additional support for this ferryl structure is found in the recent preparation and preliminary structural characterization of a new HRP-II model, Fe(IV)(O)(THF)(TPivP), **6** [26]. Although this structure suffers from disorder, it indicates that the ferryl Fe=O bond length is 1.604±0.019Å. The bond length increase of 0.04 Å between **6** and HRP-I or HRP-II is only marginally significant, however this may reflect the greater strength of the distal Fe-N(imidazole) bond in HRP relative to the Fe-O(THF) bond in **6**.

XANES Spectra

The XANES spectra for resting HRP, HRP-I, **1**, **2**, and **3** are shown in Figure 2. In each case, the principal absorption discontinuity occurs at a higher energy than that found for Fe(III) porphyrins. This suggests [13] that the metal oxidation state in all four of these species is Fe(IV) or greater. The spectra for HRP-I and HRP-II (not shown) are superimposable within the noise level of the data, and those for **1** and **2** are quite similar. This similarity confirms the EXAFS finding (above) that the metal sites in all four species have very similar structures. The extremely weak 1s→3d transition in **3** is consistent with the symmetric bis-methoxy structure proposed above.

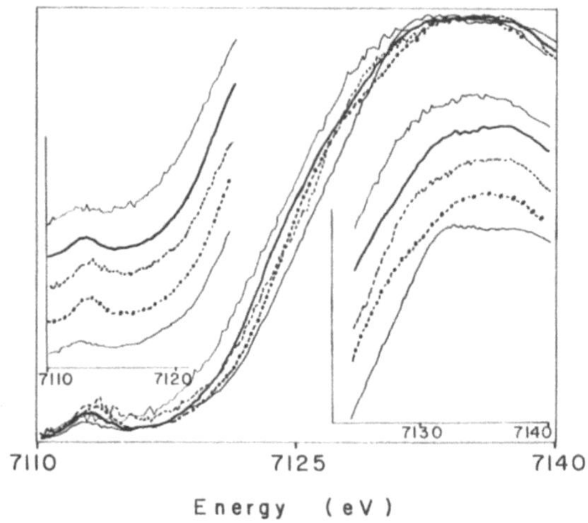

Fig. 2. XANES spectra for resting HRP (---, top-left), 1 (---), HRP-I (---), 2 (-), and 3 (---, bottom-right). Spectrum for HRP-II (not shown) is identical to that for HRP-I, within the noise of the data. [Reproduced from Ref. 13].

One of the most surprising results of these XANES studies is the finding that the putative ferryl species have only very weak pre-edge transitions. This stands in contrast with the XANES spectra of the earlier transiton-metal oxo porphyrins, e.g. (por)Ti=O, (por)V=O, and (por)Cr=O, which are characterized by an intense pre-edge transition. A powerful approach to understanding XANES spectra is via polarization dependent measurements, however the lack of suitable crystals precludes use of this technique to characterize the XANES spectra of ferryl porphyrins. We have instead relied on multiple-scattered-wave (MSW) Xα calculations. In order to test the validity of these calculations, we have also studied the oxidized Cr porphyrins 4 and 5, since these compounds are amenable to polarized XANES studies. The polarized XANES spectra for 4 are shown in Fig. 3. These spectra, and the analogous spectra for 5 (not shown) have an extremely intense pre-edge transition which is essentially completely polarized in the Cr=O (or Cr≡N) direction.

Fig. 3. Polarized XANES spectra for 4. Solid curve is for e||Cr=O; dashed curve is e⊥Cr=O. Both spectra have been normalized to give an edge jump of 1.0.

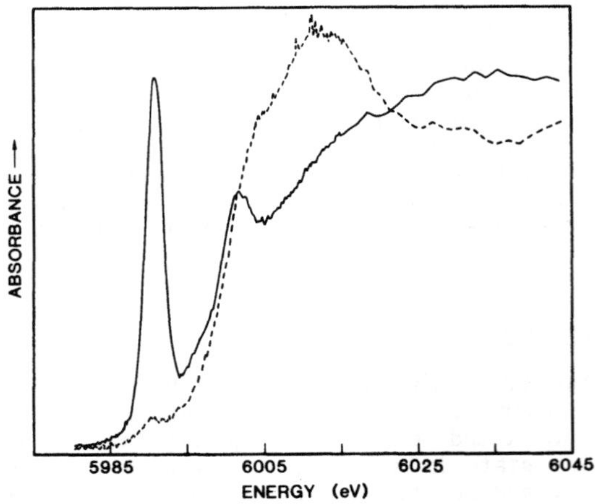

We have performed [17] MSW Xα calculations for 4 and 5 using the "extended-continuum" method [27] with a 6 atom cluster (Cr+O+4N). These calculations (shown for 4 in Fig. 4a) reproduce quite well the principal features of the polarized XANES spectra. On the basis of these calculations, the intense pre-edge transition is interpreted as a bound→bound transition with a final state composed predominantly of Cr-d and O-p orbitals. This interpretation is consistent with the results reported by Templeton and Templeton for the V=O and O=U=O units [28].

When analogous calculations are performed with the Cr(IV) replaced by Fe(IV), the results shown in Fig. 4b are obtained. These spectra clearly show that, in the context of MSA Xα calculations, the intensity of the pre-edge transition will decrease dramatically on going from (por)Cr=O to (por)Fe=O. The most likely explanation for this decrease is that the two 3d electrons which are added on going from Cr(IV) to Fe(IV) are placed in anti-bonding orbitals, thus

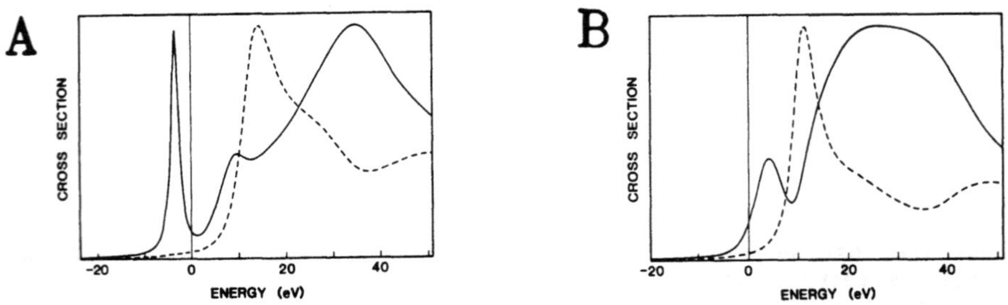

Fig. 4. Calculated XANES spectra. Solid curve is e||M=O, dashed curve is e⊥M=O. All spectra calculated in extended-continuum formalism using a 6 atom cluster as described in Ref. 17. (a) Spectra for 4. (b) Spectra for hypothetical Fe=O analog of 4.

destabilizing the Fe=O bond. This results in a longer Fe=O bond and consequently less overlap between the metal d and the oxygen p orbitals [17]. The EXAFS data (above) also show evidence for an increase in Fe=O bond length relative to Cr=O. While these calculations do not prove that the Fe=O structure proposed for HRP-I and HRP-II is correct, they do demonstrate that the absence of an intense pre-edge transition is not inconsistent with a metal-oxo structure.

DISCUSSION

Our finding that HRP-I and HRP-II are very similar structurally is at odds with the findings of Chance et al. [14] who suggest that HRP-II has a much longer, 1.93 Å, Fe-O single bond. The two proposed structures for HRP-I are very similar, thus ruling out different data analysis procedures as accounting for the discrepancy in HRP-II structures. We note that the EXAFS spectra for HRP-I and HRP-II shown in Fig. 4 of Ref. 14 are noticeably different, in contrast to the similarity which we observe (Fig. 1). The reasons for this difference remain to be resolved, however we note that our results are consistent with the suggestion (vide supra) that the principle difference between HRP-I and HRP-II is the presence of a porphyrin π-cation in the former. In particular, the similarity of the Mössbauer spectra [8] and the X-ray absorption edge spectra of HRP-I and HRP-II [13] suggests that the Fe environment in these species must be very similar. Additional evidence for the similar structures is provided by resonance Raman spectroscopy, which shows that HRP-I [29] and HRP-II [12] have nearly identical Fe-O stretching frequencies. Finally we note that the only high-valent Fe porphyrin complex to have been crystallographically characterized (complex **6**, above) is in fact a model for HRP-II. The Fe-O bond length in **6** is consistent with our results and inconsistent with those of Chance et al.

The reduction of HRP-I is known to occur with concomitant uptake of one proton, however our results show that this does not result in protonation of the

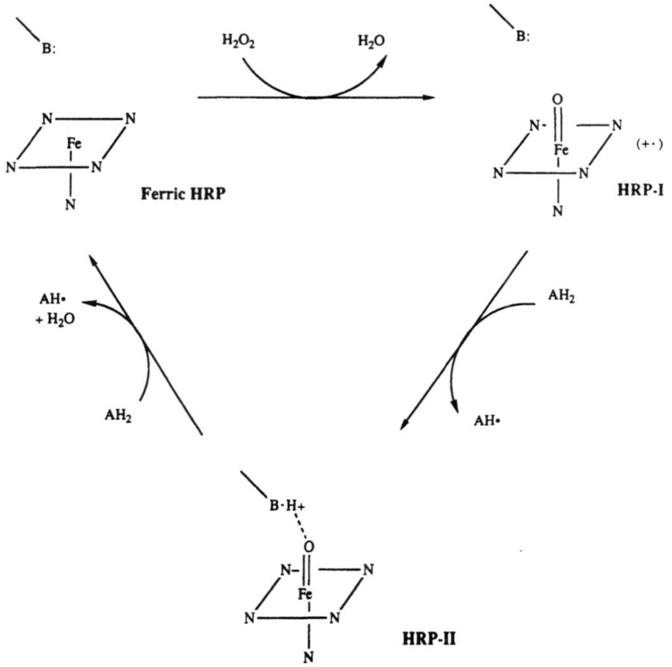

Fig. 5. Schematic illustration of HRP catalytic cycle consistent with XAS results.

oxo ligand. This is not surprising, since the Fe(IV)=O moiety is unlikely to be a strong base. A possible location for this added proton is suggested by recent resonance Raman results [30] which indicate the presence of an ionizable group (probably imidazole) within hydrogen-bonding distance of the oxo ligand. A proton in this location would then be available when HRP-II is converted, with the uptake of an electron and a second proton, to ferric HRP and water (Figure 5). The fact that HRP-I and HRP-II have similar iron-site structures may facilitate rapid reduction of HRP-I during the catalytic cycle, since there would be only a small reorganizational energy barrier to this reaction. The apparently pervasive nature of short Fe=O bonds in high valent Fe porphyrins (HRP-I, HRP-II, 1, and 2), together with the similar reactivity of HRP and P-450 on treatment with peroxide, suggests that a similar ferryl (Fe=O) complex may be the active oxygen species in P-450.

ACKNOWLEDGEMENTS

This work was supported by the National Science Foundation (CHE 85-08115 to KOH, CHE 84-06373 to JTG, and PCM 82-16799 to JHD) and a Camille and Henry Dreyfus Distinguished Young Faculty Award (to JEPH). JHD is a Camille and Henry Dreyfus Teacher/Scholar, an Alfred P. Sloan Research Fellow, and a recipient of a National Institutes of Health Research and Career Development Award (AM-01123). Synchrotron radiation was provided by the Stanford Synchrotron Radiation Laboratory which is supported by the United States Department of Energy and the National Institutes of Health through the Biotechnology Resource Program in the Division of Research Resources. We wish to thank Professor A.L. Balch for providing model compounds and Professor G.T. Babcock for communication of results prior to publication.

REFERENCES

1. (a) Dunford, H.B.; Stillman, J.S. Coord. Chem. Rev. **1976**, 19, 187-251. (b) Hewson, W.D.; Hager, L.P. In The Porphyrins; Dolphin, D., Ed.; Academic Press: New York, 1979; Vol. 7, pp 295-332. (c) Frew, J.E.; Jones, P. Adv. Inorg. Bioinorg. Mech., **1984**, 3, 175.
2. Poulos, T.G.; Finzel, B.C.; Gunsalus, I.C.; Wagner, G.C.; Kraut, J. Biol. Chem., **1985**, 260, 16122-16130.
3. Theorell, H.; Ehrenberg, A. Arch. Biochem. Biophys., **1952**, 41, 442-461.
4. Dolphin, D.; Forman, A.; Borg, D.C.; Fajer, J.; Felton, R.H. Proc. Natl. Acad. Sci., U.S.A. **1971**, 68, 614-618.
5. LaMar, G.; deRopp, J.S.; Smith, K.M.; Langry, K.C. J. Biol. Chem., **1981**, 256, 237-243.
6. Roberts, J.E.; Hoffman, B.M.; Rutter, R.; Hager, L.P. J. Biol. Chem. **1981**, 256, 2118-2121.
7. Hanson, L.K.,Chang, C.K.; Davis, M.S.; Fajer, J. J. Am. Chem. Soc. **1981**, 103, 663-670.
8. (a) Schulz, C.E.; Devaney, P.W.; Winkler, H.; Debrunner, P.G.; Doan, N.; Chiang, R.; Rutter, R.; Hager, L.P. FEBS Lett. **1979**, 103, 102-105. (b) Moss, T.H.; Ehrenberg, A.; Bearden, A.J. Biochemistry, **1969**, 8, 4159-4162.
9. (a) Schonbaum, G.R.; Lo, S. J. Biol. Chem., **1972**, 247, 3353-3360. (b) Adediran, S.A.; Dunford, H.B. Eur. J. Biochem. **1983**, 132, 147-50.
10. Roberts, J.E.; Hoffman, B.M.; Rutter, R.; Hager, L.P. J. Am. Chem. Soc. **1981**, 103, 7654-7656.
11. LaMar, G.N.; de Ropp, J.S.; Latos-Grazynski, L.; Balch, A.L.; Johnson, R.B.; Smith, K.M.; Parish, D.W.; Cheng, R.-J. J. Am. Chem. Soc. **1983**, 105, 782-787.
12. Terner, J.; Sitter, A.J.; Reczek, C.M. Biochim. Biophys. Acta **1985**, 828, 73-80.
13. Penner-Hahn, J.E.; McMurry, T.; Renner, M.; Latos-Grazynski, L.; Eble, K.S.; Davis, I.M.; Balch, A.; Groves, J.T.; Dawson, J.H.; Hodgson, K.O. J. Biol. Chem. **1983**, 258, 12761-12764.

14. Chance, B.; Powers, L.; Ching, Y.; Poulos, T.; Schonbaum, G.R.; Yamazaki, I.; Paul, K.G. Arch. Biochem. Biophys. **1984**, 235, 596-611.
15. Penner-Hahn, J.E.; Eble, K.S.; Renner, M.; Balch, A.L.; McMurry, T.J.; Groves, J.T.; Dawson, J.H.; Hodgson, K.O. J. Am. Chem. Soc., **1986**, in press.
16. Hanson, L.K., personal communication.
17. Penner-Hahn, J.E.; Benfatto, M.; Hedman, B.; Takahashi, T.; Doniach, S.; Groves, J.T.; Hodgson, K.O. Inorg. Chem., **1986**, 25, 2255-2259.
18. The HRP-II model **2** was prepared by A.L. Balch and coworkers.
19. (a) Chin, D.H.; LaMar, G.N.; Balch, A.L. J. Am. Chem. Soc. **1980**, 102,4344-4350. (b) Chin, D.H.; Balch, A.L.; LaMar, G.N. J. Am. Chem. Soc. **1980**, 102, 1446-1448. (c) Groves, J.T.; Haushalter, R.C.; Nakamura, M.; Nemo, T.E.; Evans, B.J. J. Am. Chem. Soc. **1981**, 103, 2884-2886. (d) Groves, J.T.; McMurry, T.J. Revista Portuguesa de Quimica **1985**, 27, 102-103.
20. Hahn, J.E.; Hodgson, K.O.; Andersson, L.A.; Dawson, J.H. J. Biol. Chem., **1982**, 257, 10934-10941.
21. Cramer, S.P.; Hodgson, K.O. Prog. Inorg. Chem., **1979**, 25, 1-39.
22. Teo, B.K.; Antonio, M.R.; Averill, B.A. J. Am. Chem. Soc. **1983**, 105, 3751-3762.
23. (a) Groves, J.T.; Quinn, R.; McMurry, T.J.; Nakamura, M.; Lang, G.; Boso, B. J. Am. Chem. Soc. **1985**, 107, 354-360. (b) Groves, J.T.; Gilbert, J.A. Inorg. Chem. **1986**, 25, 125-127.
24. Groves, J.T.; Kruper, W.J. Jr.; Haushalter, R.C.; Butler, W.M. Inorg. Chem. **1982**, 21, 1363-1368.
25. Srinivasan, K.; Kochi, J. Inorg. Chem. **1985**, 24, 4671-4679.
26. Schappacher, M.; Weiss, R.; Montiel-Montoya, R.; Trautwein, A.; Tabard, A. J. Am. Chem. Soc. **1985**, 107, 3736-3638.
27. Berding, M.B.; Doniach, S., manuscript in preparation.
28. Templeton, D.K.; Templeton, L.K. Acta Cryst **1980**, A36, 237-41. Templeton, D.K.; Templeton, L.K. ibid., **1982**, A38, 62-67.
29. Babcock, G.T., personal communication.
30. Sitter, A.J.; Reczek, C.M.; Terner, J. J. Biol. Chem., **1985**, 260, 7515-7522.

Energy Dispersive X-Ray Absorption Spectroscopy: Strengths and Limitations for Time-Resolved Studies of Biostructures

I. Ascone*, A.Fontaine
LURE (Lab. CNRS, CEA, MEN) Bat. 2090, 91405 Orsay Cedex, France

A.Bianconi, A.Congiu Castellano, A.Giovannelli, S.Della Longa
Dipartimento di Fisica, Università di Roma "La Sapienza", 00185 Roma, Italy

M.Momenteau
Institut Curie, Section de Biologie, U.219 INSERM, Bat. 112, 91405 Orsay Cedex, France

INTRODUCTION

Fast energy dispersive x-ray spectroscopy has been used for in-situ time dependent observations of irreversible structural changes of carboxy-hemoglobin in solution. For the first time one is able to proceed with both in situ and time-dependent investigations of the changes of the local structure by XANES (x-ray absorption near edge structure) of of the active site of a protein during a chemical process.

The x-ray energy dispersive optics takes advantage of the unique characteristics of synchrotron radiation sources. The combination of a position-sensitive detector able to work under high x-ray flux conditions, characteristic of synchrotron radiation sources, and of the x-ray energy dispersive optics allows to record x-ray absorption spectra in a very short period. The aim of this paper is to show that the energy dispersive method can be applied to the studies of bio-reactions of proteins in diluted solution. In a first part we discuss the advantages and the limitations of the dispersive scheme and its perspectives using the future high brillance synchrotron radiation sources. In the second part we report the XANES study of the denaturation of carboxy-hemoglobin.

ENERGY DISPERSIVE SCHEME

Dispersive x-ray absorption spectroscopy(1) has been used for time dependent structural studies(2-6) of solids and concentrated solutions.12 ms only are neded to collect a full x-ray absorption spectrum of a metal foil spread over 500 eV when a triangle-shaped Si(111) bent crystal is used as a focusing dispersive optics.

The energy resolution is as good as it is with a step by step scanning double crystal monochromator. Using Si(311) monochromator the resolution is increased. Because of the lack of mechanical movements the energy scale is absolutely stable, once the optics has been tuned for the absorption edge of a selected atomic species. Most of the current optics are working in the vertical plane. The horizontal optical plane of the dispersive scheme adds an advantage for focusing synchrotron radiation and the ultimate energy resolution can be even better than that obtained by the scannig method. In figure 1 the schematic set up of the dispersive x-ray absorption system is shown.

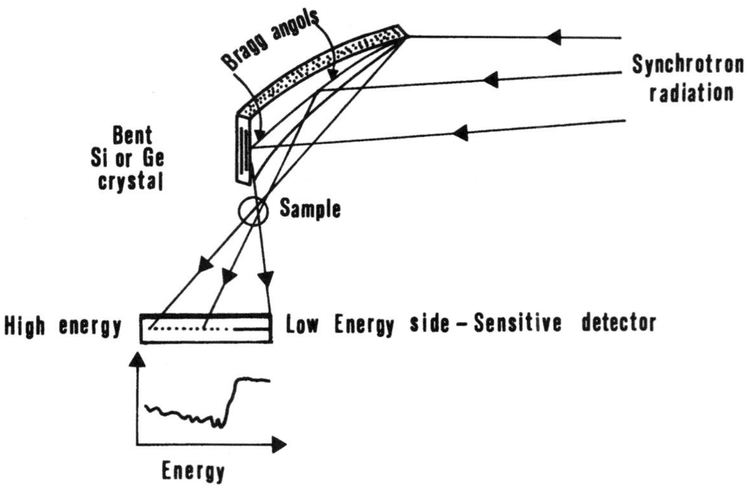

Fig. 1 Schematic layout of the dispersive x-ray absorption apparatus.

The linear detector is a cooled photodiode array which gives a good spatial resolution and a large dynamics(1). Each of the 1024 sensing elements of the photodiode array transforms an average of 10^5 x-ray photons in $8.8x10^7$ electron-hole pairs. Each frame is made of 10^8 photons which can be repeated at 550 Hz or 220 Hz according to the used analog-digital converter (10 bits or 12 bits). The linearity and the dynamics of the photodiode array gives very good signal/noise ratio.

In fig.2 the Fe XANES spectrum of the Fe foil obtained in 29 msec using the Si(311) dispersive monochromator with a resolution as good as it is with a step-by-step scan monochromator is shown.

In order to measure the XANES of proteins in solution the integration time has to be increased. In fact the thickness of sample has to be increased and the water solvent strongly attenuates the transmitted signal. The signal is essentially statistically limited as proved by comparing the XANES spectra of 50 mM iron concentration in aqueous solutions in fig.2 with the 5 mM concentration carboxy-hemoglobin in fig.3. In the present status a 1000 times more intense flux will be acceptable for the detector and the integration time can be reduced by the same factor for measuring XANES of protein solutions.

The fall down of the quantum efficiency for high energy x-ray photons of the silicon detector is an advantage for the measurements of diluted system in aqueous solution at the Fe K-edge. In fact it gives a relative unsensitivity for higher harmonic contamination which allows to work under high absorbance.

According to the length of the bent crystal (7 cm - 23 cm) the size of the focus has been measured to be 350 μ up to 800 μ. This increase is due mainly to aberration because the cylindrical shape departs from the ideal elliptical one.

However the overall reduced size of the focus beam allows to work with small volumes of solutions.

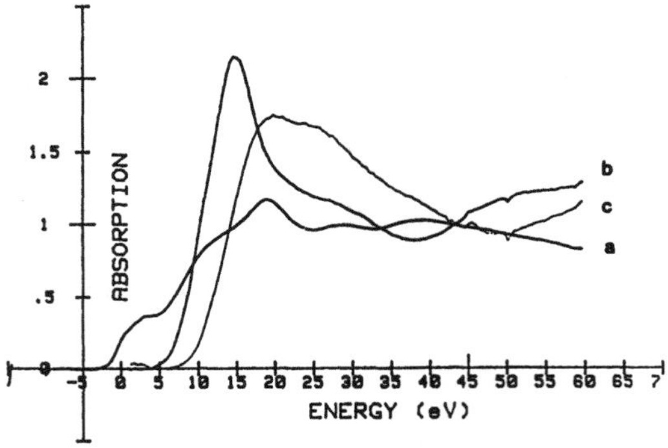

Fig. 2 The Fe K-XANES spectra of a) the Fe foil b) the Fe^{2+} in water solution and c) Fe^{3+} in water solution at 50 mM concentration.

The image of the source does not contain an homogeneous energy distribution within the focus point. The energy-position correlation is U-shaped. Working with solution gives no problem since the thickness is constant within the breadth of the image. In case of precipitation the chemical reaction may induce percolated fibrils which creates a non linear background.

An injected ingredient does not merge easily in thin liquid samples. Consequently it is an obligation to work with a stop-flow apparatus[6] if one's interest is in real characteristic time which is possible to achieve with a focused probing beam.

One has to stress once more that the dispersive optics does not reflect more photons than a single flat crystal. The main feature which differentiates a step-by-step monochromator from the dispersive optics is that the rocking curve of the bent crystal (almost constant in width) is scanning a broad energy range when one crosses the horizontal beam from one side to the other. On the contrary for the step-by-step spectrometer each elementary beam shines the crystal at the same glancing angle and then all the Bragg reflected beams come from a single rocking curve which remains at the same position on the energy scale.

XANES OF HEMOGLOBIN BY DISPERSIVE EXAFS

The absolute value of the edge jump is 0.015 in the spectra of 5 mM HbCO shown in fig. 3 which is almost 200 times lower than the value optimized for getting superior signal/noise ratio.

The measurement time for XANES spectroscopy is determined by the the overall number of photons necessary to build one spectrum with a resonable signal/noise ratio. This is determined by the product of the integration time for each frame by the number of summed frames. The spectra of HbCO in fig.3 are made by summing up to 32 frames each one collected in 2 sec.

The sensitivity is photon limited. Our detection system can work at the ms scale. Hence spectra might be collected even with a 1000 times more intense source.

XANES has been used successfully for determination of molecular binding of diatomic molecules in haem protein crystals as well as solutions[7]. Recent advances in XANES analysis by multiple scattering theory have allowed to identify a peak in the Fe K-XANES of carbonyl derivatives at 16 eV above K threshold as due to a multiple scattering resonance mainly determined by CO and its intensity vatiation is a good probe of the Fe-C-O bonding angle. It is this resonance which will be used to probe the variation of the CO bonding structure with the pH denaturation of hemoglobin.

As deduced from comparison of pH=7 and pH=1.8 spectra of fig. 3 the total change in intensity of the 16 eV CO resonance is 30% of the edge jump i.e. 4.5 10^{-4} and the error bar is estimated to be about 10% of that last value (4.5 10^{-4}).

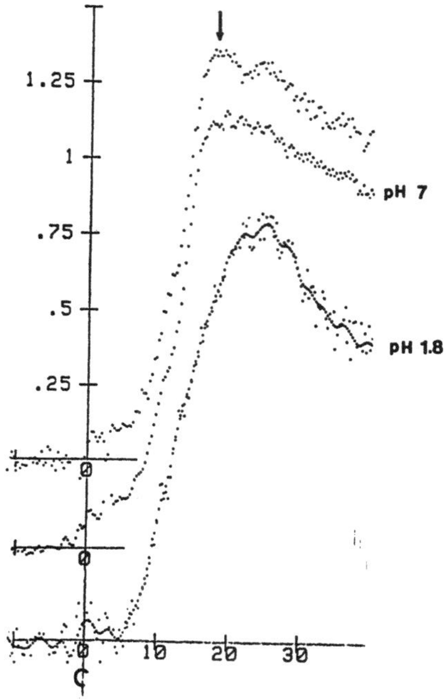

Fig. 3 Fe K-edge XANES spectra of unhindered carbonyl basket handle porphyrin with linear Fe-C-O configuration upper curve, of carboxy-hemoglobin at pH 7 and pH 1.8 and at 5 mM concentration, lower curves. The sample thickness was 2 mm. The curves were obtained with total integration time of 74 sec. The arrow in the upper curve indicates the CO multiple scattering resonance.

In the upper part of fig 3 the XANES spectrum of a synthetic iron II basket-handled complex synthetized by M.Momenteau[8]. XANES analysis has demonstrated that the Fe-C-O configuration is linear in this complex in solution [8] in agreement with x-ray diffraction data of the crystal[9]. The iron II derivative of this complex bearing a ten-membered polymethylene handle shows a great CO affinity in the presence of 1-methyl imidazole in toluene. The arrow in fig. 3 indicates the CO resonance and its intensity gives a measure of the linear Fe-C-O configuration . This resonance is observed in the spectrum of HbCO at pH 7 but it dissapears at pH 1.8. The intensity of the 16 eV CO resonance is 20% larger than the total variation observed between pH=7 and pH=1.8 in fig. 3. This overall change in the resonance intensity observed between these solutions agrees with the detailed XANES comparison of a polarization-dependant study of single crystal MbCO and the HbCO solution reported previously[7].

TIME DEPENDENT XANES STUDY OF HbCO DENATURATION

Our goal is to determine the different steps of the change of the iron active site structure which may exist during protein denaturation. Starting from the protein solution at pH=7 to a final state at pH=1.8 the protein is denatured with a complete CO detachment. This evolution includes protein unfolding which may affect the CO bending/tilting via changes in the environment of the porphyrin cycle. The reduction of pH is obtained by adding lactate to the pH 7 solution which in about one hour reduces the solution up to pH 1.8.

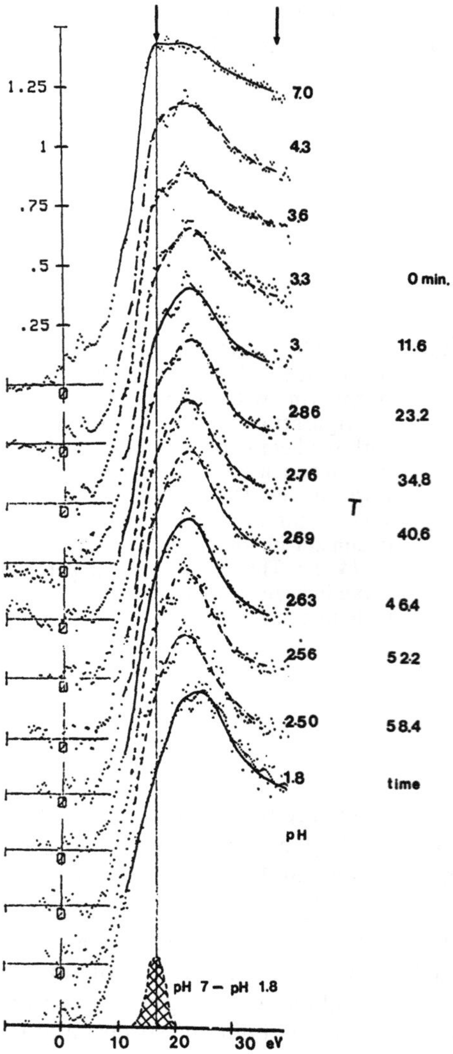

Fig.4 XANES spectra of HbCO at 5mM concentration collected duringthe action of lactate which reduces the pH from the initial state at pH 7. About 60 spectra in one hour were collected.Some of them with the corresponding value of pH are shown

The time-dependent study is used to ramp down the pH value during the slow action of the acid. According to initial the amount of lactate injected in the 5 mM solution, we measured two sets of spectra which overlap on the pH axis. Just after lactate injection the solution was splitted in two volume, one for pH measurements with a standard µ-electrode, the other in a 3 mm-thick cell with two capton windows. This cell is left without any dead volume and sealed in order to avoid oxygenation.

Measurements took about 70 s for each spectrum made of 32 frames each of them taking 2.2 s. The capability to collect all data of a full spectrum at one in a short time allows to investigate in situ such systems evolving under one or several constraints.

Fig. 4 displays the progressive change of the iron K-edge XANES versus the pH values. A guide line has been drawn within data of each spectrum to help the reader.

The pH-evolution has been quantified looking at the fractional change of the 16 eV CO resonance which is reported in fig.5. Hundred percent has been assigned to the total change found between pH=7 and pH=1.8. Whether this study is pH-dependent or time-dependent would be a relevant question if the typical time-constant of such evolution under constant pH was not small.

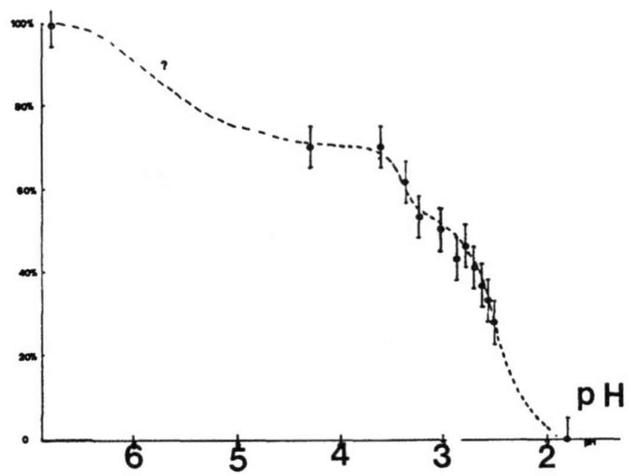

Fig.5 Relative variation of the intensity of the CO resonance at 16 eV going from pH 7 to pH 1.8

Allis and Steinhardt[10] studied both human and horse carboxy-hemoglobin denaturation. The denaturing agent was HCl and the Soret Cotton effect (400-440 nm), as well as the optical rotatory dispersion (220-240 nm) the probes. The involved time constant are within one second or less and then out of the scope of the present study. This two authors distinguished different pH ranges.

i) pH ≈3.4 where protonation of native HbCO occurs.

ii) 3<pH<3.4 where protein unfolding occurs and the heme group retains its close association with its normal imidazole binding site on the protein, although the environment of the heme has been modified. At pH>3 data from all wavelengths give first order kinetics with pH-dependant rate constant.

iii) pH<3 A second-order reaction is superimposed. This is interpreted as a heme dimerization. This second rate is pH-dependent down to pH=2.1.

iv) pH<2.1 Below this value, essentially all the heme is detached from its normal site and the pH dependance disappears.

In fig. 5 the intensity variation of the CO resonance during the action of the lactate is shown. The intensity of the multiple scattering resonance exhibits a rapid decrease at values of pH near 3.5 and 2.5 with disappearing of the CO resonance at pH less 2.1. This decrease can be due both to CO bending as well as to partial dissociation.

We associate the first step with protein unfolding as shown by quenching of the Soret band at the same pH.

A last and fast intensity decrease occurs at pH=2.6 which leads to the disappearance of the CO resonance. This can be associated with CO dissociation and with the suggested dimerization.

A decrease of the CO resonance from pH 7 to pH 4.5 is shown which may correspond to a progressive CO bending as a consequence of the HbCO protonation. But the lack of data in the range pH 5-6 , because of the first rapid pH decrease (due

to the action of lactate), does not allow to give any pH value for the transition.

CONCLUSION

The main result of this study is to show that it is possible to measure the XANES spectrum of a protein in the time scale of seconds. Before the advent of synchrotron radiation the measuring time with standard sources was so long that was impossible to measure x-ray absorption in biological molecules. About ten years ago the first XANES and EXAFS spectra of proteins were measured by synchrotron radiation using the scannig mode requiring about one hour[11-14]. Reversible CO recombination was studied by laser induced photolysis obtaining a time scale of about 20 μsecs[15]. Now the dispersive method allows the in-situ time dependent XANES measurements of the irreversible biological reactions.

This kind of experiment involves small changes of the optical density and is limited by the total transmitted photon flux.The new high brillance sychrotron radiation sources such as the 6 GeV European source should give a flux 1000 times more intense reducing the measurement time for each spectrum by the same factor.

ACKNOWLEDGMENTS

This research was supported by the European Economic Community (CEE) through the Commission of European Communities , contract ST2P-0040-1-I (CD).

The present work was performed when one of us, dr. Isabella Ascone, was at Laboratoire de Chimie Théorique, University of Nancy, France and she would like to express her thanks to Nancy University.

*Present address: Dipartimento di Fisica, Universita` di Roma "La Sapienza", 00185 Roma , Italy

REFERENCES

1) E.Dartyge, C.Depautex, J.M.Dubuisson, A.Fontaine, A.Jucha, P.Leboucher G.Tourillon, (1986) *Nuclear Instrument and Methods, A* **246** , 452
2) G.Tourillon, E.Dartyge, A.Fontaine, and A.Jucha, and C.Andrieux (1986) *J. Phys. (Paris)* **47** , C8-551.
3) E.Dartyge, A.Fontaine, G.Tourillon, R.Cortes, and A.Jucha, (1986) *Phys. Letters A,* **113A**, 38
4) E.Dartyge, A.Fontaine, G.Tourillon, and A.Jucha, (1986) *J. Phys. (Paris)* **47** , C8-607
5) T.Matsushita, H.Oyanagi, S.Saigo, H.Kihara, and U.Kaminaga, (1984) *EXAFS and Near Edge Structure III,* eds. K.O.Hodgson,B.Hedman, J.E.Penner-Hahn (Springer,Berlin) 476
6) S.Saigo, H.Oyanagi, T.Matsushita, H.Hashimoto, N.Yoshida, M.Fujimoto, and T.Nagamura (1986) *J. Phys. (Paris)* **47** , C8-555.

7) A.Bianconi, A.Congiu-Castellano, P.J.Durham, S.S.Hasnain, and S.Phillips, (1985) *Nature* **318**, 685
8) I.Ascone, A.Bianconi, E.Dartyge, S.Della Longa, A.Fontaine, and M.Momenteau (1987) *Biochem. Biophys. Acta* to be published
9) L.Richard, R.Weiss, and M.Momenteau, (1986) *J. of Chem. Soc. Chem. Com.*, **1986** ,819
10) J.W.Allis,and J.Steinhardt, (1970) *Biochemistry* **9**, 2286
11) B.M.Kincaid, P.Eisenberger, K.O.Hodgson, and S.Doniach (1975) *Proc. Natl. Acad. Sci. USA* **72**, 2340
12) R.G.Shulman, P.Eisenberger, W.E.Blumberg and N.A.Stombaugh, (1975) *Proc. Natl. Acad. Sci. USA* **72**, 4003
13) P.Eisenberger, R.G.Shulman and S.Ogawa (1976) *Proc. Natl. Acad. Sci. USA* **73**, 491
14) V.W.Hu, S.I.Chan, and G.S.Brown (1977) *Proc. Natl. Acad. Sci. USA* **74**, 3821
15) D.M.Mills, A.Lewis, A.Horootunian, J.Huang, and B.Smith (1984) *Science,* **223**, 811,

A Rapid-Freeze EXAFS Method for the Studies of Intermediates in Fast Enzymatic Reactions

S. Saigo* and N. Sone
Jichi Medical School, Minamikawachi, Tochigi 329-04, Japan.
T. Nagamura
Unisoku Scientific Instruments, Hirakata, Osaka 573-01. Japan.
H. Oyanagi
Electrotechnical Laboratory, Sakura-mura, Ibaraki 573-01, Japan.
T. Iizuka
Keio University, Shinano-machi, Tokyo 160, Japan.
M. Kusunoki
Meiji University. Higashimata, Kawasaki 214, Japan.
T. Matsushita
Photon Factory, National Laboratory for High Energy Physics, Oho-machi, Ibaraki 305, Japan.

Bray's method [R.C. Bray, Biochem. J. 81, 189-193 (1961)] has been widely used for EPR measurements of freeze trapped enzyme intermediates at low temperature. We have developed a rapid freezing apparatus for XANES and EXAFS measurement in fluorescence mode by applying this method. Solution after mixing is ejected from a nozzle into isopentane kept at -140°C. The ejected solution forms fine ice particles, which are then packed into an observation cell connected with an isopentane vessel by extracting isopentane through a filter. Performance of the apparatus was evaluated on reduction of cytochrome c and deoxygenation of hemoglobin. The packing ratio of frozen samples in the cell was found to be 50 ± 10%, but the intensity of the fluorescence x-ray decreased only by 30%. The quality of the x-ray absorption spectra of rapidly frozen samples was comparable to that of the spectra of samples in the liquid state. The quenching time of the apparatus was estimated to be 5.0 ms. These results demonstrate the feasibility of the rapid-freeze EXAFS method for characterization of intermediates in fast enzymatic reactions.

1. Introduction

Structural characterization of intermediates is essential for the basic understanding of the mechanism and efficiency of enzyme reactions. Several different approaches are possible in order to follow time-dependent phenomena by x-ray absorption spectroscopy. The dispersive method, first developed by Matsushita, *et al*. [1,2], enables one to take the entire spectrum simultaneously and has been applied to kinetic investigations of chemical reactions [3,4]. In the dispersive method, however, we cannot take advantage of the high sensitivity of fluorescence detection mode. Thus applications of this method to the studies of dilute samples such as enzymes are limited by the sensitivity of transmission detection mode. Mills *et al*. [5] have succeeded in measuring time-dependent x-ray absorption spectra in fluorescence mode during recombination of carbon monoxide

* Address before March 31, 1987: Department of Biochemistry and Biophysics, School of Medicine G4, University of Pennsylvania, Philadelphia, PA 19104, USA

with myoglobin. In this case, the ligand molecule was repetitively photodissociated and the photon energy of x-ray was scanned point by point in order to take the entire time-resolved spectra.

The rapid freezing method was originally developed by Bray [6] to characterize enzyme intermediates by EPR spectroscopy. We have developed a rapid-freezing apparatus for x-ray absorption spectroscopy in fluorescence mode by applying Bray's method. The major features of the apparatus and evaluation of its performance are described.

2. Description of the Apparatus

A schematic diagram of the apparatus is shown in Fig. 1. The apparatus is composed of a sample mixer, a low-temperature cell block, its temperature controlling unit and a liquid nitrogen container. Sample solution after mixing is rapidly frozen by Bray's method, *i.e.*, by ejecting a reacting solution into cold isopentane.

The low-temperature cell block is made of copper and gold-plated. The inner diameter and depth of the cylindrical isopentane vessel is 30 mm and 150 mm, respectively. The lower end of the vessel is shaped like a funnel and connected with the upper inlet of the observation cell. The observation cell is 5 mm thick and sandwiched by 100μm thick Mylar windows of 7 x 15 mm^2 in size. The temperature of isopentane in the vessel is kept at -140°C by regulating the flow of liquid nitrogen running through a path in the inner part of the block. The temperature of the block is detected by a silicon diode and the flow of liquid nitrogen is regulated by an electromagnetic valve. The cell block has two 25μm

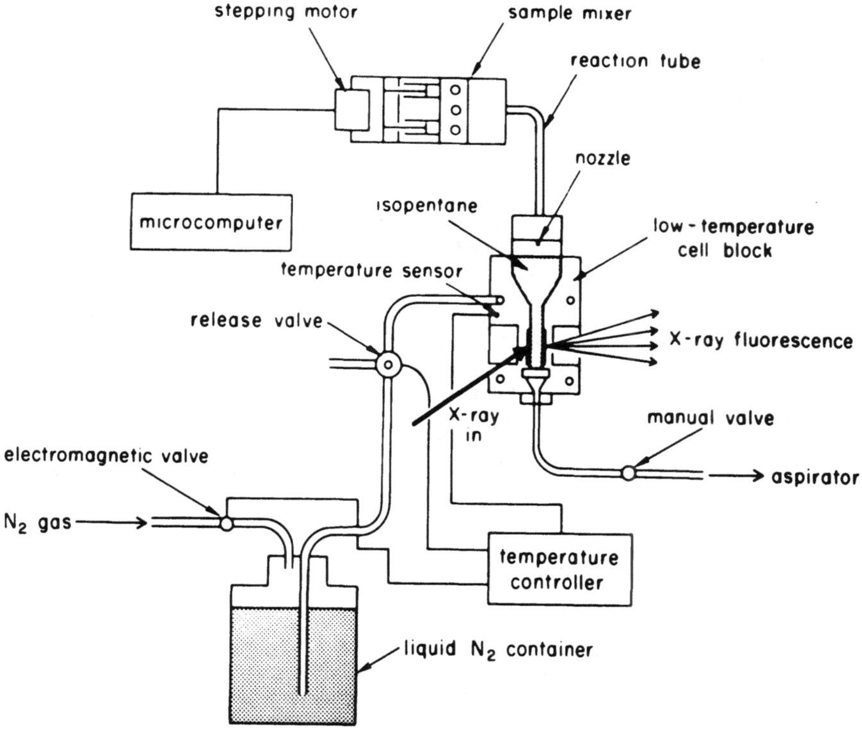

Fig. 1 Schematic diagram of the rapid freezing apparatus for EXAFS measurement.

thick double Mylar windows of 15 mm diameter for incoming and outgoing x-rays and a 25μm thick double Mylar window of 45 mm diameter for detection of fluorescent x-rays. Dry nitrogen gas is continuously passed into the inner space of the cell block to prevent frosting on the cell windows.

The syringe pistons of the sample mixer are driven by a stepping motor which is controlled by a microcomputer. Sample solutions are mixed in a double 2-jet mixer and the mixed solution is ejected from a nozzle into cold isopentane in order to quench the reaction of the solution. The reaction time, i.e., the period between mixing and quenching, is varied by changing the length of the reaction tube between the mixer and the nozzle. Double push mode can be used to produce reaction time longer than 0.6 s. The inner diameter of the reaction tube and the nozzle are 1.0 mm and 0.15 mm, respectively.

The important problem in the development of the apparatus was how to pack a frozen solution into an observation cell. We solved this problem by finding out the conditions under which the ejected solution forms fine ice particles in cold isopentane. The ice particles can then be readily packed into the cell by extracting isopentane through a mesh filter which is placed at the lower end of the cell.

X-ray absorption spectra of rapidly frozen samples were measured on Beamline 4C of the Photon Factory at Tsukuba, Japan using a fluorescence EXAFS apparatus equipped with a multidetector [7]. The sample was oriented so that the normal of the front face of the sample cell was at 45° with respect to both the direction of the x-ray beam and the detector.

3. Evaluation of Performance

a) Test on Decomposition of Carbonic Acid

To assess the performance of the apparatus by visual inspection, the reaction $H_2CO_3 \longrightarrow H_2O + CO_2$ was quenched with the apparatus [6]. We mixed 0.02 M $NaHCO_3$ with 0.01 M HCl. Both of the solutions contained 0.02% bromophenol blue as a pH indicator. The mixture is initially acidic, but becomes alkaline as the reaction proceeds. A stopped-flow study showed that the half-time of the reaction was about 0.1 s. The mixed solution was rapidly frozen with the present apparatus and ice particles formed were packed into the cell. The temperature of isopentane was kept at -140°C and the velocity of the solution coming out from the mixer was 50 cm/s. Calculation showed that the velocity of the jet ejected from a nozzle was 22 m/s. The nozzle was positioned at 10 ± 1 mm above the surface of cold isopentane. The color of the solution frozen just after mixing was completely yellow. When the frozen solution was warmed up to -50°C, the color changed slowly to purple. This shows that the quenching time of the apparatus is much less than 0.1 s. Inspection with an optical microscope showed that the average size of the frozen ice particle was 0.05 mm. 0.3 ml of each reactant solution was necessary to fill up 1.2 ml of the cell volume, so that the packing ratio of the frozen sample in the cell was estimated to be 50 ± 10%.

b) Test on Reduction of Cytochrome c

The reaction of ferricytochrome c with ascorbate was quenched after various reaction times in a similar way. The optical spectra of frozen samples packed in the cell were measured on a Unisoku computer-controlled spectrophotometer with optical fibers. Fig. 2a shows the variation of the absorbance spectra of cytochrome c with reaction time. In this measurement, the reference was a rapidly frozen buffer solution packed in the same cell. It is evident that the peak intensity at 550 nm increases with reaction time, suggesting that the reduction of cytochrome c with ascorbate was appropriately quenched with present apparatus. This was quantitatively verified by making a plot of $(A_{550} - A_{556.5})/(A_{526.5}$

$- A_{556.5})$ against reaction time, where 526.5 nm and 556.6 nm are the isosbestic points of oxidized and reduced cytochrome c. The plot is shown in Fig. 2b, where the smooth curve is an exponential function obtained by nonlinear least-squares fitting. The fitting yields the apparent rate constant of 0.13 s^{-1}, which is in reasonable agreement with the value obtained by the stopped-flow method.

Fig. 3 compares the x-ray absorption spectra of cytochrome c rapidly frozen at 0.15 s and 25 s after mixing with ascorbate. A change in XANES as well as a shift of 2 eV in the edge position is seen on reduction. The differences in the spectra are comparable to those observed by an ordinary static method between oxidized and reduced cytochrome c [8,9].

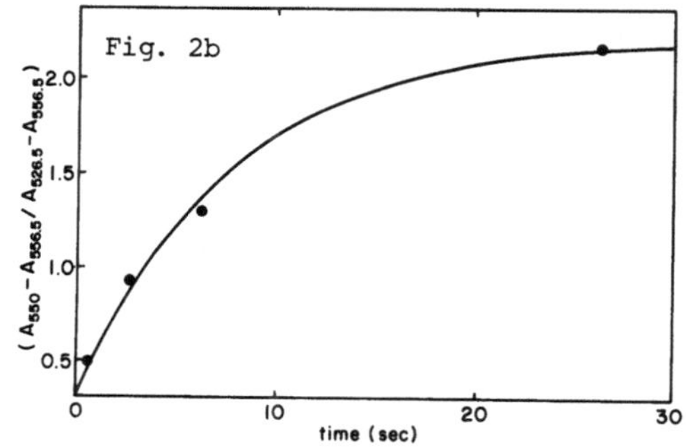

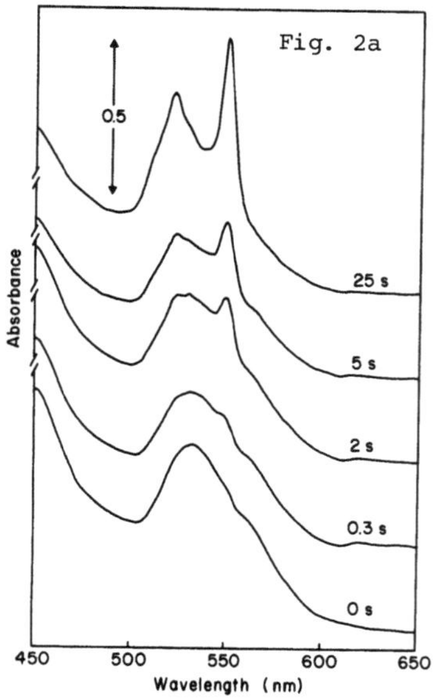

Fig. 2b Plot of $(A_{550} - A_{556.5})/(A_{526.5} - A_{556.5})$ against the time of quenching. The wavelengths 526.5 nm and 556.5 nm are the isobestic points of oxidized and reduced cytochrome c. The values of absorbance were obtained from Fig. 2a.

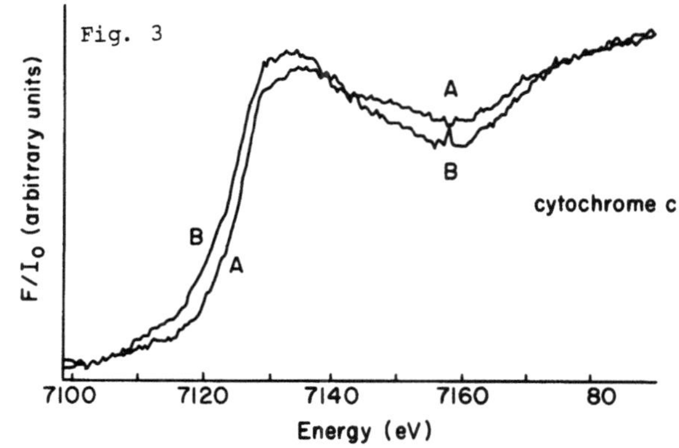

Fig. 2a Optical absorption spectra of cytochrome c rapidly frozen at various times after mixing with ascorbate. 100μM oxidized cytochrome c in 0.2 M sodium phosphate buffer, pH 6.8 was mixed with 50 mM sodium ascorbate in the same buffer at 1:1 ratio and rapidly frozen by ejecting into -100°C isopentane at 0.3 s, 1.5 s, 4 s and 25 s after mixing.

Fig. 3 X-ray absorption spectra of cytochrome c rapidly frozen after mixing with ascorbate. 10 mM cytochrome c in 0.2 M sodium phosphate, pH 6.8 was mixed with 50 mM sodium ascorbate in the same buffer and ejected into -100°C isopentane at 0.15 s (A) and 25 s (B) after mixing. The spectra of the frozen samples were measured at the same temperature.

c) Test on Deoxygenation of Hemoglobin

To evaluate the performance of the apparatus in earlier time region after mixing, we tested the apparatus on deoxygenation of hemoglobin. Fig. 4 shows the stopped-flow trace of hemoglobin after mixing with dithionite. The concentration of hemoglobin after mixing was 5 mM with respect to heme. A specially designed stopped-flow cell of 0.2 mm optical path was used in this measurement. The stopped-flow experiment was carried out under the same conditions as the rapid-freeze experiments described below. The trace yields the apparent rate constant of 75.0 s^{-1}.

Fig. 5 compares the normalized x-ray absorption spectra of oxyhemoglobin in the liquid state, oxyhemoglobin frozen just after mixing with excess dithionite and completely deoxygenated hemoglobin. A shift of 4 eV in the edge position is seen upon deoxygenation. The quenching time t_q of the apparatus is approximately given by

$$t_q = k_{app}^{-1} \ln [\Delta E_t/(\Delta E_t - \Delta E_o)] \qquad (1)$$

where k_{app} is the apparent rate constant for the deoxygenation reaction, ΔE_t is the total shift of the edge position upon deoxygenation, and ΔE_o is the difference in the edge position between oxyhemoglobin and hemoglobin just after mixing with dithionite. The quenching time was estimated to be 5.0 ms with Equation 1.

It is noticeable in Fig. 5 that the quality of the spectra of the rapidly frozen hemoglobin solutions is comparable to that of the spectrum of the hemoglobin solution in the liquid state. The magnitude of the edge jumps for the rapidly frozen oxyhemoglobin solutions was 70% of that of the edge jump for the oxyhemoglobin solution in the liquid state. In these measurements, the concentrations of the former hemoglobin solutions before freezing were made to be equal with that of the latter solution, and the spectra of the frozen and

Fig. 4

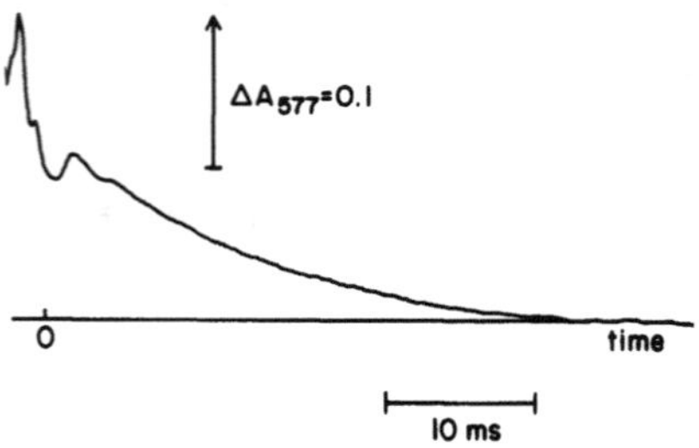

Fig. 5

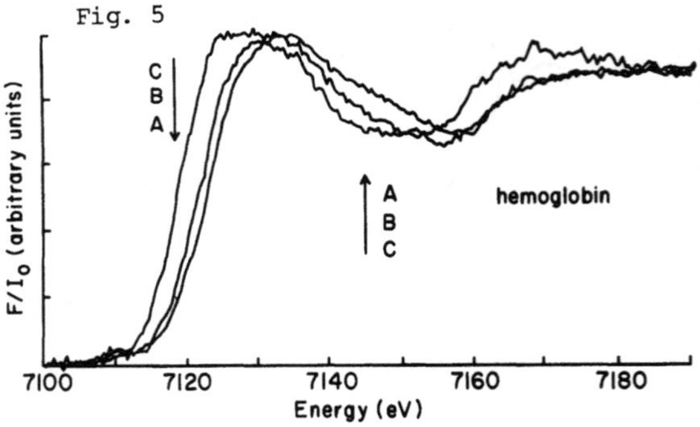

Fig. 4 Stopped-flow trace of the absorbance change of hemoglobin after mixing with dithionite. 10 mM hemoglobin in 0.1 M sodium phosphate, pH 6.8 was mixed with 100 mM sodium dithionite in the same buffer at 1:1 ratio in a stopped-flow apparatus equipped with an observation cell of 0.2 mm optical path.

Fig. 5 X-ray absorption spectra of oxyhemoglobin in the liquid state and hemoglobin rapidly frozen at different times after mixing with dithionite. 10 mM oxyhemoglobin in 0.1 M sodium phosphate buffer, pH 6.8 was mixed with 0.1 M sodium dithionite in the same buffer at 1:1 ratio and rapidly frozen just after mixing (B) and at 0.58 s after mixing (C). The mixtures were packed into the cell of the freezing apparatus and their x-ray absorption spectra were measured at -120°C. 5 mM oxyhemoglobin in 0.1 M phosphate buffer was filled into the same cell and the spectrum of the sample was measured in the liquid state at room temperature (A).

liquid solutions were measured in the same cell.

4. Concluding Remarks

Feasibility of the rapid-freeze EXAFS method has been demonstrated. This method could be applied to characterization of intermediates in a variety of reactions of enzymes and biological model compounds.

The quenching time was estimated to be 5.0 ms, which is comparable to the values reported previously for rapid freeze EPR apparatus [10,11]. The dead time calculated from the velocity of mixed solution and the volume of the flow path between the mixer and nozzle is about 3.0 ms. This value is comparable to the experimentally estimated quenching time, so that the freezing time or cooling time of the ejected solution is not the major component for the quenching time.

Mixing of reactants at subzero temperature in the presence of an antifreeze would slow down the reaction rate and accumulate reaction intermediates to a high extent [12,13]. This greatly enlarges the applicability of the rapid freezing method for EXAFS measurement.

Acknowledgement

The authors are grateful to Prof. T. Shiga of Ehime University and Prof. T. Wakabayashi of the University of Tokyo for helpful suggestions and discussions. Special thanks are due to Dr. A. Naqui of the University of Pennsylvania for his critical reading of the manuscript. This work was supported in part by a grant of the Ministry of Education, Science and Culture of Japan.

References

1. T. Matsushita and R.P. Phizackerley. Jpn. J. Appl. Phys. 20, 2223 (1981).
2. R.P. Phizackerley, Z.U. Rek, G.B. Stephenson, S.D. Conradson, K.O. Hodgson, T. Matsushita, and H. Oyanagi, J. Appl. Cryst. 16. 220 (1983).
3. T. Matsushita, H. Oyanagi, S. Saigo, H. Kihara. and U. Kaminaga, EXAFS and Near Edge Structure III, Proceedings of 3rd Int. EXAFS Conf. ed. K.O. Hodgson, B. Hedman, and J.E. Penner-Hahn, Springer-Verlag, 1984, pp. 476-478.
4. A. Fontaine, E. Dartyge and Tourillon, ibid., pp. 403-408.
5. D.M. Mills, A. Lewis, A. Harootunian, J. Huang, and B. Smith. Science 223, 811 (1984).
6. R.C. Bray. Biochem. J. 81. 189 (1961).
7. H. Oyanagi, Y. Matsushita, H. Tanoue. Y. Ishiguro, and K. Kohra, Jpn. J. Appl. Phys. 24, 610 (1985).
8. C. Yuen. M. Weissbluth, and A.M. Labhardt, SSRP Report No. 77106 (1977).
9. A. Labhardt and C. Yuen, Nature 277, 150 (1979).
10. D.P. Ballou and G.A. Palmer. Anal. Chem. 4b, 1248 (1974).
11. H. Beinert, R. Raymond, E. Hansen, and C.R. Hartzell, Biochim. Biophys. Acta 423, 339 (1979).
12. A.L. Fink and M.A. Geeves, Methods Enzymol. 63. 336 (1979).
13. P. Douzou, Cryobiochemistry, an Introduction, Academic Press, New York, 1977.

EXAFS and Structure-Function Relationships in Zinc Proteins

C. David Garner and Martinus C. Feiters

University of Manchester, Department of Chemistry, Manchester
M13 9PL U.K.
Daresbury Laboratory, Daresbury, Warrington WA4 4AD U.K.

INTRODUCTION

Zinc is known to be associated with a large variety of proteins, and is present in enzymes in each of the six categories designed by the International Union of Biochemistry [1]. In enzymes, zinc is often thought to play a catalytic role, i.e. to be essential for and directly involved in the catalysis, but also purely non-catalytic (i.e. structural or regulatory) roles have been implied, and in some cases its role is undefined [1]. Unlike manganese, iron, copper, or molybdenum, zinc exists in only one oxidation state, viz. Zn(II). Therefore, it is notably absent from the classes of dioxygen-binding proteins or electron-transfer proteins, and in the oxidoreductases where it is found either catalyses hydride transfer, as in alcohol dehydrogenase [2], or zinc assists another metal in its catalytic action, like copper in Cu, Zn-superoxide dismutase [3], rather than temporarily accommodating an electron, or having dioxygen as a ligand. The general concept of the catalytic role for zinc in enzymes appears to be the polarisation of molecules by virtue of its Lewis acidity. This may be the activation of water in its coordination sphere, e.g. in carbonic anhydrase to release a proton at physiological pH [4], or of a carbonyl function for nucleophilic attack on carbon, e.g. in carboxypeptidase [5], or of an alcohol for proton release and subsequent reactions, e.g. alcohol dehydrogenase [2]. The presence of a catalytic zinc site is typically signalled by the presence in its coordination sphere of a resting water molecule which may be activated or displaced. There is little evidence that zinc in enzymes allows favourable orientation of two substrates both coordinated to the metal, e.g., hydroxide anion and carbon dioxide in carbonic anhydrase, but this is not **a priori** excluded as the crystal structure would be consistent with coordination of two water molecules to zinc, in addition to the three protein ligands [6].

Some crystal structures of zinc proteins are available [7-11]. However, because of its filled 3d-shell, Zn is not accessible by spectroscopic techniques like optical absorption and EPR (electron paramagnetic resonance). Zinc enzymes may be indirectly studied by spectroscopic techniques if the endogenous metal is substituted by a spectroscopically more effective one. In particular, substitutions with cobalt will retain enzymic activity to a large extent and may even enhance it [12]. From studies of the optical absorption, MCD (magnetic circular dichroism), and the magnetic susceptibility of cobalt-substituted proteins, coordination numbers and ligand geometries may be inferred [13,14], and sulphur coordination may be identified by the presence of the cobalt-sulphur charge transfer band at approx. 350 nm [15]. Sometimes this information is complementary to that provided by X-ray diffraction, e.g. for carboxypeptidase, where the electron density map was originally interpreted tetracoordination for zinc [16], but refinement after a cobalt substitution study pointing to pentacoordination [8] led to the conclusion that glutamic acid coordinates to zinc with both carboxylate oxygens [10]. However, "foreign" metal ions may not always reflect the true chemical environment of zinc in an enzyme.

A relatively new technique in the study of zinc enzymes is EXAFS (Extended X-ray Absorption Fine Structure), the biological aspects of which have been reviewed [17-19]. Thus, the wiggles after the zinc K-edge step in the X-ray absorption spectrum can be described by backscattering, of the photoelectron wave leaving the excited zinc atom, by the zinc ligands leading to interference effects on the central absorber zinc atom and hence corresponding oscillations in the absorption coefficient [20]. These oscillations may be represented by the photoelectrons wave-vector k, defined as $\sqrt{(2m(E-E_o)/\hbar^2)}$ and are usually expressed in $Å^{-1}$. The phase-corrected Fourier transform of the EXAFS with respect to e^{-i2kR}, where R is the distance from backscatter to absorber, represents the radial distribution around zinc [21]. Typically the EXAFS can be analysed to determine ligand distances with high accuracy (± 0.02 Å) [22] and first shell coordination numbers to within 20% [23]. Sulphur first shell ligands can be distinguished from nitrogen and oxygen, but it is difficult to distinguish between the latter two donor atoms. Imidazole ligands give characteristic EXAFS due to additional backscattering contributions of ring atoms other than the coordinating nitrogen [24]. **EXAFS offers a unique and direct spectroscopic probe of zinc sites in proteins.** Multi-zinc proteins may be studied by selective substitution [25]; for example, cobalt-substituted sites may be studied at the cobalt K-edge, and compared with their native counterparts [25,26].

In this paper, we consider the impact of EXAFS on the understanding of zinc proteins, and investigate whether it is possible to give a meaningful classification on the basis of the zinc coordination sphere in relation to protein function. For copper proteins, a useful classification of the copper sites as Type-1, -2, or -3 exists based on spectroscopic features [27], and the different types are associated with specific coordination spheres and functions, e.g. the Type-1 site, having mixed imidazole/sulphur coordination [28] is associated with electron transfer, and the Type-3 bi-copper site, having imidazole coordination [29] is associated with dioxygen binding.

So far, only three amino acid residues have been identified as ligands of zinc in a metalloprotein; an imidazole nitrogen of histidine, the carboxylate oxygen of glutamic or aspartic acid, and the sulphur of cysteine. A tentative classification of zinc enzymes, based on features distinguishable in EXAFS, the only spectroscopic technique applicable to all zinc proteins, is:

- Type-A zinc: coordination by sulphur
- Type-B zinc: coordination by sulphur and nitrogen and/or oxygen
- Type-C zinc: coordination by nitrogen and/or oxygen

On the basis of the known chemistry of zinc, ligation of zinc by sulphur alone would lead to 4-coordination in a tetrahedral arrangement, leaving no space for additional ligands, whereas ligation by nitrogen and/or oxygen donor ligands may lead to four, five, or six-coordination.

Some zinc proteins of known coordination have been classified in Table 1. The potential of EXAFS to distinguish between clases A, B, and C derives from the facts that: (i) the backscattering amplitude of sulphur and nitrogen (or oxygen) are approximately π out of phase when placed at the same distance from the absorber atom with the backscattering amplitude from sulphur being significantly greater than that from nitrogen (or oxygen) (Fig. 1) **and** (ii) Zn-S bonds are typically **ca.** 2.3 Å in length whereas Zn-N(O) bonds are **ca.** 0.3 Å shorter; see Figure 2, which also shows the backscattering contributions from the outer shells of the imidazole ligands of zinc in superoxide dismutase. Therefore, Type-A and Type-C proteins can be readily distinguished. Because of the destructive interference of the back-scattered waves from N(O) and S at similar ligand distances, the EXAFS of Type-B zinc is also characteristic, but it may be difficult to determine the respective occupancies of the two shells as the contributions are typically not resolved in the Fourier transform.

Type-A Zinc

The zinc sites classified as Type-A are non-catalytic without exception. In view of the general concept of the catalytic role for zinc in enzymes, this is not unexpected, as the coordination sphere is saturated by sulphur and there is no resting water to be activated or displaced.

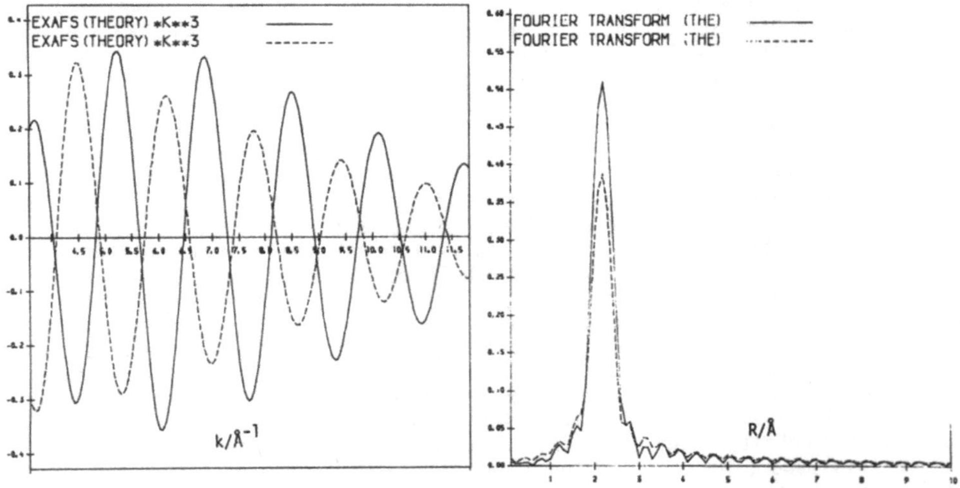

Fig. 1 Backscattering waves from one sulphur atom (——) and two nitrogen atoms (----) at 2.2Å from a Zn atom.

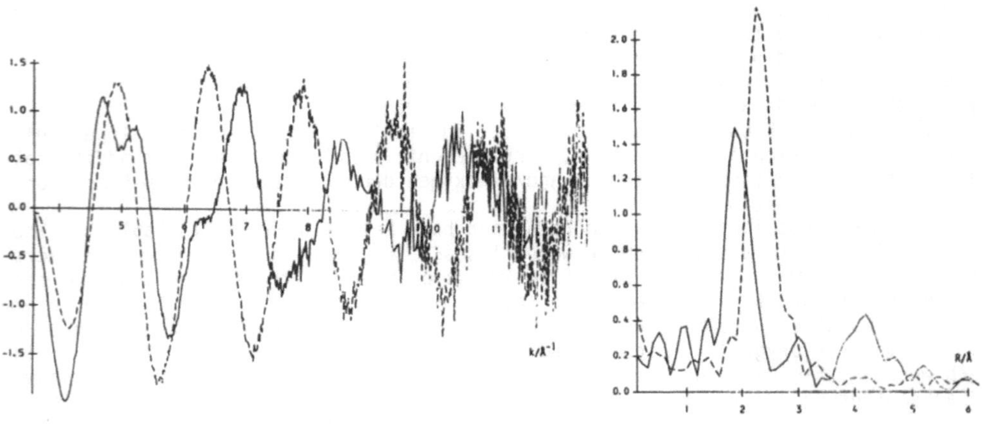

Fig. 2 Zn K-edge EXAFS (xk^3) and Fourier transform for Cu,Zn-Superoxide Dismutase (——) and Zn_7-Metallothionein (----)

Table 1 The Nature of Zinc Sites in Proteins

Protein	Coordination Sphere: Crystallography	EXAFS	Other	Function
Type-A				
Liver alcohol dehydrogenase (structural site)	4 S [9]			$NAD^+ + CH_3CH_2OH \rightarrow$ $NADH + H^+ + CH_3CHO$
Aspartate transcarbamylase		4 S [32]		$NH_2C(O)OPO_3^{2-} + {}^-O_2CCH_2CH(\overset{+}{N}H_3)CO_2^- \rightarrow$ $NH_2C(O)NHCH(CO_2^-)CH_2CO_2^- + H^+ + HPO_4^{2-}$
Metallothionein		4 S [33]		zinc storage and transport ?
Type-B				
Liver alcohol dehydrogenase (catalytic site)	2 S; 1 N; 1 H_2O [9]			$NAD^+ + CH_3CH_2OH \rightarrow$ $NADH + H^+ + CH_3CHO$
Sorbitol dehydrogenase		4-6 N/O; 1S [25]	$1H_2O$,1O; 1N,1S [37]	$NAD^+ + CH_2(OH)(CHOH)_3CH_2(OH)CH_2OH \rightarrow$ $NADH + H^+ + CH_2OH(CHOH)_3C(O)CH_2OH$
β-Lactamase (catalytic site)		4-6 N/O; 1S [25]	$1H_2O$,3N; 1S[15,35,36]	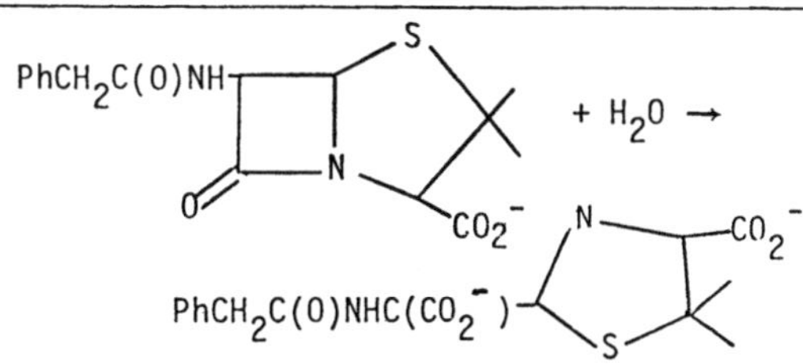

Table 1/Continued.

Type-C			
Superoxide dismutase	3 N;1,0 [10]	3 N;1 0 [24]	$2O_2^- + 2H^+ \rightarrow H_2O_2 + O_2$
Carbonic anhydrase	3 N;1,H_2O [7]	3 N;2 H_2O [26]	$OH^- + CO_2 \rightarrow HCO_3^-$
Carboxypeptidase } Thermolysin }	2 N;1,0; 1 H_2O [8,11]		$H_3\overset{+}{N}C(R_1)HC(O)NHCH(R_2)C(O)R_3 + H_2O \rightarrow H_3\overset{+}{N}CH(R_1)CO_2^- + H_3\overset{+}{N}CH(R_2)C(O)R_3$
Calmodulin		6 0	
" + Ca		5 0 [52]	saccharide binding
2 Zn-insulin	3 N; 2 0 [53]	3 N; 2 0 [53]	assists transport of glucose through liver cell membrane
Glyoxalase		2-4, N/0 [56]	$CH_3C(O)CH(OH)$S-glutathione $\rightarrow$ $CH_3C(OH)C(O)$S-glutathione
Phospholipase-C			
catalytic site		5 N/0 [25]	$R_1CH_2CH(R_2)CH_2OPO_3^{2-} + H_2O \rightarrow$
structural site		5 N/0 [25]	$R_1CH_2CH(R_2)CH_2OH + HPO_4^{2-}$

In the crystal structure of horse liver alcohol dehydrogenase (LADH), which contains 4 zinc atoms per dimer, the two zinc sites with a 4 S(Cys) coordination have been termed structural in order to distinguish them from the two other (Type-B) zinc sites in whose proximity the organic cofactor, nicotinamide adenine dinucleotide (NAD), and substrate analogues are found to bind [2]. However, there appear to be subtle differences (**ca.** 0.1 Å) in the various Zn-S distances in the non-catalytic site upon the binding of the cofactor to give the so-called closed form [2]. As the changes are on the error limit of crystallography the significance of this effect would be well worth investigating with EXAFS. In the related zinc enzyme sorbitol dehydrogenase, which contains only 1 zinc per monomer [30], the amino acid sequence shows homologies with that of LADH, but the structural site is not conserved [31].

An EXAFS study of aspartate transcarbamylase identified 4 sulphur ligands per zinc atoms at an average distance of 2.34 ± 0.03 Å, and a function for the zinc as a rivet, holding together the otherwise disordered bonding region of the regulatory chains of the enzyme in a flexible manner, which permits conformational change, was proposed [32]. EXAFS has established that Type-A sites occur in metallothionein [33], a protein of unknown function, but which has been implicated in zinc transport and storage [34].

Type-B Zinc

By definition, this category covers a range of zinc proteins with varying ratio of sulphur to nitrogen/oxygen ligands. Apart from horse liver alcohol dehydrogenase [9], no crystal structures are known, and we rely on chemical modification and spectroscopic studies, combined with EXAFS, or EXAFS alone, for the structures. We also find a large diversity in the functions implied for the Type-B zinc sites, and it is difficult to indicate clear structure-function relationship.

It is perhaps worth considering what effect different coordination spheres may have on the aspect of the zinc ion that determines its catalytic effectiveness, viz. its Lewis acidity. We note that N(His) and O(H_2O) ligands bear no formal charge (unless deprotonated), whereas carboxylate oxygen and S(Cys) ligands bear negative charges. Also, since sulphur is a softer ligand than oxygen, i.e. its electron cloud is more polarizable, it may be expected to temper the Lewis acidity of the zinc atom more than carboxylic oxygen. The accurate measurement of the position of the zinc K-edge may allow some quantitation of the relative Lewis acidity of zinc sites in proteins.

For **Bacillus cereus** β-lactamase II, evidence exists for three histidine ligands [33] and a sulphur ligand [15] at the active site zinc, forming a distorted tetrahedron. In a cryoenzymological study of the cobalt-substituted enzyme, an additional water ligand has been implied [36]. However, it has been noted that the molar extinction coefficient of the 348 nm absorption band found typical of Co-S charge transfer is low in β-lactamase compared to that in other cobalt-substituted proteins with sulphur ligands [15]. This may be due to the ease of deformation of this bond since the EXAFS indicates a sulphur ligand at a normal distance, viz. 2.34 Å, but with a relatively high Debye-Waller factor [25]. The reaction catalysed by the enzyme is hydrolysis, like that catalysed by some Type-C zinc enzymes (see below). The substrates for β-lactamase II, the penicillins, are probably susceptible to hydrolysis, because of the strain in the four-membered ring which is opened in this reaction. The function of the sulphur ligand might be to decrease the Lewis acidity of the zinc, compared to that of Type-C zinc, making the enzyme specific for hydrolysis of the lactam ring regardless of the side chains. It should be noted that this enzyme is unique among β-lactamases in requiring zinc for activity.

Recently, a structure has been proposed for sheep liver sorbitol dehydrogenase [37], based on its amino acid sequence and homologies and analogies to horse liver alcohol dehydrogenase, of which the crystal structure is known [2,9]. The best overall agreement with the alcohol dehydrogenase was found if one of the cysteine ligands to the catalytic zinc atom was changed to a glutamic acid ligand in sorbitol dehydrogenase. It is unlikely that this change would bring about a large change in Lewis acidity. The ligand change could account for the oxidation at C2 rather than

Cl of sorbitol, as the non-coordinating carboxyl oxygen of the glutamic acid ligand may be involved in a hydrogen bridge with the Cl-hydroxyl of sorbitol, leaving the C2-hydroxyl to displace water on the zinc [37]. EXAFS studies are in progress in order to test the validity of the protein structure prediction [25] and the current EXAFS interpretation is consistent with the proposed model. However, one needs to be cautious with coordination numbers in EXAFS studies of Type-B zinc, as model compound studies on the EXAFS of bis(4-toluenethiolato)(1,10-phenantroline)zinc(II), of which the crystal structure is known [38], have shown that the sulphur coordination number is underestimated, and that of nitrogen overestimated, if the EXAFS is analysed **a priori** rather than directly modelled. The alcohol dehydrogenase catalytic site, when subsituted by Cu(II), has the spectroscopic characteristics of Type-1 copper [39], and the EXAFS of such sites is notoriously difficult to analyse [40-43]. It is not well understood why the zinc sites of the dehydrogenase family require two negatively charged ligands. The resulting net formal charge of zero of the zinc site may be favourable for the hydride transfer of substrate to organic cofactor, but it is difficult to see how it helps the deprotonation of the substrate which precedes the transfer [2].

Conflicting opinions exist on the role of zinc in 5-aminolaevulinate dehydratase. In an overview of zinc enzymes, it has been classified as catalytic [1] but the results of recent EXAFS studies show a zinc site with coordination by 3 sulphurs and one nitrogen/oxygen atom, and little influence of the substrate on the zinc coordination, suggesting only a structural role [44]. It could be argued, however, that the binding of the substrate might have been affected by the fact that it was added to the air-oxidized, inactive [45], enzyme form, and that the inhibitor 2-bromo-3(imidazol-5-yl)propionic acid (against which the substrate protects [46]) does coordinate to zinc [44], suggesting that the site is catalytic. On the other hand, even though it is tempting to describe the mechanism of the reaction of two 5-aminolaevulinic acid molecules to yield porphobilinogen and two water molecules by a sequence of steps including an aldol condensation [47] which could be acid-catalysed, it is difficult to see how zinc coordinated by three sulphur ligands could act as the acid catalyst. The matter is complicated by the fact that, although the octameric enzyme molecule will bind a maximum of eight Zn(II), full enzymic acitivity is expressed with four Zn(II) per octamer [48]. This implies the existence of different binding sites, and studies on native liver alcohol dehydrogenase (M.C.Feiters and J.Jeffery, unpublished results) show that if one of them were a Type-A and the other a Type-B site, the former would dominate the EXAFS.

Type-C Zinc

It is well established that the zinc ions in carbonic anhydrase [4] and carboxypeptidase [5] are catalytic sites, and their modes of action have already been briefly discussed in the Introduction. Thermolysin, the endopeptidase from **Bacillus thermoproteolyticus**, has a Type-C site very similar to that in carboxypeptidase [11]. As far as the zinc ion in superoxide dismutase (SOD) is concerned, there is no indication of a role other than stabilizing the position of and the negative charge on the histidine residue which bridges between copper and zinc in this enzyme [10]; this bridge probably breaks at the copper end when this metal is reduced during the catalytic cycle when this metal changes its coordination number [3]. Even if other roles **in vivo** or **in vitro** for this protein become established, the crystal structure [10] suggests that the zinc is not accessible to water, and it is therefore unlikely that it is in any way more directly involved in catalysis.

Carbonic anhydrase has been the subject of two interesting EXAFS studies. The first one established that iodide anion coordinates to zinc at a normal Zn-I distance of ca. 2.65 Å [49] as opposed to the crystallographic results suggesting 3.5-3.7 Å [50]. The second study addressed the question of the coordination numbers of native and cobalt-substituted carbonic anhydrase at varying pH and in the presence of various anions [26]. Native (zinc) carbonic anhydrase was found to remain five-coordinated in the pH-range 5.5-9.7 and with the inhibitors CN^-, HCO^-_3, and NO^-_3, the average ligand distances varying between 1.99 and 2.03 Å.

Interestingly, the analysis for the complex with chloride gave anomalous results with regard to the coordination numbers, probably due to the fact that the chlorine backscattering amplitude, like that of sulphur, is approximately π out of phase with that of nitrogen/oxygen and the resulting destructive interference effects, as in Type-B zinc and Type-1 copper EXAFS, are difficult to unravel. For cobalt-substituted carbonic anhydrase, definite trends for lower coordination number (5.0 to 4.3) and shorter average ligand distance (2.07 and 2.00 Å) with increasing pH were observed. Some anions, e.g. CN^-, gave short average ligand distance and low coordination number, viz. 1.95 Å and 4.2, respectively, when added at neutral pH, whereas others, e.g. acetate, gave long average ligand distance and high coordination number, viz. 2.11 Å and 5.2. The XANES (X-ray Absorption Near Edge Structure) matched the EXAFS in this respect. The conclusion is that cobalt, though restoring approx. 50% enzymic activity to the apoprotein, is more reluctant to accept five-coordination in the enzyme than zinc. The EXAFS results are in agreement with the interpretation of optical absorption spectra as far as the cobalt is concerned [51], but extrapolation of the latter results to the native enzyme could be invalid.

Carbonic anhydrase represents a Type-C zinc site with five-coordination and a relatively short average zinc-ligand distance. There is increasing evidence, partly from EXAFS, that a subclass of Type-C sites exists with significantly longer average metal-ligand distances. Such a site exists in the jack bean protein, concanavalin A, the ability of which to bind saccharide reversibly depends on the occupation of a calcium binding site, which in turn depends on the binding of one of a variety of divalent metal ions at another binding site. The zinc EXAFS of this protein with and without calcium have been studied using ZnO as a reference [52]. It was found that the zinc coordination number changes from 6 to 5 upon calcium binding with a concomitant change in average ligand distance from 2.06 to 2.02 Å. A model for the zinc binding site, involving coordination by 1 N(His), 1 O(Glu), 1 $O(H_2O)$, 1 O(Asp) and 2 O from one other aspartic acid carboxylate group, was proposed. Upon calcium binding, the zinc ligation by one of the latter oxygens appears to be broken and the two aspartic acid ligands are shared by calcium and zinc.

2-Zn insulin hexamers from pig have been the subject of a comparative X-ray crystallography and EXAFS study [53]. The crystallography indicated octahedral zinc coordination by three nitrogens from histidines of different protein chains and three oxygens from water molecules. EXAFS simulation based on the crystallographic model gave average ligand distances for three nitrogens at 2.07 Å and two oxygens at 2.28 Å after refinement, both in reasonable agreement with the diffraction study, but failed to detect the third water at approx. 3.0 Å observed in the crystal structure and gave anomalous distances for the outer imidazole ring atoms.

'Octahedral' sites occur in the zinc enzymes glyoxalase and **B.cereus** phospholipase C. Upon substitution with cobalt, optical and MCD spectra pointing to distorted octahedral ligand geometry are observed [54,55]. In an X-ray absorption study of glyoxalase [56], the edge structure and the average ligand distance (2.07 Å) have been interpreted as indicating distorted octahedral zinc coordination but a coordination number of only 2-4 was determined from the EXAFS. The ligands not observed in EXAFS were assumed to be water oxygens at $\geqslant$ 2.3 Å and, in the Mn substituted glyoxalase [57], two of the ligands are fast-exchanging water molecules. **B.cereus** phospholipase C has two 'octahedral' sites, the relatively tightly bound 'structural' zinc and exchange-labile but catalytically essential 'catalytic' site [58,59]. Recent spectroscopic studies on selectively Co-substituted enzyme species indicate that sites are very similar, have distorted octahedral ligand geometry and are in close contact with each other[55]. EXAFS data have been collected on the native ZnZn-enzyme and on the enzyme species with the catalytic site zinc substituted by cobalt. The Zn EXAFS of the latter represents the structural site, whilesubstraction of the later from the former gives information on the catalytic site. The catalytic site Co EXAFS can be directly related to the spectroscopic studies. The analyses gave 5 $\pm$ 1 nitrogen/oxygen ligands for both sites, which refined to 3 ligands at a relatively short distance, and 2-3 at a longer distance, the positions of te minor imidazole shells suggesting that the coordinating nitrogens of these ligands are at the shorter distance. The coordination number from the Co-EXAFS of the catalytic site data is 7 $\pm$ 2. This suggests that the cobalt coordination number is higher than that of zinc in the same site, and that the cobalt spectroscopic results cannot be extrapolated to the native zinc site.

No evidence for the direct participation of zinc in the catalysis of the Type-C zinc enzymes with relatively long average ligand distances has yet been found and the designation of the 'structural' and 'catalytic' zinc sites of phospholipase C should not be taken as a literal analogy of those in horse liver alcohol dehydrogenase. For glyoxalase, the strong competitive inhibitor 5-p-bromobenzylglutathione was found to affect the EPR spectrum of the cobalt-substituted enzyme, indicating a conformational change in the metal ion environment [54] but to have little effect on the Zn EXAFS, apart form a possible rotation of the imidazole ligands [56]. In a mechanism of carbonyl polarisation, zinc might operate by hydrogen bonding via an intervening ligand [56]. For phospholipase C, inhibition by monovalent anions has been reported [60] but the iodide anion was found to interact with an essential arginine [61] rather than with the zinc as in carbonic anhydrase [49].

Conclusion

EXAFS has already contributed to the elucidation of coordination spheres in zinc enzymes and clearly many more studies are possible and should be accomplished. The results already obtained show that care has to be taken in extrapolating results from cobalt substituted proteins to the native proteins. The classification of zinc sites tentatively proposed on the basis of the limited information presently available appears to have some merit but, apart from the Type-A zinc, no strict structure-function relationships have emerged. It is unfortunate that EXAFS cannot distinguish between carboxylate oxygens and water oxygens as ligands, since differences in the population of these groups will affect the ease and degree of substitution and the Lewis acidity of the zinc site and, therefore, its catalytic activity.

References

1) Vallee, B.L. (1983) in Metal Ions in Biology (Spiro, T.G., editor) Vol. V: Zinc Enzymes, Wiley, pp. 1-24

2) Eklund, H. and Brändén, C.-I. (1983) in Ref. 1, pp. 123-152

3) Blackburn, N.J., Hasnain, S.S., Binsted, N., Diakun, G.P., Garner, C.D., and Knowles, P.F. (1984) Biochem. J. 219, 985-990.

4) Lindskog, S. (1983) in Ref. 1, pp. 77-122.

5) Vallee, B.L., Galdes, A., Auld, D.S., and Riordan, J.F. (1983) in Ref. 1, pp. 25-76.

6) Clementi, E., Corongiu, G., Jönsson, B., and Romano, S. (1979) FEBS Lett. 100, 313-317.

7) Liljas, A., Kannan, K.K., Bergstén, P.C., Waara, I., Fridborg, B., Strandberg, B., Carlbom, U., Järup, L., Lövgren, S., and Petef, M. (1972) Nature 235, 131-137.

8) Rees, D.C., Lewis, M., Konzatko, R.B., Lipscomb, W.N., and Hardman, K.D. (1981) Proc. Natl. Acad. Sci. USA, 78, 3408-3412.

9) Eklund, H., Nordström, B., Zeppezauer, E., Söderlund, G., Ohlsson, I., Boiwe, T., Soderberg, B.-O., Tapia, O., Brändén, C.-I., and Åkeson, Å. (1976) J. Mol. Biol. 102, 27-59.

10) Richardson, J.S., Thomas, K.A., Rubin, B.H., and Richardson, D.C. (1975) Proc. Natl. Acad. Sci. USA 72, 1349-1353.

11) Colman, P.M., Jansonius, J.N., and Matthews, B.W. (1972) J. Mol. Biol. 70, 701-724.

12) Prescott, J.M., Wagner, F.W., Holmquist, B., and Vallee, B.L. (1985) Biochemistry 24, 5350-5356.

13) Holmquist, B., Kaden, T.A., and Vallee, B.L. (1975) Biochemistry 14, 1454-1461.

14) Rosenberg, R.C., Root, C.A., Wang, R.-H., Cerdonio, M., and Gray, H.B.

(1973) Proc. Natl. Acad. Sci. USA 70, 161.
15) Baldwin, G.S., Galdes, A., Hill, H.A.O., Waley, S.G., and Abraham, E.P. (1980) J. Inorg. Biochem. 13, 189-204.
16) Lipscomb, W.N., Hartsuck, J.A., Reeke, G.N., Quiocho, F.A., Bethge, P.H., Ludwig, M.L., Steitz, T.A., Muirhead, H., and Coppola, J.C. (1968) Brookhaven Symp. Biol. 21, 24-28.
17) Cramer, S.P., and Hodgson, K.O. (1979) Prog. Inorg. Chem. 25, 1-39.
18) Powers, L. (1982) Biochim. Biophys. Acta 683, 1-38.
19) Bordas, J. (1982) in Uses of Synchrotron Radiation in Biology (Stuhrmann, H.B., ed.), Academic Press, London, pp. 107-143.
20) Stern, E.A. (1974) Phys. Rev. B. 10, 3027.
21) Sayers, D.E., Stern, E.A., and Lytle, F.W. (1971) Phys. Rev. Lett. 27, 1204.
22) Citrin, P.H., Eisenberger, P., and Kincaid, B.M. (1976) Phys. Rev. Lett. 36, 1346.
23) Eisenberger, P., and Lengeler, B. (1980) Phys. Rev. B, 3551-3561.
24) Blackburn, N.J., Hasnain, S.S., Diakun, G.P., Knowles, P.F., Binsted, N., and Garner, C.D. (1983) Biochem. J. 213, 765-768.
25) Feiters, M.C., Little, C., Waley, S.G., and Jeffery, J. (1986) J. Phys., in the press.
26) Yachandria, V., Powers, L., and Spiro, T.G. (1983) J. Am. Chem. Soc. 105, 6596-6604.
27) Malkin, R., and Malmström, B.G. (1970) Adv. Enzymol. 33, 177-244.
28) Colman, P.M., Freeman, H.C., Guss, J.M., Murata, M., Norris, V.A., Ramshaw, J.A.M., and Venkatappa, M.P. (1978) Nature 272, 319-324.
29) Gaykema, W.P.J., Hol, W.G.J., Vereijken, J.M., Soeter, N.M., Bak, H.J., and Beintema, J.J. (1984) Nature 309, 23-29.
30) Jeffery, J., Chesters, J., Mills, C., Sadler, P.J., and Jörnvall, H. (1984) EMBO Journal 3, 357-360.
31) Jörnvall, H., von Bahr-Lindström, H. and Jeffery, J. (1984) Eur. J. Biochem. 140, 17-23.
32) Phillips, J.C., Bordas, J., Foote, A.M., Koch, M.H.J., and Moody, M.F. (1982) Biochemistry 21, 830-854.
33) Abrahams, I.L., Bremner, I., Diakun, G.P., Garner, C.D., Hasnain, S.S., Ross, I., and Vasak, M., (1986) Biochem. J., 236, 585-589.
34) "Metallothionein" (Kägi, J.H.R. and Nordberg, M., eds.) Birkhauser Verlag, Basel (1979) and references therein.
35) Hill, H.A.O., Sammes, P.G., and Waley, S.G. (1980) Phil. Trans. R. Soc. Lond. B 289, 333-344.
36) Bicknell, R., and Waley S.G. (1985) Bicohemistry 24, 6876-6887.
37) Eklund, H., Horjales, E., Jörnvall, H., Brändén, C.-I., and Jeffery, J. (1985) Biochemistry 24, 8005-8012.
38) Cremers, T.L., Bloomquist, D.R., Willett, R.D. and Crosby, G.A. (1980) Acta Cryst. B36, 3097-3099.
39) Maret, W., Dietrich, H., Ruf, H.-H., and Zeppezauer, M. (1980) J. Inorg. Biochem. 12, 241-252.
40) Tullius, T.D., Frank, P., and Hodgson, K.O. (1978) Proc. Natl. Acad. Sci. USA 75, 4069-4073.
41) Scott, R.A., Hahn, J.E., Doniach, S., Freeman, H.C., and Hodgson, K.O. (1982) J. Am. Chem. Soc. 104, 5364-5369.

42) Groeneveld, C.M., Feiters, M.C., Hasnain, S.S., van Rijn, J., Reedijk, J., and Canters, G.W. (1986) Biochim. Biophys. Acta, accepted for publication.
43) Peisach, J., Powers, L., Blumberg, W.E., and Chance, B. (1982) Biophys. J. 38, 277-285.
44) Hasnain, S.S., Wardell, E.M., Garner, C.D., Schloesser, M., and Beyersmann, D. (1985) Biochem. J. 230, 625-633.
45) Batlle, A.M. del C., Ferramola, A.M. and Grinstein, M., (1967) Biochem. J. 104, 244-249.
46) Beyersmann, D., and Cox, M. (1984) Biochim. Biophys. Acta 788, 162-166,
47) Cheh, A.M., and Neilands, J.B. (1976) Structure and Bonding 29, 123-169,

48) Bevan, D.R., Bodlaender, P., and Shemin, D. (1980) J. Biol. Chem. 255, 2030-2035.
49) Brown, G.S., Navon, G., and Shulman, R.G. (1977) Proc. Natl. Acad. Sci. USA 74, 1794-1797.
50) Bergstén, P.-C., Waara, I., Lövgren, S., Liljas, A., Kannan, K.K., and Bengtsson, U. (1972) in Oxygen Affinity of Hemoglobin and Red Cell Acid-Base Status, Alfred Benzon Symp. IV, pp. 363-383, Academic Press, New York.
51) Bertini, I., Canti, G., Luchinat, C., and Scozzafava, A. (1978) J. Am. Chem. Soc. 100, 4873-4877.
52) Kalb, A.J., Stern, E.A., and Heald, S.M. (1979) J. Mol. Biol. 135, 501-506.
53) Bordas, J., Dodson, G.G., Grewe, H., Koch, M.H.J., Krebs, B., and Randall, J. (1983) Proc. R. Soc. Lond. B219, 21-39.
54) Sellin, S., Eriksson, L.E.G., Aronsson, A.-C., and Mannervik, B. (1983) J. Biol. Chem. 258, 2091-2093.
55) Bicknell, R., Hanson, G.R., Holmquist, B., and Little, C. (1986) Biochemistry, accepted for publication.
56) Garcia-Iniguez, L., Powers, L., Chance, B., Sellin, S., Mannervik, B., and Mildvan, A.S. (1984) Biochemistry 23, 685-689.
57) Sellin, S., Eriksson, L.E.G., and Mannervik, B. (1982) Biochemistry, 21, 4850-4857.
58) Little, C., and Otnaess, A.-B. (1975) Biochim. Biophys. Acta 391, 326-333.
59) Little, C. (1981) Acta Chem. Scand. B35, 39-44.
60) Aakre, S.-E., and Little, C. (1982) Biochem. J. 203, 799-801.
61) Aalmo, K., Hansen, L., Hough, E., Jynge, K., Krane, J., Little, C., and Storm, C.B. (1984) Biochem. Int. 8, 27-33.

EXAFS and XANES Studies of Copper Proteins

S. Samar Hasnain
Synchrotron Radiation Source, SERC Daresbury Laboratory, Warrington WA4 4AD, U.K.

Introduction

A number of copper proteins have been investigated using the EXAFS and XANES techniques. Some of this work is reviewed here to demonstrate the contribution of these techniques in specific structure determination. I also report here some of the recent advances in theoretical calculations for inter- and intra-ligand multiple scattering effects present in the EXAFS and XANES region. These developments offer an exciting possibility of investigating accurately the sterochemical details around a metal site of a protein in solution.

Theoretical background

The simplified plane-wave approximation [1] for EXAFS,

$$\chi(k) = \frac{\Delta\mu}{\mu_o} = \Sigma_j \frac{N_j}{kR_j^2} \left|f_j(\pi)\right| \sin(2kR_j + 2\delta_1 + \phi_j) \exp(-2\sigma_j^2 k^2) \exp(-2R_j/\lambda) \qquad (1)$$

is valid at high photoelectron energies. At lower electron energies, the curvature of the electron wave cannot be neglected and thus the plane-wave approximation breaks down and leads to errors in the calculated phase, which in turn can result in incorrect determination of the interatomic distances. Furthermore, it is this low energy part of the EXAFS spectrum which contains most of the information, particularly from the more distant shells. This range is particularly important in the case of metalloproteins, where the range of the EXAFS data is limited in view of the weaker scattering from the low-Z atoms, which are typically the metal ligands (e.g. nitrogen, oxygen). The low energy part of the EXAFS spectrum can be analysed by use of the exact theory given by Lee and Pendry [2] which takes account of the curvature of the electron wave and thus has been named the 'spherical wave method'. Unfortunately, this 'exact' theory has not been used in a majority of metalloprotein studies owing to its mathematical complexity and requirements for large computational time. Recently, this exact theory has been simplified by performing an average over the angular positions of the scattering atoms relative to the x-ray beam direction [1]. This simplification does not compromise the exact nature of the theory for polycrystalline or amorphous samples and thus is well suited for studies of metal ions in a protein solution. This simplification has resulted in a major improvement in the analysis programme EXCURVE to the extent that it is as quick to run as the approximate plane-wave theory. In general this single scattering theory is sufficient for interpretation of EXAFS data. However, strong multiple scattering effects can result from geometrical arrangement of ligands, e.g. in an imidazole ligand strong multiple scattering exists due to an almost collinear geometry of $Cu-N_\alpha-C_\gamma$. This has been referred to as a 'shadowing' or sometimes a 'focusing' effect and has been suggested to cause a shortening of the $Cu-C_\gamma$ distance when the low energy part of the EXAFS spectrum is fitted [3]. This bond shortening results from a phase modification due to the intervening atom [4]. The photoelectron is forward scattered by the intervening atom which is then backscattered by the C_γ atom. Forward scattering is generally strong at lower electron energies (E < 200 eV) over a significant angular range (0-70°). At higher electron energies the strong forward scattering is increasingly confined to a narrower cone and thus in the high energy part of the EXAFS spectrum, multiple scattering contributions become unimportant except when an approximately collinear geometry exists. Recently the single scattering curved-wave EXAFS function has been modified [5] to include the double- and triple-scattering contributions, which are expected to be the only other contributions of significance in the EXAFS region.

These multiple scattering contributions and higher order scattering become even more important in the x-ray absorption near-edge region (XANES) [6]. In the XANES, multiple scattering contributions are strong even in the open structure and thus offer a possibility of obtaining a fuller stereochemical picture of the metal site in proteins. Recent theoretical developments have made this region of the spectrum tractable to theoretical interpretation and offer a unique method of obtaining higher correlation functions in aqueous protein samples.

Copper environment in metallothionein

Metallothioneins are a ubiquitous class of proteins presumed to have a function of metal storage (Cu and Zn) and are thought to play an important role in the detoxification of metals such as cadmium and mercury [7,8]. A number of structural studies have been carried out to elucidate the structure of metal binding sites of these low molecular weight (~ 6000), cysteine-rich proteins. EXAFS is ideally suited to probe the environment of these different metal atoms. Hasnain et al [9] have recently reviewed most of the EXAFS and XANES work on metallothionein.

Abrahams et al [10], have recently reported an EXAFS study of two Cu-Zn metallothioneins, 6 Cu:3 Zn MT (77 K) and 5 Cu:2 Zn MT (at 300 K) from pig liver. The two EXAFS spectra, fig. 1, are very similar except that the spectrum of the

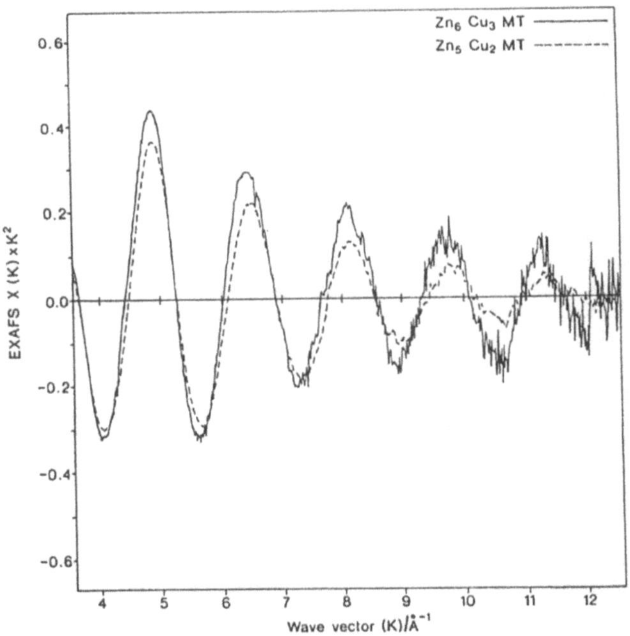

Figure 1 Cu K-edge spectra for 6 Cu:3 Zn MT at 77 K (——) and 5 Cu:2 Zn MT at 300 K (----) from pig liver.

6 Cu:3 Zn MT sample shows an increase in amplitude typical for a decrease in the Debye-Waller factor σ^2. For both data sets the best simulation is obtained with three Cu-S separations of 2.25 ± 0.01 Å with σ^2 = 0.007 Å^2 and 0.005 Å^2 for 5 Cu:2 Zn MT and 6 Cu:3 Zn MT, respectively. Abrahams et al also showed that the data range limited by the presence of Zn K-edge (K = 12 Å^{-1}) does not allow a distinction to be made between (i) three S atoms at the same distance 2.25 Å, (ii) one S atom at 2.17 Å and two S atoms at 2.29 Å, and (iii) two S atoms at 2.21 Å and one S atom at 2.32 Å. Figure 2 compares simulations based on models (i) and (ii), which shows that the differences occur in the k-range 9.5-13 Å^{-1}. Abrahams et al therefore question the validity of Bordas et al [11] analysis of yeast Cu-Thionein, who had suggested 2 Cu-S at 2.16 Å and two at 2.28 Å from the analysis of a much limited data range. However, it is important to remember that Cu-Zn MT and yeast Cu Thionein are quite different proteins with different amino acid compositions. Winge and co-workers (private communication) have recently studied the β-domain of rat liver copper metallothionein and found that the copper is co-ordinated by 2-3 sulphur ligands at 2.25 ± 0.03 Å, which agrees well with the Cu-Zn MT results. No evidence for higher shells, e.g. Cu-Cu or Cu-Zn approaches was found in these studies. However, Freedman et al [12] have analysed their spectra from $Cu_{7.8}$-MT from canine liver to suggest that clustering occurs beyond the inner co-ordination sphere of 4 sulphur atoms at 2.27 Å. These higher shell contributions are small and extreme care is required in Fourier filtering and their subsequent analysis. As noted by Freedman et al, the data range, limited signal-to-noise at higher k and back-scattering amplitude transferability can all seriously affect the final analysis of these shells. In order to establish the structure beyond the first shell, a more rigorous ab initio approach to EXAFS data analysis may be more appropriate. The number of cysteinyl residues coordinated to copper in metallothionein is most likely to be three, as Cu-S bond lengths appear to correlate with the co-ordination number in chemical systems. Thus, $Cu(SR)_2$, $Cu(SR)_3$ and $Cu(SR)_4$ have Cu-S bond lengths of 2.15 Å, 2.24-2.29 Å and 2.35 Å, respectively [13].

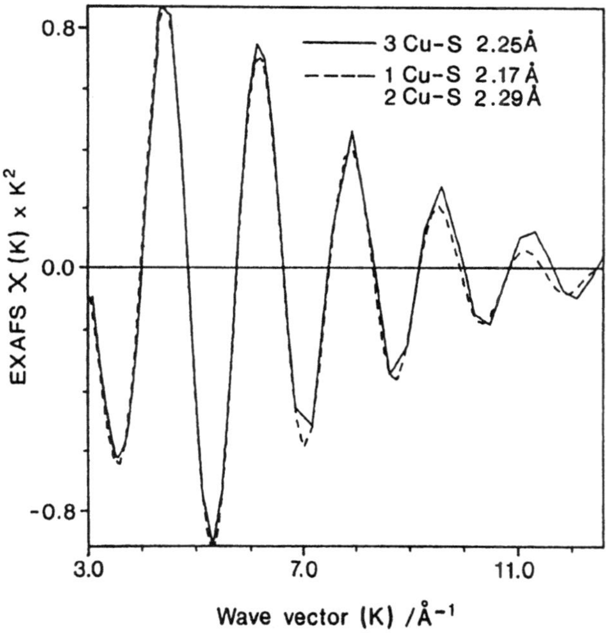

Figure 2 Comparison of best theoretical simulation obtained with models with three Cu-S separation at 2.25 Å (——) and with one Cu-S separation at 2.17 Å and two Cu-S at 2.29 Å (----).

Copper in Azurin

Azurin and plastocyanin are the most well characterised single 'blue' or type-1 copper proteins. They are implicated as electron carriers. A number of other single 'blue' copper proteins, whose functions are still uncertain, which have attracted attention include stellacyanin, umecyanin, rusticyanin and amicyanin. Detailed structural information has been provided by the crystal structure studies on oxidized and reduced plastocyanin [14,15] at different pH; and on oxidized azurin from *alcaligenes denitificans* [16] and *pseudomonas aeruginosa* [17]. In all these cases, copper is coordinated by two histidine nitrogens, one cysteine sulphur and one long distance (~ 3 Å) methionine sulphur in a distorted tetrahedral geometry. An EXAFS study of freeze-dried powders [18] of plastocyanin (and more recently of a single crystal) [19] and azurin from *pseudomonas aeruginosa* has yielded data in reasonable agreement with the high resolution crystal structure determinations of these proteins; however, the long Cu-S(met) distance could not be determined. This and the work on the oxidized and reduced stellacyanin has already been reviewed by powers [20].

Groenveld et al [21] have recently carried out a pH-dependent EXAFS study of *pseudomonas aeruginosa* azurin for the oxidized and reduced protein solution. This work is briefly reviewed here. Table 1 summarises the main results of the EXAFS analysis.

Table 1. Summary of the main results from an EXAFS study [21] of *pseudomonas aeruginosa*. Distances in Å, Debye Waller term $2\sigma^2$ in $Å^2$.

	Oxidised			Reduced		
	Crystal structure	EXAFS			EXAFS	
	pH 6.0 [16,17]	pH 9.1	pH 4.1		pH 9.2	pH 5.5
2 N_{His}	2.03	1.95±0.02 ($2\sigma^2$=0.003)	1.95±0.02 (0.004)	1 N_{His}	1.89±0.02 ($2\sigma^2$=0.005)	1.89 (0.005)
				1 N_{His}	2.05±0.02 ($2\sigma^2$=0.001)	2.04 (0.001)
1 S_{Cys}	2.12	2.23±0.02 ($2\sigma^2$=0.012)	2.19±0.02 (0.006)		2.21±0.02 (0.011)	2.24±0.02 (0.01)
1 S_{met}	3.1	3.4 ± 0.2 ($2\sigma^2$=0.03)	3.4 ± 0.2 (0.01)		2.74±0.05 (0.03)	2.70±0.05 (0.01)

Effect of pH

From T-jump studies of the electron transfer reaction between azurin and cytochrome-C551, it has been concluded that azurin may exist in two forms [22], while the equilibrium between these forms is effected by pH. The ^{1}H-NMR spectrum of azurin shows that protein can exist in two forms, depending on the state of the protonation of His-35 [23]. This residue itself is not a ligand of the copper, but is situated approximately 4 Å away from His-46 and 8 Å from the copper atom. The protonation/deprotonation of this amino acid appears to be connected with a conformational change of the protein, which affects, among others, the NMR signals of a number of copper ligands. Thus, the hypothesis was advanced that conformational changes are induced around the copper which would in turn alter the redox activity of the azurin. Groeneveld et al show conclusively from their EXAFS data, fig. 3, that the coordination sphere of copper is hardly affected by pH variations, the only effect being an increased disorder of the Cu-S(met) bond at

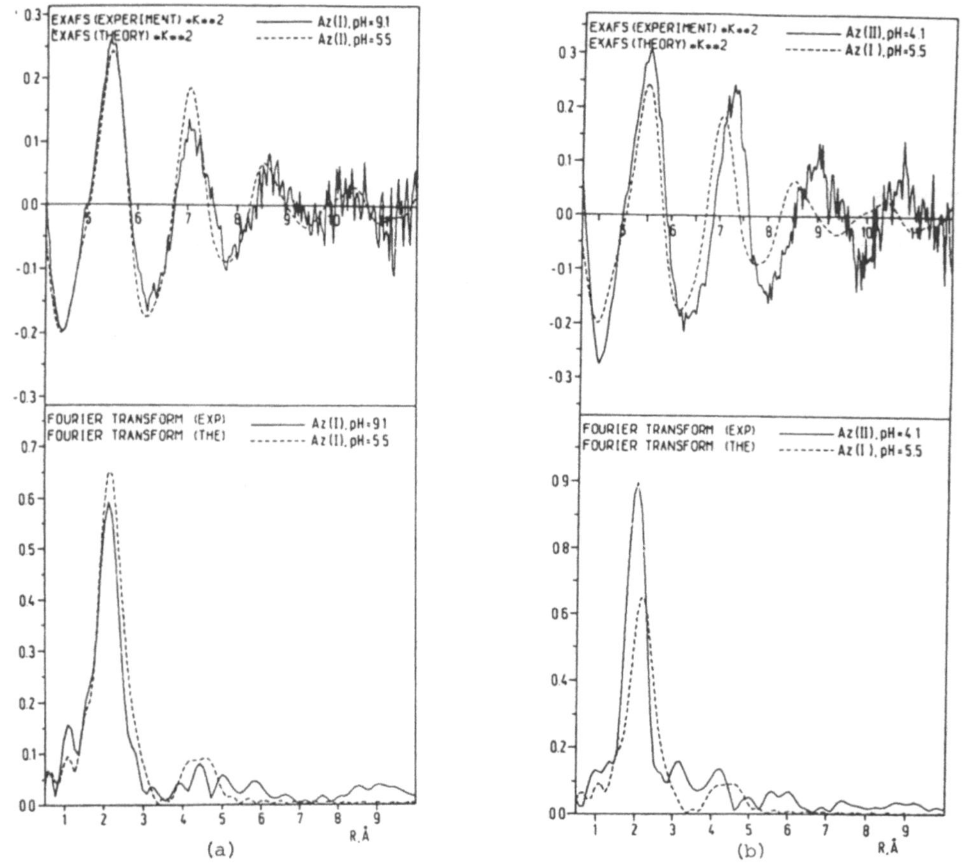

Figure 3 EXAFS spectra at the Cu K-edge of azurin solution from Pseudomonas aeruginosa. For clarity the effect of pH and reduction is shown by comparison with the simulated spectrum of the low pH reduced protein (which was indistinguishable from the raw experimental spectrum except for the statistical variation). (a) pH effect; (b) oxidised vs reduced.

high pH, indicating an increased labile nature of this contact. They argue that the pH effects in the ^{1}H-NMR spectrum of azurin must therefore be due to small rotations of ligands or to changes in the secondary coordination sphere. Unfortunately, the XANES spectra, which should contain such information, were not recorded by these authors with high energy resolution and thus they were not able to comment on changes in this region of the spectrum.

These results show that the effect of pH on the copper site in azurin is different from that in plastocyanin, which has been studied crystallographically [15] in the reduced state at $3 < pH < 8$. At $pH \leqslant 5$ in reduced plastocyanin, one of the histidine ligands is protonated, resulting in a breaking of one Cu-N(His) bond and a shortening of the Cu-S(Met) bond.

Effect of reduction

Reduction is accompanied by a considerable decrease in amplitude and a shift in phase. This results from an increased spread in the $Cu\text{-}N_{His}$ distances and is due to the appearance of a stronger contribution from the methionein S at a substantially shorter Cu-S distance at 2.70 Å. The average Cu-N_{His} distance is the same as in the case of the oxidized protein but with a much increased Debye-Waller factor, 0.02 $Å^2$. A detailed analysis has shown that the Cu-N_{His} distances are split by 0.15 Å in the case of the reduced protein. However, the Cu-S_{cys} is essentially unaltered. Thus, the EXAFS results have shown that the copper site is only slghtly re-arranged, providing support for the concept of an 'entatic state' where the copper co-ordination is energetically well suited for an efficient electron transfer.

For the reduced protein at low pH (5.5), Groeneveld et al [21] have provided unambiguous evidence for the Cu-S_{met} contribution at 2.70 ± 0.05 Å. In order to demonstrate this clearly, they have employed an interesting difference method which could be of use in other similar cases. The Cu-S_{met} contribution appears as a shoulder of the major peak in the Fourier transform, therefore it could not be readily isolated. They Fourier filtered the major shell including the shoulder

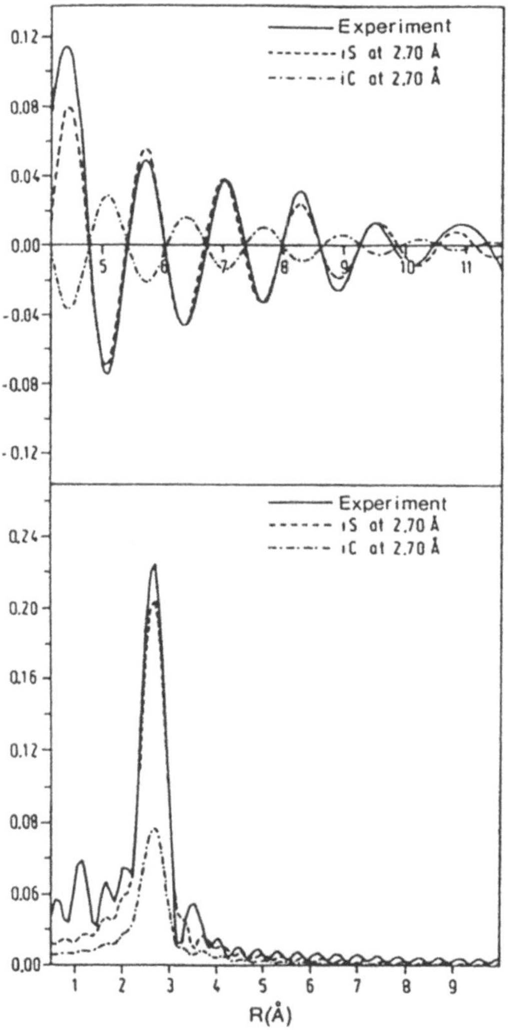

Figure 4 Simulation of the difference spectrum (Cu-S_{met} contribution, see text) with a Cu-S (---) and a Cu-C (-·-·-·) wave.

and subsequently the simulated Cu-N(His) and Cu-S(cys) contributions [the contributions responsible for the main peak in the Fourier transform] were substracted. Analysis of this difference spectrum clearly shows that it represents a sulphur atom at 2.70 ± 0.05 Å rather than the backscattering from carbon atoms, fig. 4.

Superoxide dismutase (Cu-Zn)

Copper-zinc superoxide dismutase (Cu,Zn-SOD) was first isolated from bovin erythrocytes and liver in 1938 by Mann and Keilin [24]. It was after 31 years, in 1969, that McCord and Fridovich [25] discovered the superoxide dismutase activity. In addition to the Cu,Zn-SOD, two other classes of Fe or Mn containing superoxide dismutase have been found in mitochondria [26]. It has been suggested that these superoxide dismutases are part of a defence mechanism to protect aerobic organisms from dioxygen activity.

A great deal of structural information is available for bovine Cu,Zn-SOD, including its amino acid sequence, and the crystal structure determination at high resolution has been achieved [27]. The human and the yeast proteins show a high degree of homology; 82% and ~ 50% residues are conserved, respectively. These are all dimeric proteins, of molecular weight ~ 32,000 dalton, with each of the subunits containing a Cu and a Zn atom. The crystal structure of the bovine superoxide dismutase suggests that the subunits are held together by noncovalent interactions. There is some evidence for communication between the metal binding sites of the subunits, which are separated by ~ 34 Å. Within the subunit, the Cu(II) and Zn(II) ions are observed to be separated by ~ 6 Å and bridged by the imidazole ring of a histidine (His-61) residue, which is believed to be deprotonated. The Zn ion is coordinated to the nitrogen atoms of each of the imidazole rings of histidine residues 61, 69 and 78 and an oxygen atom (carboxylate) of the aspartate 81. All of the Zn ligands are derived from one continuous piece of the polypeptide chain, which is not the case for the Cu(II) ion. The crystal structure shows that Cu(II) ion is coordinated to four imidazole rings from histidines 44, 46, 61 and 118. There is considerable spectroscopic evidence that copper is five coordinate; the fifth coordination position is taken by a water molecule.

Blackburn et al have used the EXAFS technique for studying the Cu and Zn sites of the freeze-dried powder [3] and solution [28] Cu-Zn bovine superoxide dismutase. They were able to show that the 4 Cu-N distance for the protein was 2.00 ± 0.02 Å, which is typical for $[Cu(imid)_4]^{2+}$ complexes. At the time the 2 Å crystal structure data with residual factor R ≈ 30% showed Cu-N to be 2.11 Å. This has recently been determined as 2.05 ± 0.05 Å [29] where the R factor has been improved to 16%. Blackburn et al also were able to show that spectra changed on freeze drying and this change was most compatible with the loss of a coordinated water, which was found at 2.24 ± 0.03 Å. For the Zn site, 3 Zn-N were established at 2.01 ± 0.02 Å and Zn-O as 2.04 ± 0.03 Å. We note that, given the data range the distinction between Zn-N and Zn-O contributions is not possible.

Blumberg et al [30] used the Zn-edge XANES of the native and the reduced protein to show that the near edge spectrum of the two were very similar, thus suggesting that the bridging imidazole from His-61 is coordinated to Zn in both cases. Blackburn et al [28] extended this study to show that while the Zn-edge spectrum did not change, the Cu-edge XANES and EXAFS changed.

Comparison of the Cu K-edge EXAFS for the two forms showed that the overall amplitude decreased by ~ 20% in the reduced protein, thus suggeseting a decrease in the coordination number. Blackburn et al analysed their EXAFS data using a three-shell model. They suggest that Cu(I) is three coordinate, involving only imidazole coordination. In addition to a decrease in coordination number, they find that the Cu-N bond length is decreased to 1.94 ± 0.02 Å, which again is indicative of a coordination number lower than in the native protein. These results have thus provided the first direct structural information on the copper site of the reduced protein.

An interesting feature of the multishell analysis of the EXAFS of the Cu site of SOD has been the requirement to split the Cu-Cβ distances by ~ 0.2 Å. Blackburn and Hasnain [31] interpreted this splitting as primarily due to asymmetric coordination of the imidazole rings in the protein. They have used the interactive molecular graphics to assess the possibility of deriving information on the orientation of rings relative to the metal centre. They used interactive graphics to fit the EXAFS distances to the crystallographic coordinates of Tainer et al [27] by carrying out translations and rotations of the imidazole groups whilst keeping the other protein residues fixed at their crystallographic positions. Figure 5 shows a comparison of the EXAFS-generated site with the crystallographic data. This is further discussed below in relation to the XANES calculations.

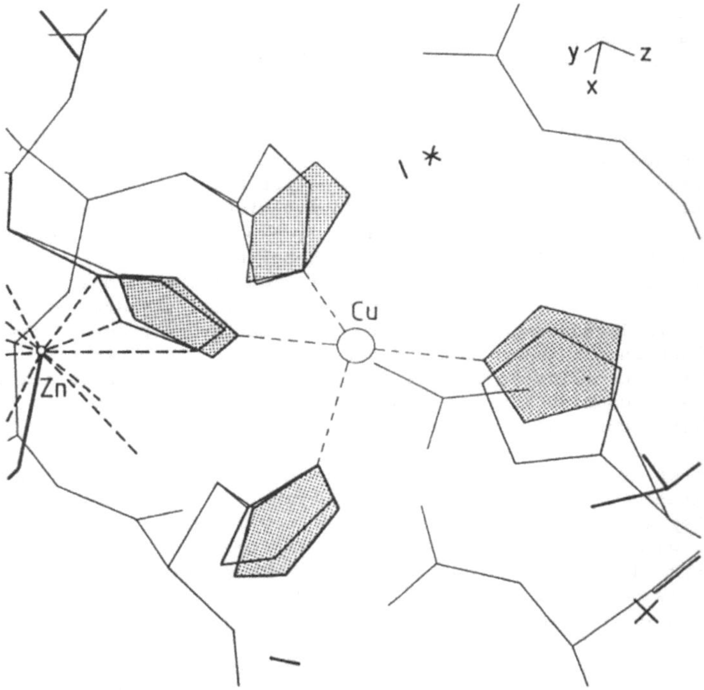

Figure 5 A comparison of the EXAFS-generated site with the crystallographic data of Cu-Zn superoxide dismutase.

Blackburn et al [3] also recognised a strong multiple scattering contribution in their EXAFS spectra due to the nearly collinear arrangement of the Cu-N_α-C_γ atoms, which resulted in an artificial shortening of the third shell atoms by ~ 0.3 Å as only a single scattering theory was employed. In the most common analysis approach of EXAFS, which uses plane-wave approximation and relies on extraction of phase shifts from chemical complexes, these effects must be carefully analysed. Strange et al [4] have recently provided the first quantitative examination of multiple scattering contributions from the atoms of an imidazole ring. They show that once the multiple scattering effects are taken into account, the distances of all the atoms of an imidazole ring can be determined accurately. In view of the strong multiple scattering contribution exhibited by the imidazole ring, these should be easily recognisable in EXAFS data and thus a distinction between an imidazole ligand and another low-Z ligand should not prove difficult.

Thus, ligands such as water, aspartate, azide, and cyanide should be distinct from histidine imidazole. In addition, a distinction may be possible between histidine and a tyrosine, even though both are ring systems, but should exhibit different multiple scattering in view of differences in the scattering pathways. These ideas need to be explored systematically by using a reasonable number of different chemical systems.

Alagna et al [32] have recently carried out detailed multiple scattering calculations to understand the XANES of a number of copper imidazoole compounds which show EXAFS and XANES spectra akin to the copper site of the freeze-dried and aqueous superoxide dismutase. Figure 6 shows the experimental XANES for three imidazole complexes, each with two axial ligands to the CuN_4 equatorial plane. Also shown are the spectra for the Cu site of superoxide dismutase. Each of the spectra show a number of spectral features, with varying degree of resolution, suggesting significant stereochemical differences between them. Alagna et al have shown that each of these features, their relative intensity and resolution are a signature of a particular aspect of the structure. For example, the peak marked 'b' arises due to the axial ligands; its intensity and position is related to the

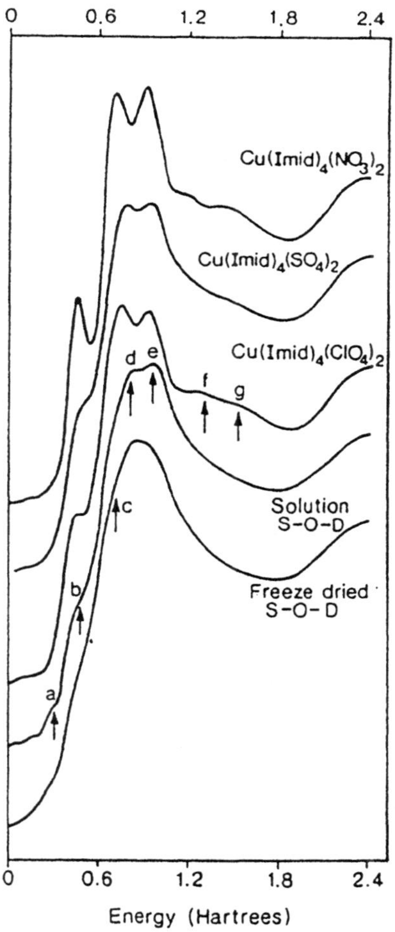

Figure 6 Experimental XANES for three copper imidazole complexes and for the Cu site of Cu-Zn superoxide dismutase in the freeze-dried and aqueous state. Energy zero (E_0) has been defined arbitrarily but corresponds to the same monochromator position in all cases. 1 Hartree = 27.2 eV.

number and position of these ligands. Similarly, features marked 'd' and 'e' (and associated features 'f' and 'g', respectively) depend on the orientation of imidazole rings with respect to the CuN_4 plane. The latter aspect is shown in fig. 7, where XANES spectra are calculated for $(CuN_4)O_2$ system. In one case all the imidazole rings are nearly perpendicular to the CuN_4 plane and in the other two are flat and two perpendicular. This is a clear demonstration of inter-ligand scattering which is potentially very exciting as it may allow the determination of the geometrical arrangement of one ligand with respect to the other, and therefore should be able to provide a detailed stereochemical picture of the metal site in aqueous protein samples.

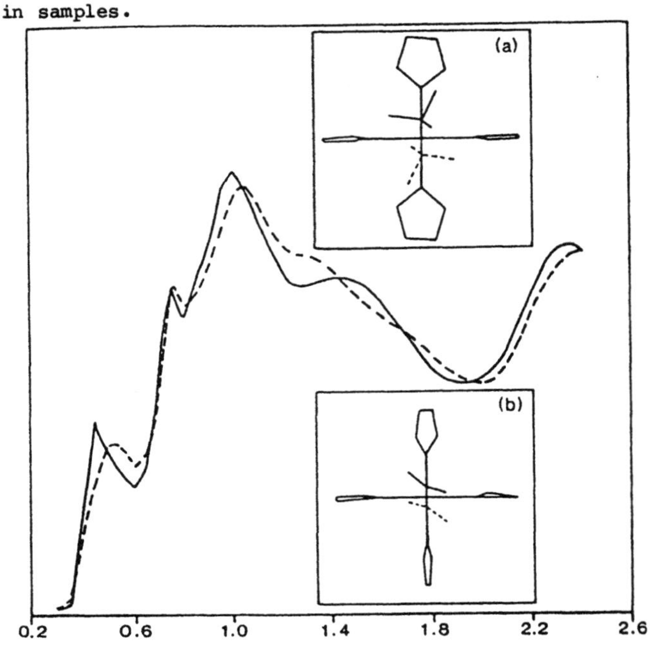

Figure 7 XANES calculations for $(CuN_4)O_2$, where N represents an imidazole group. (——) all imidazole rings are nearly perpendicular to the CuN_4- plane; (----) calculation with two rings perpendicular and two parallel to the CuN_4 plane. Differences in the 1.2-1.8 H range result from the inter-ligand scattering.

Alagna et al also attempted to simulate the XANES of superoxide dismutase. Figure 8 shows these calculations and compares them with the observed spectrum. The calculation (b) was based fon the crystal structure coordinates available in May 1984 in the Brookhaven Protein Databank, calculation (d) is from the EXAFS-modified site using the molecular graphics and calculation (c) is obtained with the revised crystal structure coordinates obtained from further crystallographic refinement (R = 14% at 2 Å), as made available in December 1985. Thus, Alagna et al were able to show that the EXAFS-modified picture of the Cu site improved the agreement with the experimental spectrum in a way similar to the better refined crystallographic data. This opens up an exciting possibility where use is made of EXAFS and XANES not only to obtain a more refined structure of a metal site in a protein, but more importantly, in obtaining this information for protein in solution and also when the protein is modified either by ligand binding, change of pH or oxidation state. Figure 9 shows the sensitivity of this region where the spectra for the Cu site of SOD in its different forms are shown and thus, the importance of understanding the XANES region cannot be over emphasised.

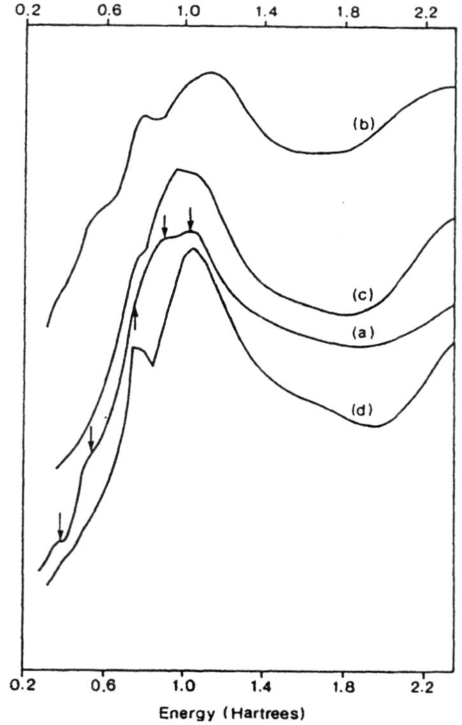

Figure 8 (a) The experimental spectrum of an oxidised solution of superoxide dismutase. (b)-(d) Three calculations with (b) crystallographic data (May 1984), (d) EXAFS-modified site using molecular graphics and (c) further refined crystallographic data.

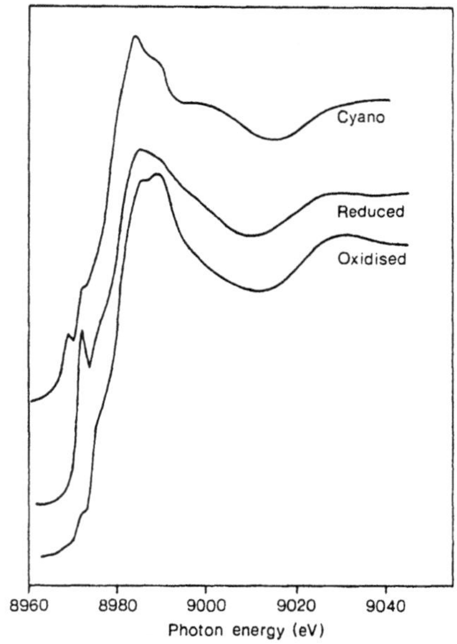

Figure 9 Cu K-edge XANES spectra of the native, dithionite-reduced and CN-bound superoxide dismutase.

Acknowledgement

I am especially grateful to the Science and Engineering Research Council for support and to the Director of the Daresbury Laboratory for the provision facilities. I am indebted to my colleagues Drs. I.L. Abrahams, L. Alagna, N.J. Blackburn, I. Bremner, G. Canters, G.P.Diakun, P. Durham, M. Feiters, R.W. Strange and Professor C.D. Garner. It is also a pleasure to acknowledge Norman Binsted for developing the EXCURVE programme package without which most of the work presented would not have been possible.

References

1. S.J. Gurman, N. Binsted and I. Ross, J. Phys. C17 143 (1984)
2. P.A. Lee and J.B. Pendry, Phys. Rev. B11 2795 (1975)
3. N.J. Blackburn, S.S. Hasnain, G.P. Diakun, P.F. Knowles, N. Binsted and C.D. Garner, Biochem. J. 213 765 (1983)
4. R.W. Strange, S.S. Hasnain, N.J. Blackburn and P.F. Knowles, Proc. IV Int. EXAFS Conf. 7-11 July 1986, J. de Phys. (In Press)
5. S.J. Gurman, N. Binsted and I. Ross, J. Phys. C19 1845 (1986)
6. P.J. Durham, J.B. Pendry and C.H. Hodges, Comput. Phys. Commun. 25 193 (1982)
7. J.H.R. Kagi and M. Norøberg (eds.) Metallothionein, Birkhäuser Verlag, Basel (1979)
8. A. Bakka and M. Webb, Biochem. Pharmacol. 30 721 (1981)
9. S.S. Hasnain, G.P. Diakun, I. Abrahams, I. Ross, C.D. Garner, I. Bremner and M. Vasak, in Proc. 2nd Int. Meet. Metallothionein, 1985 (J.H.R. Kägi, ed.) Birkhäuser, Basel. (In Press)
10. I.L. Abrahams, J. Bremner, G.P. Diakun, C.D. Garner, S.S. Hasnain, I. Ross and M. Vasak, Biochem. J. 236 585 (1986)
11. J. Bordas, M.H.J. Koch, H.J. Hartmann and U. Wesser, FEBS Lett. 140 19 (1982)
12. J.H. Freedman, L. Powers and J. Peisach, Biochemistry 25 2342 (1986)
13. C.D. Garner, Proc. IV Int. EXAFS Conf., 7-11 July 1986, J. de Phys. (In press)
14. P.M.Colman, H.C. Freeman, J.M. Guss, M. Murata, V.A. Norris, J.A.M. Ramshaw, M.P. Venkatappa, Nature 272 319 (1978)
15. H.C. Freeman, in Coordination Chemistry, Vol.21 (J.P. Laurent, ed.) Pergamon Press, Oxford (1981) p.29
16. C.E. Norris, B.F. Anderson and E.N. Baker, J. Mol. Biol. 165 501 (1983)
17. E.T. Adman and L.H. Jensen, Isr. J. Chem. 21 8 (1981)
18. T.D. Tullins, P. Frank and K.O. Hodgson, Proc. Natl. Acad. Sci. USA 75 4069 (1978)
19. R.A. Scott, J.E. Hahn, S. Doniach, H.C. Freeman and K.O. Hodgson, J. Am. Chem. Soc. 104 5364 (1982)
20. L. Powers, Biochem. Biophys. Acta 683 1 (1982)
21. C.M. Groeneveld, M.C. Feiters, S.S. Hasnain, J. Van Rijn, J. Reedijk and G.W. Canters, Biochem. Biophys. Acta (In press)
22. O. Farver, Y. Blatt and I. Pecht, Biochemistry 21 3556 (1982)
23. E.T. Adman, G.W. Canters, H.A.O. Hill and N.A. Kitchen, FEBS Lett. 143 287 (1982)
24. T. Mann and D. Keilin, Proc. R. Soc. London B126 303 (1938)
25. J.M. McCord and I. Fridovich, J. Biol. Chem. 244 6049 (1969)
26. J.S. Valentine and M.W. Pantoliano in Copper Proteins, ed. T.G. Spiro (John Wiley, N.Y.) Chapter 8 (1981)
27. J.A. Tainer, E.D. Getzoff, J.S. Richardson and D.C. Richardson, Nature 306 284 (1983)
28. N.J. Blackburn, S.S. Hasnain, N. Binsted, G.P. Diakun, C.D. Garner and P.F. Knowles, Biochem. J. 219 985 (1984)
29. J.A. Tainer (Private Communication, Dec. 1985)
30. W.E. Blumberg, J. Peisach, P.M. Eisenberger and J.A. Fee, Biochemistry 17 1842 (1978)
31. N.J. Blackburn and S.S. Hasnain, in Biological and Inorganic Copper Chemistry (K.D. Karlin and J. Zubriata, eds.) Adenine Press (1986)
32. L. Alagna, R.W. Strange, P. Durham and S.S. Hasnain, J. Am. Chem. Soc. (submitted)

The Active Site of Cu,Zn Superoxide Dismutase as Studied by EXAFS: The Binding of Chloride to the Reduced Enzyme

A. DESIDERI, S. MORANTE, G. ROTILIO
Dipartimento di Biologia e Fisica, II Università di Roma, Tor Vergata
Roma, Italia.

INTRODUCTION

Bovine erithrocyte copper-zinc superoxide dismutase Cu(II)-Zn(II)SOD is a dimer made of two equivalent subunits each containing one copper and one zinc atom. X-ray diffraction study (1) on crystals of native Cu,Zn SOD have shown the copper to be coordinated to four histidines (His 44, His 46, His 61 and His 118) and one molecule of water. One of the histidines (His 61) provides an imidazolate bridge between the copper and the zinc.
The biological role of this enzyme is to disproportionate the toxic O_2^- by the following catalytic reactions:

$$O_2^- + E\text{-}Cu^{2+} \longrightarrow O_2 + E\text{-}Cu^+$$

$$O_2^- + E\text{-}Cu^+ + 2H^+ \longrightarrow H_2O_2 + E\text{-}Cu^{2+}$$

Since O_2^- is an anion and reacts with both the oxidized and reduced form of the enzyme it has been of great interest to study the interaction of the enzyme woth other single charged anions with regard to their mode of binding at the active site and their inhibition of enzyme activity. This problem has been extensively investigated by spectroscopic means such as EPR, NMR, electronic absorption, ENDOR, on the oxidized enzyme (2-4) but not on the reduced one, as the Cu(I) atom is magnetically and optically silent. Therefore it is reasonable to undertake an investigation of the interaction of the reduced enzyme by EXAFS, the only spectroscopic technique that may probe the surrounding of the copper irrespective of its oxidation state. In particular in this work an EXAFS investigation on the reduced enzyme with the Cl^- anion is reported.
A direct binding of Cl^- to the reduced form of bovine Cu,Zn SOD was previously suggested by observing the linewidth at half height of the Cl^- resonance as monitored by $^{35}Cl^-$ NMR spectroscopy (5). On the other hand a more recent $^{35}Cl^-$ NMR investigation (6) claims that

the anion does not bind to the copper atom but to a positively charged chain, namely Arg 141, which lies at close distance (7 Å) from the copper (1). The limit of the NMR technique for this problem resides in the fact that the probe is the Cl^- anion and not the Cu(I) ion, making difficult to detect any perturbation affecting the metal atom.

Materials and Methods

Bovine superoxide dismutase was freshly prepared as previously described (1). Reduction of the enzyme was performed by addition of a 5-fold excess of solid dithionite. The Cl^- anion was added to the enzyme solution as a powder to reach a concentration of 2 M. The enzyme concentration of the samples used for X-ray absorption measurement was between 2 and 3 mM. All the samples were carefully examined during and after the data collection to check the occurrence of any photo-redox reaction or inactivation during measurement. No such changes were found. The Cs_2 $CuCl_4$ complex, of known X-ray structure (8), was used as model compound to obtain the phase and amplitude function of the copper-chloride pair.
X-ray absorption spectra were recorded at the Frascati Synchrotron Radiation Source, in the fluorescence mode for the enzyme solution and in the absorption mode for the model compound. Data analysis was made on spectra that were an average of 10-15 different scans (time integration for each point was approximately 2 sec).

Results and discussion

The X ray absorption spectrum of the reduced enzyme in the presence of 2 M Cl^- is reported in Fig. 1. The EXAFS region of the spectrum, indicative of imidazole coordination (9), is reported in Fig. 2. Four imidazoles are known to be in the first coordination sphere of the oxidized copper in the native enzyme (1), whilst several lines of evidence suggest that one of this ligand is lost upon reduction (10, 11). The backfourier transform of the first shell of the copper of the reduced enzyme in the presence of Cl^- is reported in Fig. 3 (solid line). The best fit of experimental data was first attempted by considering just nitrogen in the first coordination sphere but a poor fit was obtained. A much better simulation was obtained by splitting the first shell into two subshells, one consisting of nitrogens and the other containing one chloride ion at 1.9 Å from the copper atom.
A similar result was obtained for the second shell (Fig. 4). The best fit of the experimental data was obtained by inserting beside the carbon atoms a new subshell consisting of one Cl^- at 32 Å distance from the metal.
These data indicate that the Cl^- anion is able to reach the first coordination sphere of the copper in the reduced state, and that another stable position at $\sim$ 3 Å from the metal is also available. On the basis of the aminoacid sequence and the crystallographic distance of the different aminoacid residues the positively charged chain of Arg 141 seems to be the most suitable binding site of the anion. In fact an hydrogen bond of the anion with such protonated

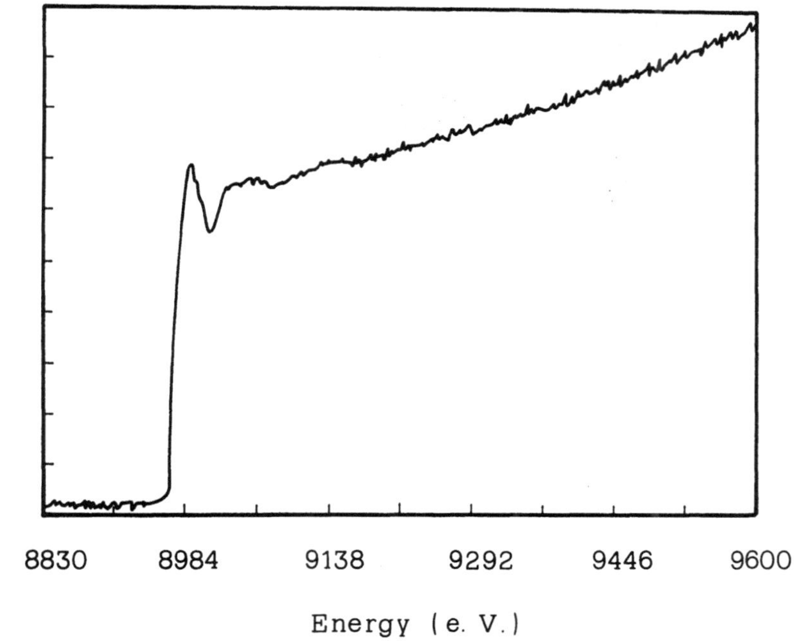

Fig. 1. Cu X ray absorption spectrum of Cu(I) SOD in presence of 2M Cl^-.

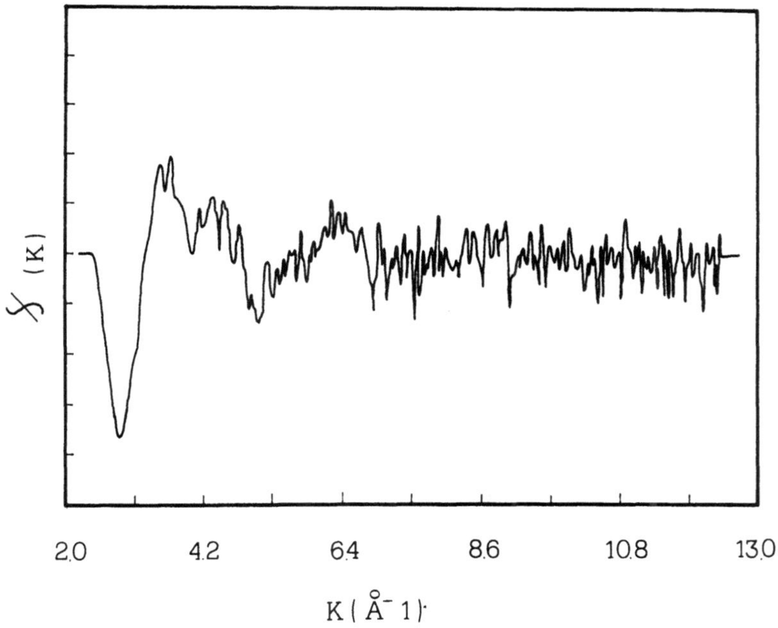

Fig. 2. EXAFS modulation of the sample of Fig. 1.

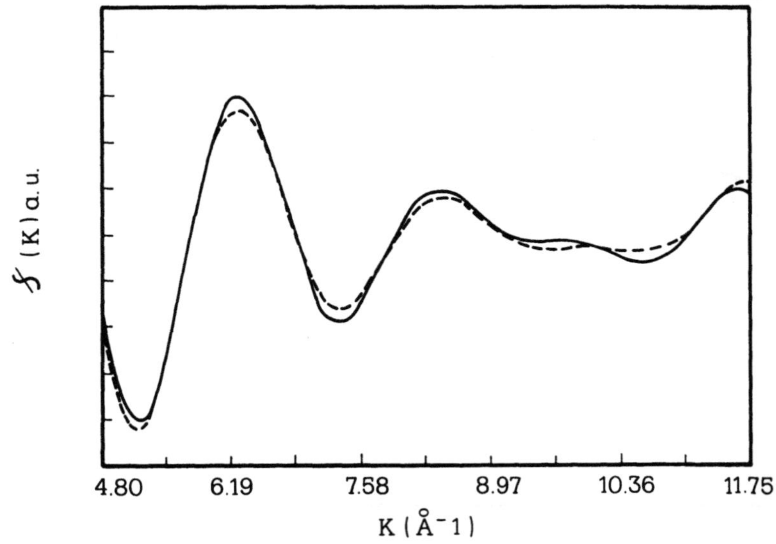

<u>Fig. 3.</u> Backfourier of the I shell of the sample of Fig. 1 (solid line); the dashed line represents the best fit obtained including, beside the nitrogen contribution, the presence of a Cl^- al 1.9 Å distance.

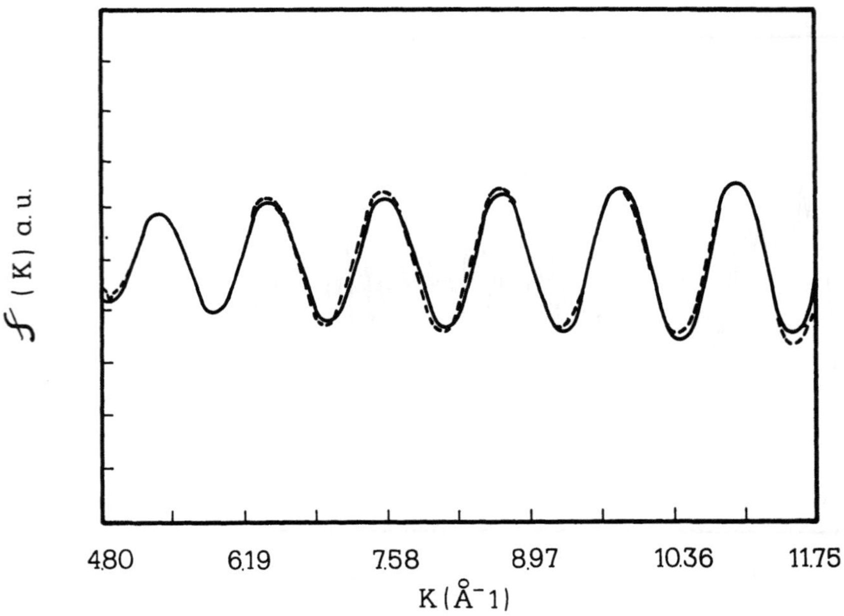

<u>Fig. 4.</u> Backfourier of the II shell of the sample of Fig. 1 (solid line); the dashed line represents the best fit obtained including, beside the carbon contribution, the presence of Cl^- at 32 Å distance.

residue would result in an anion-copper distance of 3 Å as obtained from the EXAFS spectra.
An extensive analysis of the interaction of the oxidized and reduced form of the enzyme with all the halide series is under progress.

Acknowledgement. This work is part of a program supported by the Progetto Strategico Biotecnologie of the Consiglio Nazionale delle Ricerche, Roma, Italy.

REFERENCES

1) Tainer J.A., Getzoff E.D., Beem K.M., Richardson D.C., Richardson J. (1982) J. Mol. Biol. 160, 181.
2) Rotilio G., Morpurgo L., Giovagnoli C., Calabrese L., Mondovì B. (1972) Biochemistry 11, 2187.
3) Rigo A., Stevenato R., Viglino P., Rotilio G. (1977) Biochem. Biophys. Res. Commun. 79, 776.
4) Van Camp H.C., Sands R.H., Fee J.A. (1982) Biochim. Biophys.Acta 704, 75.
5) Fee J.A., Ward R.L. (1976) Biochem. Biophys. Res. Commun. 71,427.
6) Valentine J. and Mota De Freitas D. in "Superoxide and Superoxide Dismutase in Chemistry, Biology and Medicine" G. Rotilio ed., Elsevier Amsterdam (1986) pag. 149-154.
7) Mc Cord J.M. and Fridovich I. (1969) J. Biol. Chem. 244, 6049.
8) Helmholn L. and Kruh R.F. (1952) J. Am. Chem. Soc. 74, 1176.
9) Co M.J., Scott R.A., Hodgson K.O. (1981) J. Am. Chem. Soc. 103, 986.
10) Rotilio G., Calabrese L., Mondovì B., Blumberg W.E. (1974) J.Biol. Chem. 249, 3157.
11) Blackburn N.J., Hasnain S.S., Binsted N., Diakun G.P., Garner C.D. and Knowles P.F. (1984) Biochem. J. 219, 985.
12) Tainer J.A., Getroff E.D., Richardson J.S. and Richardson P.C. (1983) Nature 306, 284.

Structural Changes in Hemerythrin and Error Analysis

Edward A.Stern
Dept. of Physics FM-15, University of Washington, Seattle, WA 98195

Abstract

Results are presented on the structure of the active site of hemerythrins as determined by EXAFS measurements. The μ-oxo bridge between the two Fe atoms in the oxy-form is converted to a μ-OH bridge as O_2 is released, forming deoxyhemerythrin. In order to draw such a conclusion the uncertainties in the structure parameters must be small enough. A careful analysis of the sources of errors indicates that the major source is from the nontransferability of amplitude and phase of the backscatters between the standards and the unknown. This source is usually neglected. A new function is presented that should be minimized to obtain the best fit between assumed models and the experimental data. Using this new function a definite prescription is given to obtain the errors in the fitting parameters.

Introduction

All aerobic organisms use oxygen and therefore require respiratory proteins to carry out the binding, storage, and transport of O_2. Based on the nature of the oxygen binding site, respiratory proteins may be hemoglobins, hemerythrins, or hemocyanins. Hemerythrins are found in sedentary creatures and in most cases have neither homotropic nor natural heterotropic effectors. The animals that use hemerythrins are generally less advanced than those using hemoglobins or hemocyanins, because they have a minimal vascular system [1].

Historically, the hemoglobins have been the most extensively investigated among the respiratory proteins. However, in the last few decades much progress has been made in understanding the structure of hemerythrins, and it appears reasonable to expect that with the correct effort the complete molecular basis of the respiratory function of hemerythrin will soon be understood in great detail. Comparison of the functioning of hemerythrin and the other respiratory proteins on a molecule basis may give further insight on why such quite different proteins evolved to serve similar functions and how nature uses different strategies to find mechanisms to reversibly bind O_2.

Hemerythrins

The active site of hemerythrin (Hr) contains two iron atoms which reversibily bind O_2. In the oxy- form the two iron atoms are the ferric state. while in the deoxy- form they are in the ferrous state. X-ray diffraction measurements [2] have been made to high resolution on azidomet-Hr and met-Hr. However, measurements on the biologically active forms of oxy-Hr and deoxy-Hr have not been made yet to a resolution high enough to solve the structure because of the rapid deterioration of the samples.

In this report the results of EXAFS measurements on both the biologically active forms of Hr and the met- forms are summarized. In order to draw conclusions about these results, estimates of the errors in the parameters of the structure need to be assessed. A discussion of error estimation is presented which details a new and more reliable way to estimate errors in the fitting parameters of modeling to the data.

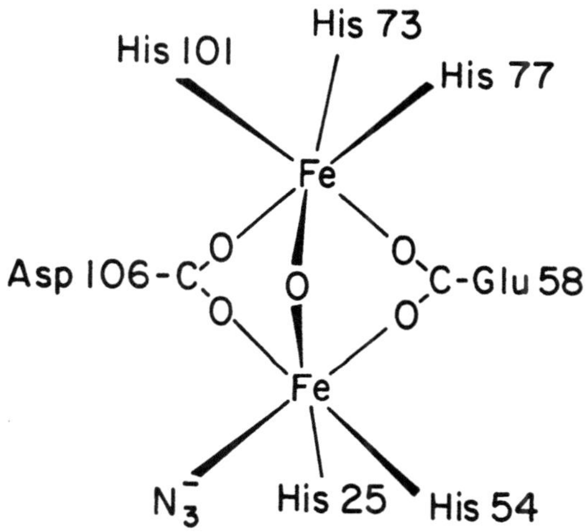

Fig. 1 - Schematic of the active site in azidomethemerythrin as determined by x-ray diffraction

EXAFS Results

The structure of the active site of azidomet-Hr as determined by x-ray diffraction [2] is presented in Fig.1. The two 6-coordinated Fe atoms are bridged by a μ-oxo atom and two carboxyl groups. Five of the remaining ligands are nitrogen atoms of histidines, while the sixth is an inorganic nitrogen of the azide. X-ray diffraction [3], EXAFS [4,5], and other studies indicate that the oxy-form of Hr is similar to the azidomet- form, with the azide replaced by O_2..

EXAFS measurements [5] on deoxy-Hr are compared with those made on oxy-Hr, met-Hr, and azidomet-Hr. The Hr samples used in the EXAFS measurements are in solution, while the diffraction measurements are on single crystals. Table I

lists the distances from the iron atoms and the Fe-O-Fe angle for the various forms of Hr as determined by EXAFS. The conclusions from the EXAFS measurements are as follows:

1. The met-forms in solution and in crystals have the same structure about the active site.
2. The oxy-form is similar to the met- forms, with O_2 replacing the inorganic ligand binding group.
3. On release of O_2 the μ-oxy bridge with Fe-O distances of 1.8 Å is converted to a μ-OH bridge with Fe-O distances of 2.0 Å. Accompanying this change is a large increase in vibrational amplitude around the active site.The entropy increase associated with this vibrational change constitutes most of the entropy change driving the transition.
4. The iron atom that had the O_2 attached probably is penta-coordinated with an empty site where the O_2 was located.

Error Analysis

To draw these conclusions requires an accurate assessment of the errors in the parameters. For example, when the μ-oxo bridge is replaced by a μ-OH bridge, a coordination of one oxygen atom per iron atom changes its distance from 1.8 Å to 2.0 Å. If the error in this coordination change is $\Delta N > 0.5$, then one cannot conclude that the change occurred. For instance, if the μ-oxo bridge is broken, adds an H^+ iron, and remains attached to only one of the Fe atoms, the result would be a coordination of $N = 1$ at a distance of 1.8 A, changing to $N = 0.5$ at 2.0 A. Our error analysis shows that the coordination is $N = 1$ for both these bonds, with an uncertainty of $\Delta N = \pm 0.3$, so it can be concluded that the deoxy- form has a μ-OH bridge.

Table I. Distance (in angstroms) from Fe atoms and the Fe-O-Fe angle of the active site of hemerythrin, as determined by EXAFS. The column headings O,N and N,O indicate that the EXAFS cannot uniquely distinguish between oxygen and nitrogen atoms, but the more likely atom is listed first. The numbers in parentheses in the second rows for the met-forms are x-ray diffraction results. The μ-O heading gives the μ-oxy distance for all forms except deoxy, which has a μ-OH distance.

Form	μ-0	0,N	N,0	Fe-Fe	Fe-0-Fe Angle
Azidomet	1.80(3) (1.77)	2.08(4) (2.23)	2.18(4) (2.25)	3.25(5) (3.25)	127°(4) (133°)
Met	1.82(3) (1.80)	2.07(4) (2.11)	2.14(4) (2.21)	3.14(5) (3.21)	117°(4) (126°)
Oxy	1.82(3)	2.11(4)	2.22(4)	3.24(5)	125°(5)
Deoxy	1.97(3)	2.12(4)	2.24(4)	3.50(6)	125°(5)

In a similar manner, in order to conclude that the deoxy- form has one Fe hexa-coordinated and the other penta-coordinated instead of both hexa-coordinated, it is necessary to be able to distinguish between an average coordination per Fe atoms of 5.5 and 6.0. Our error analysis [5] indicates that the uncertainty in the total coordination determination is $\Delta N = \pm 0.5$, so that such a distinction cannot be made from the EXAFS results. However, additional information is available. The strenght in the pre-edge 3-d pip suggests that one Fe in deoxy is penta-coordinated [5].

The usual discussion of errors in EXAFS determination in proteins assumes that $\Delta N \geq 1$ and the uncertainty in distances is $\Delta R \sim 0.02$ A. However, such statements are usually generic and are not based on an actual assessment of the specific data and analysis technique at hand. In this paper the results of a careful analysis of errors are summarized for the general case of fitting a model to EXAFS data on biological samples.

There are three types of errors:

1. *Errors in the data.* These are both random and systematic. The random errors are from the shot noise of the finite number of photons, from beam noise of the synchrotron source, and from electronics noise. The systematic errors are glitches from the monochromator, harmonic contamination of the x-ray beam, and sample contamination or degeneration.

2. *Data analysis errors.* These are systematic errors introduced by inadequacies in

the background subtraction and window effects in isolating a particular shell.

3. *Nontransferability of phase and amplitude*. The phase and amplitude obtained from theory and standards are not strictly transferrable to the unknown because of differences in bond distances and chemical state [6].

Another important consideration is the number of fitting parameters that can be obtained from the data. When a particular coordination shell is isolated a window of width δR is employed. This limits the rapidity with which k can vary and thus the number of independent points in k-space. The relation is [7]

$$\delta R \delta k = \frac{\pi}{2} \tag{1}$$

where δk is the spacing between independent k-points. The number of independent k-values is thus

$$m = \frac{k_2 - k_1}{\delta k} = \frac{2(k_2 - k_1)}{\pi} \delta R, \tag{2}$$

where k_1 and k_2 are the minimum and maximum k-values in the data, respectively. Here k_1 is determined by the smallest value at which the EXAFS can be separated from the rapidly varying background near the edge, while k_2 is determined by the limitation of the noise becoming comparable to the signal. For each independent k-value two experimental quantities are determined: the amplitude $A(k)$ and the phase $\phi(k)$ of the EXAFS signal $\chi(k)$. Thus, the number of independent data points is

$$N_D = 2m = \frac{4(k_2 - k_1)\delta R}{\pi}. \tag{3}$$

The fit should be made to $A(k)$ and $\phi(k)$, not $\chi(k) = \mathrm{Im} A(k) \exp[i\phi(k)]$, because these are the functions which are limited in their k-variation, as given by Eq.(1).

A standard statistical analysis [5,8] indicates that the function that should be minimized to determine the fitting parameters is

$$\chi_\nu^2 = \frac{1}{\nu} \sum_i \left[\frac{A^2(k_i)[\phi_f(k_i) - \phi(k_i)]^2}{\sigma^2(k_i)} + \frac{[A_f(k_i) - A(k_i)]^2}{\sigma^2(k_i)} \right] \tag{4}$$

Here k_i is an independent k-data point, $\sigma^2(k_i)$ is the rms error in the EXAFS $\chi(k_i)$, the subscripts f is the fitted result while the absence of a subscript denotes the measured quantity, and ν is the number of degrees of freedom given by

$$\nu = N_D - P \,, \qquad (5)$$

where P is the number of parameters used in the fitting model. An acceptable fit occurs when $\chi_\nu^2 \leq 1$. The errors in a given parameter can be determined by keeping all other parameters fixed and varying the one under consideration to change χ_ν^2 by unity. This assumes that the correlation between the parameters is small. Note that in order to be able to define the errors in the parameters, $P < N_D$; i.e., the maximum number of parameters that can be determined is less than N_D. In fact, to obtain a reliable value of the error, it should be at least 3 less than N_D.

The error σ^2 are the sum of the three types of errors discussed above:

$$\sigma^2 = \sigma^2{}_1 + \sigma^2{}_2 + \sigma^2{}_3 \,, \qquad (6)$$

where the subscripts relate to the respective numbered sources of errors.

In the case of Hr it is found that $\sigma_1{}^2$ $\sigma_2{}^2$ 0.03 for oxy-Hr and 0.04 for deoxy-Hr, and $\sigma_3{}^2$ 0.09 and 0.08, respectively. Thus the errors are dominated by the nontransferability of phase and amplitude because the standards employed were not close enough to the environment of Hr. The problem is caused by presence of the μ-oxo and μ-OH bridges in Hr and not in the standards. The standards, by necessity, have neighbors of only one type of atom localized around one distance, while the unknown has a distribution of distances and both N and O atoms. The quite different distance of the μ-oxo bond and the differente charge transfer introduced by the bridging atoms are the main causes of the $\sigma_3{}^2$ error.

Using a model compound with a similar mixture of liganding and bridging atoms should decrease the $\sigma_3{}^2$ error, but the problem is to separate the individual contributions from each ligand. These is no unique way to do it. A way around this problem is to take a difference of the $\chi(k)$ between the unknow and the model compound. If this difference $\Delta\chi(k)$ is $\leq$ $1/2\chi$ (k), then σ_3 can be reduced by fitting to $\Delta\chi(k)$ using the original standards. It must be remembered that σ^2 of $\Delta\chi(k)$ is $\approx 2\sigma^2$ of $\chi(k)$, but this can ben compensated for if $\Delta\chi$ is small enough, since the $\sigma_3{}^2$ errors should decrease approximately in the ratio $\Delta\chi/\chi$.

Conclusions

The EXAFS determination of the structure around the active site of Hr is limited by the nontransferability of amplitude and phase of backscattering atoms between standards and Hr. This error is typically not accounted for, implying that error estimates usually are too optimistic in EXAFS structure determinatin of metalloproteins.

A new method for estimating errors in fitting parameters is presented which should be more reliable than one currently employed. Employing the new method, error estimates are made for EXAFS studies on deoxy-Hr and oxy-Hr. It can be concluded that the μ-oxo bridge in oxy-Hr converts to a μ-OH bridge in deoxy-Hr. However, the experimental errors of EXAFS preclude a definite conclusion as to whether one of the Fe atoms is penta-coordinated in the deoxy-form. Other data are necessary to resolve this question, and the strength of the 3-d pip pre-edge absorption suggests that one Fe is penta-coordinated.

Acknowledgments

The hemerythrin sample were kindly supplied by Prof. Joann Loehr. The hemerythrin measurements were performed in conjunction with Dr. W.Timothy Elam, Prof. John Ellis, and Mr.Ke Zhang, who also analyzed the data. The research was partially supported by NSF grant PCM82-04234.

References

[1] I.M. Klotz and D.M. Kurtz, Acc. Chem. Res 17. 16 (1984)

[2] R.E.Stenkamp, L.C. Sieker, L.H. Jensen and J. Sanders-Loehr. Nature 291, 263 (1981); R.E. Stenkamp, L.C. Sieker, and L.H. Jensen, J. Am. Chem. Soc. 106, 618 (1984); J. Inorg. Biochem. 19, 247 (1983).

[3] R.E. Stenkamp L.C. Sieker, L.H. Jensen, J.D McCallum, and J. Sanders-Loehr, Proc. Nat. Acad Sci. U.S.A. 32, 713 (1983).

[4] W. T. Elam, E. A. Stern, J. D. McCallum, and J.Sanders-Loehr, J.Am. Chem. Soc. 104, 6369 (1982).

[5] K. Zhang, E.A. Stern, and J.Sanders-Loehr (unpublished).

[6] E. A. Stern, B.Bunker, and S.M. Heald, Phys. Rev. B 21, 5521 (1980); B.A. Bunker and E.A. Stern, Phys. Rev. B 27, 1017 (1983).

[7] L.Brillouin, *Science and Information Theory,* 2nd ed. (Academic Press, N.Y., 1962).

[8] P.R. Bevington, *Data Reduction and Error Analysis for Physical Sciences* (McGraw-Hill, N.Y.,1969).

The Interaction of Iron With Transferrins

Richard Garratt *§, Robert Evans §, Samar Hasnain + and Peter Lindley *

* *Dept. Crystallography, Birkbeck College, Malet St., London, WC1E 7HX,UK.*
§*Division of Biochemistry, UMDS (Guy's), London, SE1 9RT,UK.*
+ *SERC Daresbury Laboratory, Warrington, WA4 4AD, UK.*

INTRODUCTION

In this work we show that 1.) X-ray absorption studies of diferric chicken ovotransferrin (Fe_2COT) provide evidence for direct water coordination at the metal binding sites. 2.) The immediate environment of the iron atoms appears unaltered as a function of pH when spectra are recorded in the non-chelating buffer MES. 3.) Studies on human serum transferrin provide no evidence for structural changes at the metal coordination sites as a consequence of proposed interdomain cooperativity.

THE TRANSFERRINS

Ovotransferrin and serum transferrin are homologous iron-binding glycoproteins found in avian egg white and vertebrate blood plasma respectively. At present the precise natura of the iron-binding sites is unknown. Evidence from spectroscopic studies on transferrins isolated from several sources suggests the iron-binding sites to be similar in all cases and implicates tyrosine phenolate and histidine imidazole groups as the protein-derived ligands. A unique feature of the interaction of iron with the transferrins is the obligate requirement for a synergistic anion ((bi)carbonate) for the formation of a stable complex (1), and together with a water molecule (or OH^-) this is thought to complete the primary coordiantion sphere (see 2).

Herein we present further evidence for the presence of coordinated water to the iron atoms and using the sensitivity of XANES spectroscopy attempt to broach two further questions of transferrin biochemistry. Firstly, the mechanism by which diferric transferrin surrenders its iron to recipient cells once inside the acidic vesicle where iron release is known to occur and secondly the possibility that structural rearrangement of the metal-binding sites occurs as a consequence of interdomain cooperativity (see 2.).

SOLVENT COORDINATION IN OVOTRANSFERRIN

Fig.1 shows a comparison of the Fe K-edge EXAFS of Fe_2COT recorded in 0.1M $NaHCO_3$ buffer and as a freeze-dried powder. The reducation in the amplitude and frequency of the major shell contribution upon freeze-drying suggests the loss of a ligand from the first coordination sphere. This is evident in the Fourier transform (inset) where the first shell is shifted to lower R and is of reduced intensity.

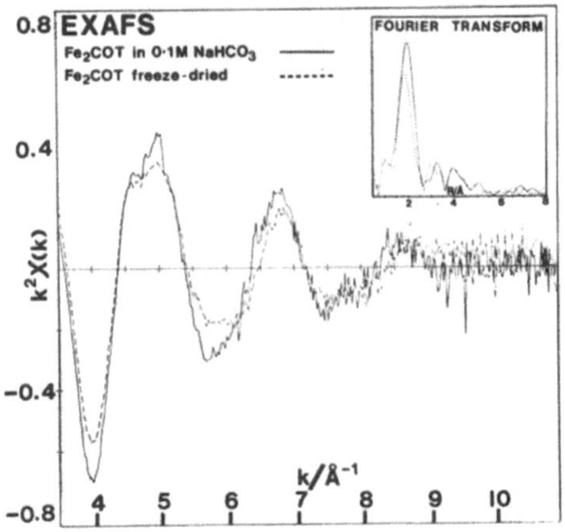

Fig.1 EXAFS and Fourier Transforms (inset) of Fe_2COT

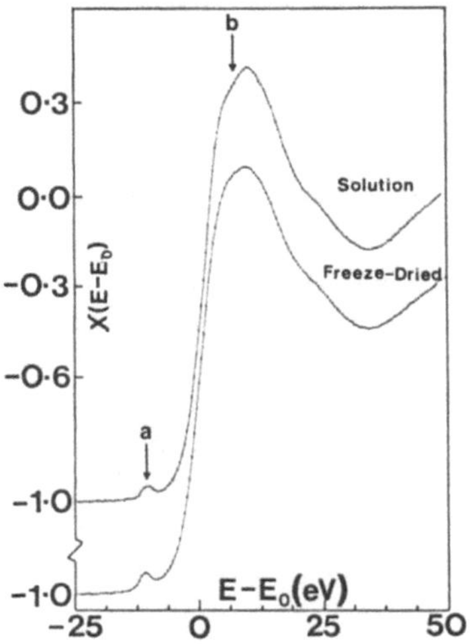

Fig. 2 XANES spectra of Fe_2COT

Changes in the R = 3 Å region and in the intensity of the features a and b in the edge structure (Fig.2) suggest that minor rearrangement of the remaining ligand geometry has also occured. Evidence (3) has been provided for a negative correlation between the intensity of the pre-edge feature a and the coordination number, consistent with a reduction in coordination for the freeze-dried protein.

Data presented elsewhere (4) has shown that the EXAFS spectrum of the freeze-dried sample can be adequately simulated by removal of on first shell atom from parameters used for simulation of the solution spectrum (5). We suggest therefore that a water molecule present in the solution structure of Fe_2COT is lost as a consequence of freeze-drying.

IRON REMOVAL FROM OVOTRANSFERRIN

Fig.3 shows XANES spectra and their first derivative for FE_2COT recorded in 0.1M 2-(N-Morpholino) ethanesulphonic acid (MES) buffer at pH7.2 and 4.8. Whilst it is often stated that iron is progressively lost from transferrins below pH6.5 this is in fact dependent on the availability of a suitable iron accepting species present in the medium (see 2). Physiologically this species has not been unequivocally identified but it is known that iron is released from transferrin inside an acidic compartment.

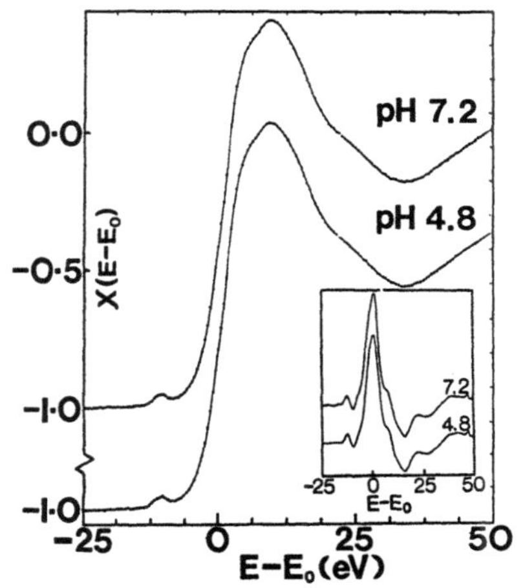

Fig. 3 XANES spectra and first derivative (inset) of Fe_2COT in 0.1M MES at pH7.2 (above) and 4.8 (below).

By using a non-chelating buffer we have attempted to mimic the pH-dependent changes which occur to the protein in vivo whilst preventing iron release. The XANES spectra, which are extremely sensitive to geometrical changes in the immediate vicinity of the probe atom clearly demonstrate that no such effects occur for Fe_2COT as a function of pH alone. It would seem that the structural changes to the molecule which facilitate iron removal intracellularly may be dependent on for example, the presence of the acceptor molecule or on protein-protein interactions between transferrin and its receptor.

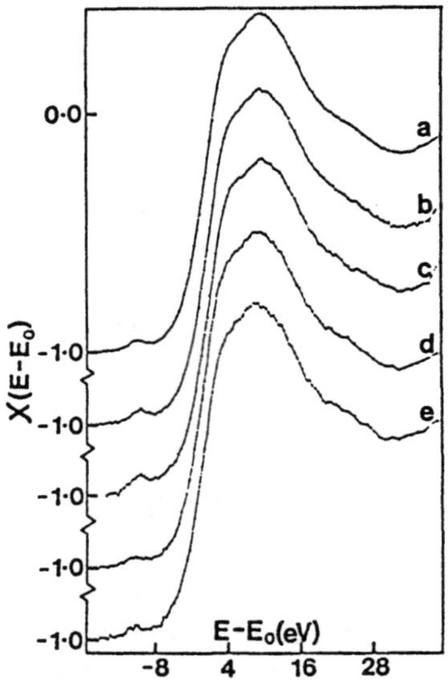

Fig.4 XANES spectra of Diferric Human Serum Transferrin (a), N-monoFe HST (b), C-monoFe HST (c) N-monoFe C-monoIn HST(d), and N-monoIn C-monoFe HST (e).

COOPERATIVITY BETWEEN THE TWO SITES OF HUMAN SERUM TRANSFERRIN

No significant differences are observed in the Fe K-XANES spectra of diferric human serum transferrin and its two monoferric counterparts (Fig.4 a,b, and c). This is consistent with EPR data on chicken ovotransferrin (6) which indicated that the signal from the diferric molecule was a composite of two distinct signals originating from the N-and C- domains of the protein. These data are readily interpreted in terms of structural independence of the two metal binding sites, indicating that no structural interpretation can be inferred for the cooperativity of metal binding (2). A similar result was observed for mixed Iron/Indium complexes where the remaining binding sites of a monoferric preparation were saturated with In^{3+} (Fig.4 d, e). Should the binding of metal to one domain of transferrin generate a conformational change at the unoccupied site, this is evidently undetectable once the second metal ion has bound.

Acknowledgements

We gratefully acknowledge the SERC for financial support and the provision of facilities. It is a pleasure also to thandk Dr. Barry Dobson for his help during many hours of data collection.

REFERENCES

(1) Schade, A.L., Reinhart, R.W. and Levy, H. (1949) *Arch. Biochem. Biophys.* **20**, 170.

(2) Brock, J. (1985) in "*Metalloproteins Part 2; Metal proteins with non-redox roles*" (Harrison, P. ed.) Mc Millan Press Basingstoke, pp.183-262.

(3) Roe, A.L., Schneider, D.J., Mayer, R.J., Pyrz, J.W., Widom, J. and Que, L., Jr. (1984) *J. AM. Chem. Soc.* **106**, 1976

(4) Lindley, P.F., Evans, R.W., Garratt, R.C. and Hasnain, S.S. , (1986) *EXAFS and Near Edge Structure J. de Phys. C 8 - 1189.*

(5) Garratt, R.C., Evans, R.W., Hasnain, S.S. and Lindley, P.F. (1986) *Biochem. J.* **233**, 479.

(6) Butterworth, M., Gibson, J.F. and Williams, J. (1975) *Biochem. J.* **149**, 559.

Structural Features of the Metal Binding Sites in Transferrins by EXAFS and Other Spectroscopic Techniques

I. Bertini[a], S. Mangani[b], L. Messori[a], S. Mobilio[c] and P. L. Orioli[a]
a) Department of Chemistry, University of Florence
b) Department of Chemistry, University of Siena
c) INFN National Laboratories of Frascati

ABOUT THE PROTEIN.

Transferrins, or siderophylins, are a group of proteins from different sources, with molecular weight around 80,000, constituted by a single polypeptide chain (1,2). The best known representatives of this class of proteins are human serum transferrin (Tf), chicken egg ovotransferrin (OTf) and human milk lactoferrin (Lf). Transferrins are involved in metal transport, particularly iron transport, through biological fluids, from absorption to storage and utilization sites. These proteins have an ellipsoidal shape, with two metal binding sites well apart each other, respectively near the C-terminal and the N-terminal protein extremes. The primary structures of Tf, OTf and Lf have been recently determined; large sequence homologies have been revealed for the three proteins as well as for the two domains of the same protein (3-5).

A low resolution X-ray structure, available for the rabbit serum transferrin at 3.5 Å refinement, shows that the protein is organized in two domains (6); each of them contains one metal binding site. Interestingly, metal binding at both sites occurs through formation of a ternary complex with the so called "synergistic anion" which *in vivo* is bicarbonate (7). This means that the affinities for the protein of both metal ions and bicarbonate are synergistically enhanced when both species are present. *In vitro* oxalate and other bifunctional ligands, bearing a carboxylate group, can be used as synergistic anions (7). The role of the synergistic anion is probably related to the metal uptake and release processes. The functional differences between the two binding sites, if any, are, at the moment, unclear.

The two binding sites can be identified through spectroscopic measurements on various metal substituted derivatives. Through the different derivatives, the two sites display very similar but not identical spectroscopic properties.

Iron(III) EPR signals, at liquid N_2 temperature, are slightly different for the two sites (8,9). EPR spectra of copper(II) and oxovanadium(IV) disubstituted derivatives are also capable of distinguishing between the two sites (10-15).

^{205}Tl NMR spectroscopy provides another tool for revealing the subtle structural differences exsisting between the N-terminal and C-terminal site; the latter technique can be used to monitor the relative occupancy of the two sites in etherodimetallic systems (16). ^{13}C NMR of the synergistic anion is also sensitive to the nature of the sites (*vide infra*)(17).

Dependence on pH of the affinities for metal ions is significantly different for the two sites (18). In general, metal ions bind transferrin and bicarbonate at pH higher than 6; the pK_a for metal ion binding ranges between 5 and 6 for tripositive metal ions and is about 7 for bipositive metal ions. However, the pH dependence of site occupancy appears to depend on more than one acidic group. At low pH the C-terminal site is definitely favoured, so that it is possible to prepare monosubstituted transferrins, whereas at higher pH (~ 8-9) both sites have comparable affinities. Despite such marked inequivalence, no significant physiological difference between the sites has been yet demonstrated.

Copper(II) and oxovanadium(IV) EPR spectra on both ovotransferrin and serum transferrin (11-15) as well as CD and ^{1}H NMR spectra on the cobalt(II)/OTf/oxalate system (19) have shown that each site displays pH dependent conformational variations; this is often referred as "heterogeneity" of the sites. There are several pK_a's determining such heterogeneity. Surely, in bipositive metal ion derivatives, a pK_a of 9.5 has been ascertained through different compounds (12-14,19). The same pK_a has been determined also for a conformational change in the Th(IV) transferrin derivative (20). The metal/protein/anion ternary complex is stable up to pH 9.5-11 through the different derivatives; above these pH values the complex breaks down, possibly in consequence of deprotonation of a key guanidinium group in the binding site, and protein denaturation takes place.

THE PROTEIN LIGANDS

Chemical modification studies on the apo and the diferric forms of transferrins have shown that seven tyrosines and four histidine residues become protected upon iron binding (21,22) as well as two arginine residues (23).

EPR studies on the copper protein provide evidence that one histidine is bound to Cu(II) since the low field hyperfine component is splitted by ^{14}N (10). The g value of the oxovanadium(IV) derivative has been interpreted as consistent with a single histidine in the plane orthogonal to the V-O bond (13-15). Spin echo studies on copper(II) transferrin reveal modulation patterns indicative of a His imidazole group coordinated to the metal (24).

Resonance Raman studies on iron(III), manganese(III), cobalt(III) and copper(II) transferrin demonstrated the binding of tyrosine to the metal ion since irradiation of the charge transfer transition in the visible region enhances the vibration modes assigned to the C-O stretching typical of tyrosinate. The charge transfer bands, that are often present in native and metal substituted transferrins, are definitely assigned to tyrosinate to metal charge transfer transitions (25,26).

One important hint in the understanding of metal tranferrins spectral properties has come from the investigation of metal complexes with the EHPG (27-29) and, more recently, with EHGS (30) ligands. They mimick quite accurately the protein metal chromophore (see Fig.1) and reproduce very closely its visible spectra (namely the intensity and position of the phenolate to metal LMCT band) through several metal substituted derivatives (29).

A

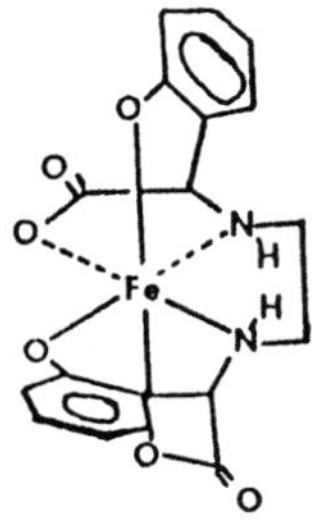

B

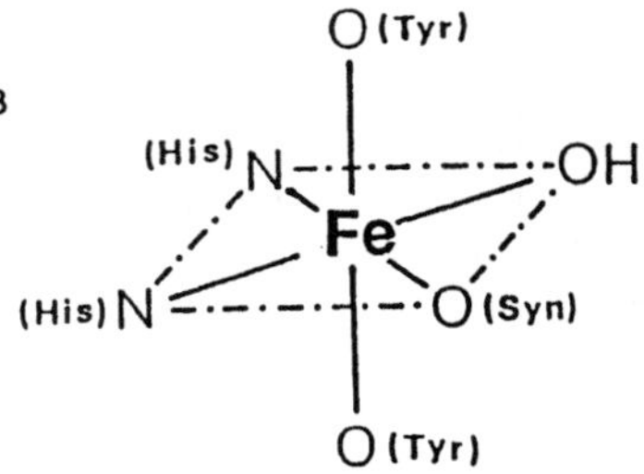

Figure 1. Structure of the $|Fe(meso\text{-}EHPG)|^-$ complex (A) compared to the proposed structure of the iron site in transferrin (B).

Furthermore the molar absorption coefficients for the internal phenolate ligand-ligand transitions at 242 nm and 290 nm, generated upon metal binding and subsequent tyrosine deprotonation, have been used to quantitate the number of coordinated tyrosines through the different metallosubstituted derivatives. It is found that each site contains two metal coordinated tyrosinates (27).

^1H NMR spectra of the Co(II) ovotransferrin derivatives show four signals for the ortho protons of two tyrosinates in each site, shifted far away upfield (31); the spectra also show six signals downfield in the 30-70 ppm region, which may originate from the protons of two histidines in each site. Two of them disappear when the spectrum is recorded in D_2O, and are therefore assigned to the two hystidine NH protons. The other signals are assigned to two ortho-like and two meta-like His protons. This may be taken as an indication of two imidazoles bound to each metal ion. An alternative assignment is that an exchangeable proton i.e. that related to the signal closer to the diamagnetic position, is due to a group like an Arg or a Lys which is not bound to the metal ion but is located in its immediate vicinities. In this case, the fourth protein ligand, which cannot be a sulphur atom (vide infra the EXAFS results), could be an oxygen from a β-CO_2 aspartate group as proposed through chemical modification studies (32).

THE SYNERGISTIC ANION

In ^{13}C NMR spectra of paramagnetic metal transferrin derivatives, prepared in presence of ^{13}C enriched bicarbonate, no signal is observed for the bound anion; this sets an upper limit of few angströms for metal to carbon distance depending on the estimated electronic relaxation time (33).

Again, no ^{13}C modulation of the spin echo is observed with ^{13}C enriched bicarbonate in the dicopper transferrin derivative (24). This is very consistent with bicarbonate directly bound to the metal. The EPR spectra of both iron(III) and oxovanadium(IV) derivatives dramatically change upon changing the synergistic anion, giving a further indication of direct binding of the synergistic anion to the metal ion (8,15). Even the LMCT bands in iron(III) transferrin are extremely sensitive to the nature of the synergistic anion (7).

^{13}C NMR spectra of enriched bicarbonate in Al(III), Ga(III) and Zn(II) derivatives of both Tf and Otf show only one signal for the bound anion; the position of the signal is close to that observed in the case of the $|Co(NH_3)_4CO_3|^+$ complex and sligthly upfield with respect to the signal of free CO_3^{2-}. These results are interpreted on the basis of a carbonate-like chemical state of the synergistic anion and of a close similarity between the anion binding environments in the two domains (17). The above data are quite different from those reported for $Co(III)_2/Tf/HCO_3^-$ which showed different ^{13}C signals for the two sites (34).

Spin echo studies on the Cu(II)/Tf/^{13}C oxalate system reveal a modulation pattern which is consistent with a carboxylate group not bound to the metal ion, i.e. with oxalate monodentate binding; indeed the spin echo technique reveals only weakly coupled nuclei (24).

^{13}C NMR studies on Al(III) and Ga(III) transferrin complexes prepared in presence of ^{13}C enriched oxalate have shown two AB quartets, one for each metal site; consequently, in each site, the two oxalate carbons are not equivalent. Monodentate binding of the oxalate anion to the metal is proposed. The magnitude of nonequivalent signals and the position of the shift relative to the free oxalate have been interpreted in terms of a scheme in which oxalate introduces some strain enhancing the structural differences between the two metal sites (17). This is also clearly shown by the proton NMR spectra on the Co(II)/OTf/oxalate derivative and by CD spectra on the same system (19).

The definitive evidence that the synergistic anion is directly bound to the iron comes from EXAFS spectroscopy; in the Fourier Transform of the EXAFS spectrum of the adduct iron(III)/transferrin/thioglycolate a second peak appears at 2.3 A which is assigned to the Fe-S distance (35).

All the data so far mentioned are consistent with a model which proposes the synergistic anion bridging the metal ion and a positively charged residue (possibly an Arg guanidinium group) ("interlocking sites model") (7).

THE SIXTH LIGAND

The electronic spectra of Co(II)/OTf show the metal to be six coordinated; furthermore it is quite feasible that Fe(III), Mn(III) and Co(III) ions are six coordinated; therefore we should look for a further ligand. A reasonable candidate is a solvent molecule.

Proton concentration variation upon tripositive metal binding seems to indicate that three protons are released upon trivalent ion binding (in the case of d-series bipositive metal ions two protons are released) (27,36). Two of them should come from deprotonation of tyrosines and one, possibly, from bicarbonate which forms carbonate upon metal binding. Oxalate derivatives, however, behave in the same way giving support to the idea that a proton comes from dissociation of a metal coordinated water (36). Whatever the results, the metal/donor atoms moiety is surely negatively charged because two tyrosinates and one carbonate (or OH^-) overcome the metal charge. Indeed the metal ion is not accessible to strong metal coordinating anionic or neutral ligands; positively charged residues, which may wrap the metal site, probably compensate the negative charge. Very recent studies have suggested the presence of clusters of positive charges near the binding sites from Lys and Arg residues (2,37); the latter residues probably constitute the binding sites for the so called "modifier" anions like perchlorate, pyrophosphate, thiocyanate and chloride (38).

Solvent ^1H nuclear magnetic relaxation dispersion (NMRD) is a powerful tool to detect water bound to paramagnetic metal ions under fast exchange conditions (39). Water ^1H NMRD of iron transferrin under various conditions provided not univocal answers on this respect; probably the proton exchange between bound solvent and bulk solvent is slow on the NMR time scale (40). Analogous experiments on copper(II) transferrin provided evidence for exchangeable protons as many as in oxovanadium(IV) transferrin (41); in the latter case no solvent molecule is expected to interact with the metal ion. Similar results are obtained with Gd(III) transferrin (42). Therefore either there is no solvent molecule interacting with the metal ion or its exchange is slow on the NMR time scale.

In favor of the latter possibility there is a change of the EPR signal linewidth for Cu(II)/Tf/carbonate on passing from H_2O to D_2O solvent (12). A significant decrease in amplitude of the EXAFS spectrum of iron(III) transferrin freeze dried samples with respect to solution samples has been suggested to be an indirect proof for the presence of a metal bound water (43); indeed, the

EXAFS spectrum of the former samples was best fit with a first coordination shell of five N/O atoms whereas six N/O atoms are needed for the fitting of the EXAFS data of the dissolved samples (see later). A restored amplitude of the EXAFS has been observed after the exposure to the air of the freeze dried samples; a similar behavior is displayed by the EPR signal of the same system.

XAS DATA

X-ray absorption spectroscopy provides further insight into the metal binding sites of transferrins. Iron K-edge spectra of diferric OTf and Tf have ruled out coordination numbers lower than six for the iron atoms (44,45). A subsequent EXAFS study on iron ovotransferrin has given direct evidence for the binding of the sulphur atom of the thioglycolate synergistic anion to the ferric center (35); the view of a six coordinate environment of light atoms for Fe(III), with a mean Fe-ligand distance of 1.96 A., has been also confirmed in the case of native diferric ovotransferrin. The Fourier Transform of the EXAFS spectrum of the latter system displays a second shell peak at 3.23 A which is similar to the ones observed in several complexes with imidazole ligands; this peak has been therefore attributed to histidine imidazoles participating in iron coordination. The quality of the data did not allow a quantitative analysis of that peak (35).

A very recent EXAFS work on the C- and N-terminal fragments of iron ovotransferrin, reporting a detailed analysis of the spectra, indicates only small differences between the two sites. In fact, theoretical calculations, performed including up to three shells of atoms to simulate the observed data, result in a first coordination shell of six low Z atoms distributed over two distances: two atoms at 1.87 Å and four atoms at 2.06 Å for the C-terminal site and two atoms at 1.83 Å and four atoms at 2.04 Å for the N-terminal site. The shortest distances are atributed to iron-tyrosinate bonds upon comparison with model compounds (46).

Independent EXAFS measurements on the N-terminal fragment of diferric human serum transferrin, analyzed with the aid of the model compounds $Fe(acac)_3$ and $|Fe(rac\text{-}EHPG)|^-$, again showed a six coordinated environment of oxygen/nitrogen atoms for the iron center (47). The quality of the data did not allow an exhaustive analysis of fine structural details; nevertheless, also in the latter case, a single distance fit did not reproduce the contributions to the EXAFS spectrum from the first coordination shell. A satisfactory simulation was obtained only by using a two distances fitting (Fig.2); the final best fitting parameters were, in this case, 3.8 O(N) atoms at 1.86 Å and 2.4 O(N) atoms at 1.98 Å from the iron center (Table I).

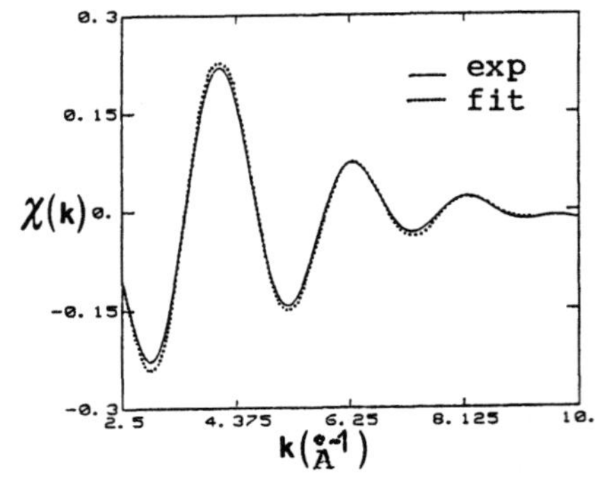

Table I

N	3.8 O(N)	2.4 O(N)
R(Å)	1.86	1.98
$\Delta\sigma^2(Å^2)$	0.0035	0.0011
Eo(eV)	0.04	9.2
Fit Index R = 0.0037		

Figure 2. Experimental and simulated backtransformed function of the EXAFS spectrum of the N-terminal fragment of iron transferrin. The best fitting parameters are also reported (Table I).

CONCLUSIONS

Despite the absence of high resolution X-ray diffraction studies, a picture of the metal binding sites in transferrins comes out from the several spectroscopic investigations. There are two high spin iron(III) centers, far apart each other, situated in very similar but not identical environments, respectively in the C-terminal and N-terminal domains; they are surrounded by two tyrosinate oxygens, the synergistic anion, a solvent molecule and two histidine nitrogens (or a His nitrogen and an Asp carboxylate oxygen) in a distorted octahedral chromophore. EXAFS studies give precise information on coordination numbers and distances. The synergistic anion is arranged in a bridging fashion and modulates site reactivity. The iron clusters lie in a "closed" conformation and are hardly solvent accessible. Only "modifier" anions labilize this conformation and make the metal ion accessible by changing the protein backbone.

REFERENCES

1 Aisen, P., and Listowsky, I. (1980) Annu. Rev. Biochem. 49, 357.
2 Chasteen, N.D. (1983) Adv. Inorg. Biochem. 5, 201.
3 MacGillivray, R.T.A., Mendez, E., Sinha, S.K., Sutton, M.R., Lineback-Zins, J., and Brew, K. (1982) Proc. Natl. Acad. Sci. USA 79, 2504.

4 Williams, J., Elleman, T.C., Kingston, I.B., Wilkins, A.G., and Kuhn, K.A. (1982) Eur. J. Biochem. 122, 297.

5 Metz-Boutiguè, M.H., Jolles, J., Mazurier, J., Schoentgen, F., Legrand, D., Spik, G.J., Montreuil, J., and Jolles, P. (1984) Eur. J. Biochem. 145, 659.

6 Gorinsky, B., Horsburg, C., Lindley, P.F., Moss, D.S., and Watson, J.L. (1979) Nature, 281, 157.

7 Schlabach, M.R., and Bates, G.W. (1975) J. Biol. Chem. 250, 2182.

8 Aisen, P., Pinkowitz, R.A., and Leibman, A. (1973) Ann. N.Y. Acad. Sci. 222, 337.

9 Zak, O., and Aisen, P. (1985) Biochim. Biophys. Acta 829, 348.

10 Zweier, J.L., and Aisen, P. (1977) J. Biol. Chem. 252, 6090.

11 Zweier, J.L. (1978) J. Biol. Chem. 253, 7616.

12 Zweier, J.L. (1980) J. Biol. Chem. 255, 2782.

13 Casey, J.D., and Chasteen, N.D. (1980) J. Inorg. Biochem. 13, 111.

14 Chasteen, N.D., White, L.K., and Campbell, R.F. (1977) Biochemistry 16, 363.

15 Campbell, R.F., and Chasteen, N.D. (1977) J. Biol. Chem. 252, 5996.

16 Bertini, I., Luchinat, C., and Messori, L. (1983) J. Am. Chem. Soc. 105, 1347.

17 Bertini, I., Luchinat, C., Messori, L., Scozzafava, A., Pellacani, G.C., and Sola, M. (1986) Inorg. Chem. 25, 1782.

18 Chasteen, N.D., and Williams, J. (1981) Biochem. J. 193, 717.

19 Bertini, I., Luchinat, C., Messori, L., Monnanni, R., and Scozzafava, A. (1986) J. Biol. Chem. 261, 1139.

20 Harris, W.R., Carrano, C.J., Pecoraro, V.L., and Raymond, K.N. (1981) J. Am. Chem. Soc. 103, 2331.

21 Rogers, T.B., Gold, R.A., and Feeney, R.E. (1977) Biochemistry 16, 2299.

22 Williams, J. (1982) Biochem. J. 201, 647.

23 Rogers, T.B., Boerresen, T., and Feeney, R.E. (1978) Biochemistry, 17, 1105.

24 Zweier, J.L., Aisen, P., Peisach, J., and Mims, W.B. (1979) J. Biol. Chem. 254, 3512.

25 Gaber, B.P., Miskowsky, V., and Spiro, T.G. (1974) J. Am. Chem. Soc. 96, 6868.

26 Tomimatsu Y., Kint, S., and Scherer, J.R. (1976) Biochemistry 15, 4918.

27 Pecoraro, V.L., Harris, W.R., Carrano, C.J., and Raymond, K.N. (1981) Biochemistry 20, 7033.

28 Patch, M.G., Simolo, K.P., and Carrano, C.J. (1983) Inorg. Chem. 22, 2630.

29 Patch, M.G., Simolo, K.P., and Carrano, C.J. (1982) Inorg. Chem. 21, 2972.

30 Carrano, C.J., Spartalian, K., Appa Rao, G.V.N., Pecoraro, V.L., and Sundaralingam, M. (1985) J. Am. Chem. Soc. 107, 1651.
31 Bertini, I., Luchinat, C., Messori, L., and Scozzafava, A. (1984) Eur. J. Biochem. 141, 375.
32 Shewale, J.G., and Brew, K. (1982) J. Biol. Chem. 257, 9406.
33 Harris, D.C., Gray, G.A., and Aisen, P. (1974) J. Biol. Chem. 249, 5261.
34 Zweier, J.L., Wooten, J.B., and Cohen, J.S. (1981) Biochemistry 20, 3505.
35 Schneider, O.J., Roe, A., Mayer, R.J., Que, L. Jr.(1984) J. Biol. Chem. 259, 9699.
36 Gelb, M.H., and Harris, D.C. (1980) Arch. Biochem. Biophys. 200, 93.
37 Thompson, C.P., McCarty, B.M., and Chasteen, N.D. (1986) Biochim. Biophys. Acta 870, 530.
38 Folajtar, D.A., and Chasteen, N.D. (1982) J. Am. Chem.Soc. 104, 5775.
39 Bertini, I., and Luchinat, C. "NMR of Paramagnetic Molecules in Biological Systems" (1986) Benjamin Cummings, Boston.
40 Koenig, S.H., and Schillinger, W.E. (1969) J. Biol. Chem. 6250.
41 Bertini, I., Briganti, F., Koenig, S.H., and Luchinat, C. (1985) Biochemistry 24, 6287.
42 O'Hara, P.B., and Koenig, S.H. (1986) Biochemistry 25, 1445.
43 Garratt, R.C., Evans, R.W., Hasnain, S.S., and Lindley, P.F. IV International Conference on EXAFS and Near Edge Structure, Fontevraud, July 7-14 1986 (France).
44 Alagna, L., Bianconi, A., Desideri, A., and Tomlinson, A.A.G. (1980) Ital. J. Biochem. 29, 73.
45 Roe, A.L., Schneider, D.J., Mayer, R.J., Pirz, J.W., Widom, J., and Que, L. Jr. (1984) J. Am. Chem. Soc. 106, 1676.
46 Garratt, R.C., Evans, R.W., Hasnain, S.S., and Lindley, P.F. (1986) Biochem. J. 233, 479.
47 Bertini, I., Mangani, S., Messori, L., Mobilio, S., and Orioli, P.L. (1986) J.Physique, in press.

Applications of X-Ray Absorption Spectroscopy to the Study of Ferritin

Sayers, D.E.*, Theil, E.C.#, Yang, C.Y.*
Departments of Physics* and Biochemistry#, North Carolina State University, Raleigh, North Carolina, U.S.A., 27695

Fontaine, A. and Dartyge, E.
Laboratoire Pour l'Utilisation du Rayonnement Electromagnetique CNRS Bat 209D, 91405 ORSAY, FRANCE

I. Introduction

Iron is required by plants, animals and microorganisms for a variety of activities, e.g. synthesizing DNA, transporting oxygen and electrons, reducing nitrogen and trapping radiant energy. To overcome the corrosive action of iron in the aerobic and saline environment of most organisms and to provide a soluble reservoir of iron, ferric ions are stored as a polynuclear complex inside a hollow protein coat made of multiple (n=24) peptide chains (subunits). The complex of iron and protein is ferritin; the protein coat is stable without the iron core and is designated apoferritin.

Ferritin has a similar structure in all organisms and among vertebrates. The amino acid sequence has been conserved throughout evolution (reviewed in refs. 1,2), although small variations occur both in apoferritin and the iron core. The function of ferritin can also vary, particularly in multicellular organisms, from providing a reservoir of iron for other cells or for other iron proteins such as the cytochromes, to protection from the toxic effects of excess iron. In general, variations in protein structure are related to variations in function and we have based our experiments on ferritin upon the same hypothesis. The simplest apoferritin is that from horse spleen; all (i.e.>90%) of the peptide chains in the protein coat are identical.

Initially we have two goals: (1) to determine the mechanism(s) by which apoferritin controls the formation of the iron core; and, (2) to define small molecules which can alter the structure of the iron core either directly, or indirectly by interacting with apoferritin. We use the simplest apoferritin, that prepared from ferritin from the horse spleen. Later, using cloned DNA encoding ferritin from other sources (e.g. ref. 3) and engineered to produce subunits of a single type in the bacterium E. coli, we can systematically study the effect of variations in apoferritin by preparing protein coats in which all the peptides are identical to each other but different from horse spleen apoferritin. Amino acids common to both apoferritins can also be altered by site-directed mutagenesis, and the different peptides can be mixed. EXAFS has been used to study the structure of model and natural iron cores formed in different environments (part A) and the role of the protein in core formation (aborted at intermediate stages by controlling the amount of iron added) (part B). Dispersive X-ray absorption spectroscopy (DXAS) studies can exploit the changes in oxidation state during iron core formation [(Fe(II) to Fe(III)] and dissolution of the iron core [Fe(III) to Fe(II)] plus the large amounts of iron that can be accommodated by each ferritin molecule (up to 4500) to study the effect of ions on dissolution of the iron core (part C) and core formation. Both applications of X-ray absorption spectroscopy will be used to study modified apoferritin produced using engineered DNA. Our first, preliminary, use of such apoferritin is described in part D.

II. Experiment

Apoferritin was prepared from horse spleen ferritin (Miles) as previously described (4) by reduction, chelation of the iron with thioglycollic acid, and dialysis against 0.05 M Hepes. Na. pH-7. The Fe(III) nucleation complex was prepared at 4·C by purging solutions of apoferritin and freshly prepared ferrous sulfate with argon, mixing at a ratio of 0.4 Fe/subunit, maintaining anerobiosis for 30-60 minutes to insure Fe(II) binding, admitting air and allowing oxidation to occur for at least 12 hours before making the measurements. The iron concentration was 2mM and all the iron was attached to the protein as judged by the absence of precipitated iron (Ks ca, 10^{-18}M). Studies on model and natural iron cores were made using horse spleen ferritin, Imferon, a therepeutic iron-dextran complex donated by Merrill Laboratories, and Blutal, a therapeutic iron-chondroitin sulfate complex donated by Dainippon Pharmaceutical.

EXAFS measurements on solids or liquids were made at SSRL or NSLS (line X-11) as previously described (5), at room temperature, using Silicon(111) monochromator crystals. Standards included $Fe(NO_3)_3$ in 0.1 N HNO_3, Fe(II)O, Fe(II) oxalate and Fe(III) oxalate. The Fe(III) apoferritin complex, the standards, ferritin, Blutal and Imferon were measured in the fluorescence mode and, when appropriate, were measured in the transmission mode as well.

For DXAS experiments either 1.4 M thioglycollic acid (TGA) 0.5M Hepes.Na, pH=7 or 1.4 M TGA/Tris, pH=7 were diluted 1:10 with ferritin in distilled water and immediately placed in the sample holder for measurements (within 10 seconds of mixing). The measurements were made at LURE as described(6).

III. RESULTS AND DISCUSSION

A. Cores and Models

We have hypothesized that the local structure of ferritin cores may vary. To determine environmental parameters which may alter core structure we examined the effect of anions, phosphate and sulfate, on the structure of the iron core of ferritin.

In the case of phosphate, a soluble complex of polynuclear Fe(III) and adenosine triphosphate (ATP) formed at concentrations up to 12mM in Fe, was found to have an X-ray absorption spectrum similar to ferritin iron cores(7). However, detailed analysis of the second shell neighbors revealed that in contrast to the mammalian ferritin that had been studied previously (5) and which contained little phosphate, the environment of the average iron atom was modelled best with an Fe-Fe interaction at 3.34 Å and an Fe-P interaction at 3.27 Å, suggesting that in the Fe(III)/ATP complex, phosphate was distributed throughout the polynuclear Fe(III) complex and thus might be distributed throughout ferritin cores. During the course of the EXAFS study, a natural ferritin from a bacterial source (A. vinelandii) was found to be rich in phosphate (Fe:P=1.4:1) (9). Whether the distribution of phosphate in such a natural ferritin will be similar to the Fe(III)/ATP model is not yet known, but the hypothesis that the local structure of ferritin cores can vary is supported by the composition of the bacterial ferritin; high resolution electron microscopy reveals the high phosphate, natural ferritin cores to be more disordered than low phosphate ferritin cores.

The effect of a second anion, sulfate, on the structure of soluble, polynuclear iron cores was tested by X-ray absorption and Mössbauer spectroscopy of mammalian ferritin and two polysaccharide compounds which provided different types of ligands for nucleation of polynuclear iron (8). The protein provided -COOH ligands, the dextran polysaccharide provided -OH and a few -COOH ligands, and the other polysaccharide, chondroitin sulfate, provided -OH, - COOH and sulfate ligands. Apparently, -COOH and -OH provide very similar environments for nucleation for the Fe(III) in mammalian ferritin and Fe(III)-dextran was very similar with only small differences were observed in the apparent particle size. In contrast, when sulfate ligands were present, the average Fe-Fe interaction was more ordered (a large negative Debye-Waller factor). Moreover, two types of magnetic domains were observed, one of which was virtually the same as the hydrated ferric oxide in ferritin and Fe(III) dextran, and the other of which appeared to be like hematite (Fe_2O_3). The results suggest that the sulfate ligand nucleated the hematite-like environment in the soluble Fe(III) chondrotin sulfate complex.

B. Fe(III) - Apoferritin Nucleation Complex

The number of iron atoms in a typical ferritin core is so high (ca 2000) that contributions from Fe(III) at the outer surface of the core, which might be in a distinctive environment due to interactions with the inner surface of the protein coat, would be undetectable. Since the protein coat is stable without the iron core, it is possible to remove the iron (to less than 0.5 Fe atoms/mole of protein less than 0.03 Fe/subunit) and reinitiate core formation by adding back small numbers of iron atoms. An amount of iron equivalent to 10 Fe atoms/molecule was added and allowed to oxidize; the stoichiometry was selected to be below the apparent saturation of the protein by Fe(II) (ca 0.5 Fe/subunit or 12/molecule(2)]. EXAFS analysis of the Fe(III) apoferritin complex(7) with protein from mammalian ferritin showed that the near neighbor environment of the Fe(III) was different from that in the full iron core and consisted of two low Z shells of atoms at 1.94 Å and 2.79 Å, respectively. The data were well modelled by carboxylate (O,C) but the possibility of N as a ligand cannot be excluded. Beyond the two low Z atoms, a third neighbor was detected at 3.57 Å which appeared to be Fe. Thus, when small numbers of Fe(II) atoms are added to the protein coat and oxidation to Fe(III) is allowed to occur (as it does during full core formation) the Fe(III) atoms are still attached to the protein but also are bridged to each other indicating migration of Fe atoms during oxidation as previously suggested (2) to form small Fe(III)-oxo clusters. Both the mononuclear character of added Fe(II) and the polynuclear character of the Fe(III) were confirmed using Mossbauer spectroscopy(8). However, the magnetic ordering in the Fe(III)-apoferritin complex was considerably weaker than in the full ferritin core; the blocking temperature was 7K compared to 35K for ferritin.

The existance of a polynuclear Fe (III) complex in which the metal atoms are essentially all attached to the protein and to each other indicates that the protein coat actually participates in nucleation of the iron core rather than passively permitting nucleation in a purely inorganic phase. Moreover, the protein may also provide sites for the binding of anions, such as phosphate and sulfate, which affect core structure. Finally, a core nucleation step, which involves the protein, provides a site of influence for the known variations in apoferritin structure.

C. The Use of Dispersive X-Ray Absorption Spectroscopy to Analyze Ferritin Iron Core Formation

Dissolution of the hydrous ferric oxide core of ferritin is usually monitored as the rate of formation of Fe(II) bipyridyl or the removal of Fe/thioglycollate by dialysis. However, the dissolution of the core involves at least two steps: (1) reduction and breaking of bridging oxygen ligands; and (2), formation of Fe(II) bipyridyl complex. Separation of the iron by dialysis can only occur after completion of both steps. By measuring the rate at which the Fe edge in the X-ray absorption shifts from the position for Fe(III) to that for Fe(II), the first step in core dissolution can be monitored. Preliminary analysis of x-ray absorption data for mammalian ferritin collected in the dispersive mode indicates that reduction of iron in the core by thioglycollate is complete in 1-2 minutes whereas in a separate experiment the use of dialysis to remove the iron reduced by thioglycollate requires days. Formation of the Fe(II) bipyridyl complex measured at 522 nm(11), with reduced flavin as the reductant, continues for hours. Thus reduction of the iron in the ferritin core and conversion to small complexes appears to be temporarily separated. The rate of reduction itself also appears to be variable and in the case of reduction by thioglycollate, to be influenced by buffer ions such as Tris and Hepes. The properties of cores with different structures has yet to be examined.

Formation of the ferritin iron core, in vitro at least, can be measured by the conversion of Fe(II) to colored, polynuclear Fe(III) species using the change in optical density at 420nm(11). However, the rate of oxidation and formation of polynuclear complexes may not be equivalent. Using the same approach for core dissolution, core formation can also be studied using dispersive X-ray absorption spectroscopy.

D. Apoferritin from engineered DNA

As part of our plan to determine the effect of variations in apoferritin structure on core formation, core structure and iron release, we have cloned cDNA for three different ferritin subunits and are subcloning them in "expression vectors", i.e. plasmids, which will direct the synthesis, in E. Coli, of the apoferritin subunits encoded in the cDNA. Initially we selected a clone, 1D8, derived from bullfrog mRNA which has the most deviant amino acid sequence, compared to horse spleen apoferritin, of any apoferritin subunit so far characterized. For example, the protein encoded in 1D8 has 16 serine residues compared to 9 in horse spleen apoferritin. Nevertheless, there are a number of residues that are potential iron binding ligands that are conserved. In addition, the mRNA and the protein encoded in 1D8 displays unusual genetic regulation by iron. The protein produced by the cloned DNA in E. Coli is clearly the bullfrog protein because it reacts with a specific monoclonal antibody to bullfrog ferritin. The addition of ferrous sulfate to a preparation of such protein, followed by oxidation to permit iron core formation produced a complex with the electrophoretic properties of ferritin and with an edge structure characteristic of a polynuclear iron core, but with distinctive higher shell interactions (Fig. 1). The added iron migrated with ferritin during electrophoresis and remained soluble at a concentration ten times above the solubility of free Fe(III). Taken together, the data suggest that the apoferritin subunits produced by the engineereed DNA assembled to form an apoferritin coat which functioned to incorporate iron into an iron core. Future EXAFS comparisons of the core and DXAS analysis of core formation for native apoferritin and protein modified by site-directed modification of the cDNA should provide a great deal of information about the mechanism(s) by which the apoferritin coat influences the iron core.

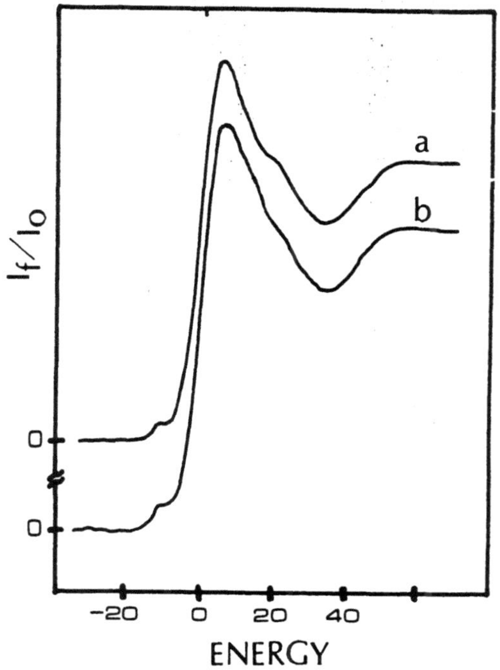

Figure 1: XANES for horse spleen ferritin (curve a) and ferritin reconstituted from Fe(II) and the protein produced by E. Coli (curve b) under the direction of cDNA for tadpole red cell ferritin subunit 1D8. Note the similarity of the edges for the native and engineered ferritins, although differences in detail are evident. Preliminary examination of the Fourier transform of the data indicate differences in the higher order interactions.

The experiments described were supported in part by NIH grants AM20251 (E.C.T. and D.E.S.) and GM24675 (E.C.T.), by DOE grant 80ER10742 (D.E.S.) and by the North Carolina Agricultural Research Service (E.C.T.). Measurements were made at the Stanford Synchrotron Radiation Laboratory (SSRL), the National Synchrotron Light Source (NSLS) at Brookhaven National Laboratory and the Laboratoire pour l'Utilisation du Rayonnement Electromagnetique (LURE) at Orsay.

REFERENCES

1. Theil, E.C. (1983) Adv. Inorg. Biochem. 5:1-29.

2. Theil, E.C. (1987) Ann. Rev. Biochem. 56 (in press).

3. Didsbury, J.R., Theil, E.C., Kaufman, R.E., and Dickey, L.F. (1986) J. Biol. Chem. 261:949-955.

4. Chasteen, N.D. and Theil, E.C. (1982) J. Biol. Chem. 257: 7672-7677.

5. Theil, E.C., Sayers, D.E., and Brown, M.A. (1979) J. Biol. Chem. 254:8132-8134.

6. Sayers, D.E., Bazin, D., Dexpert, H., Jucha, A., Dartyge, E. Fontaine, A., and Lagarde, P. (1984) in: EXAFS AND XANES NEAR EDGE STRUCTURE III (Hodgson, K.O., Hedman, B., and Penner-Hahn, J.E., eds.) pp 209-216, Springer-Verlag, New York.

7. Mansour, A.N., Thompson, J.C., Theil, E.C., Chasteen, N.D. and Sayers, D.E. (1985) J. Biol. Chem. 260: 7975-7979.

8. Yang, C.Y., Bryan, A.M., Theil, E.C., Sayers, D.E. and Bowen. L.H., J. Inorg. Biochem. (In press).

9. Mann, S., Bannister, J.V., Williams, R.J.P. (1986) J. Mol. Biol. 188:225-232.

10. Yang, C.Y., Meagher, A., Huynh, B.H., Sayers, D.E., Theil, E.C. Biochemistry (in press).

11. Mertz, J.R. and Theil, E.C. (1983) J. Biol. Chem. 258:11719-11726.

Structural Investigations of Biomimetic Complexes of Cytochrome P-450 by Difference EXAFS Spectroscopy

J. Goulon[1,2], M. Loos[1], I. Ascone[1,2], C. Goulon-Ginet[1], P. Battioni[3], J.P. Battioni[3], J.P. Mahy[3], D. Mansuy[3] and B. Meunier[4].

1. INTRODUCTION

Over the past decade, intensive efforts have been made in the synthesis and structural study of high valent (porphyrinato):metal-oxo complexes which could mimic the heme mediated activation of oxygen as performed by monooxygenase hemoproteins like cytochromes P-450 or by hemoperoxidases like the horseradish peroxidase (HRP). Various oxidants : NaOCl, PhIO, alkylhydroperoxides or even molecular oxygen have been used in the frame of competing strategies [1]. In all cases, the central question is to know whether or not these compounds are featuring a real and short Metal...Oxygen double bond because it is currently postulated that formal Fe^{V}=O species or a π-cation porphyrin Fe^{IV}=O complex might be the active centers in the catalytic cycle of alkane hydroxylation by cytochromes P-450 and because such a short Fe...O bond was also detected recently by PENNER-HAHN et al. for HRP-I [2].

Similarly there is a growing interest in iron carbenes, iron nitrenes and nitrosoalkane complexes which are the carbon and nitrogen analogues of the high valent metal oxo-complexes discussed above. The reason for this stems from the suggestion that iron metabolite complexes formed upon metabolic oxidation of several drugs by hepatic cytochromes P-450 might involve such iron carbene, iron nitrene or iron nitrosoalkane complexes [3-6]. As regards iron...carbon bonding, one should keep in mind that rather unstable σ-alkyl or σ-aryl iron(II) or iron(III) complexes might also be important intermediates as proven in the case of the metabolic oxidation of arylhydrazines by cytochromes P-450 [7]. Accurate determinations of the axial Metal...Carbon or Metal...Nitrogen distances are therefore needed here again.

As the chemical instability or reactivity of most of these compounds make it difficult, if not impossible, to grow single crystals suitable for X-ray studies, it is quite attractive to use X-ray absorption spectroscopies i.e. EXAFS and XANES in order to discriminate between such single or double bonding of the metal to carbon, nitrogen or oxygen. It is however our experience that conventional EXAFS data analyses based upon more or less sophisticated fits of only the coordination shells very often fail to give reliable results. There are mixed reasons for this failure : (i) The contributions of the axial ligands and of the equatorial nitrogen atoms are usually not well enough resolved experimentally and, even worse, can give rise to partially destructive interferences [8,9,10]. (ii) The pyrrole rings frequently take "ruffling" or "doming" conformations which are not easily detectable from a direct analysis of the EXAFS spectra and make fundamentally dubious

[1]Laboratoire de Chimie Théorique, U.A. CNRS n° 510.
Université de Nancy-I, B.P. 239, F54506 - Vandoeuvre-lès-Nancy, France.

[2]L.U.R.E., L.P. CNRS n°008, C.E.A. et M.E.N.,
Université de Paris-Sud, Bât. 209d, F91405 - Orsay Cedex, France.

[3]Laboratoire de Chimie et Biochimie Pharmacologiques et Toxicologiques,
U.A. CNRS n° 400, Université René Descartes, 45 Rue des Saints-Pères,
F75270 - Paris Cedex 06, France.

[4]Laboratoire de Chimie de Coordination du CNRS, 205 Route de Narbonne,
31400 - Toulouse, France.

any attempt to calculate the axial shift of the metal from observed variations of the Metal...N_{eq} distances [11].

In order to overcome these difficulties we have been led to develop, over these recent years, difference FT-analyses taking into account structural perturbations such as a variable axial shift of the metal, a variable contraction of the macrocyclic cavity radius and doming effects. Although the basic principle of these "perturbed" difference analyses has already been described in details elsewhere [12], we believe that it is worthwhile to review here some of the most recent developments of the method itself before we go into the details of specific applications regarding structural characterizations of iron carbene, iron nitrene or nitrosoalkane complexes.

2. DIFFERENCE FT ANALYSES WITH STRUCTURAL PERTURBATIONS

Direct difference Fourier analyses of EXAFS spectra were first introduced by CRAMER et al. [13]. These authors wanted to localize the contribution of the unbound oxygen O' in the oxy-picket fence model of oxyhemoglobin from a direct difference FT analysis of the spectra of the oxy- v.s. deoxy- complexes. The results were not really convincing and the method was then abandoned to the benefit of fitting procedures. ELAM et al. met more success in analyzing the direct difference spectrum of deoxy- v.s. oxy-hemerythrin but their investigation was restricted to the coordination shell only [14]. As regards the stereochemistry of porphyrins more specifically, a serious objection raised by CRAMER himself against direct difference FT analyses was that variations of the axial displacement of the metal were neglected. Indeed when such variations do occur, it becomes impossible to have a perfect cancellation of the successive signatures of the porphyrin macrocycle and misleading residual signals are obtained [12].

It was however shown by GOULON et al. [12] that these difficulties can easily be overcome by applying some appropriate structural perturbation to the FT spectrum $\tilde{\chi}(R)$ of the reference compound before carrying out any difference. An illustrative block-diagram of the "perturbed difference" analyses is shown in figure 1.

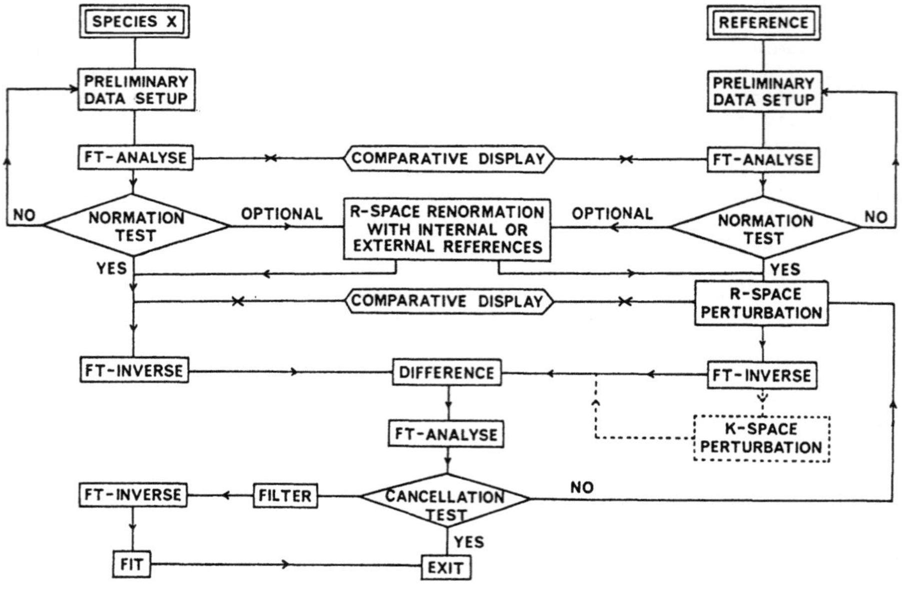

Figure 1 : Block diagram of the perturbed difference analyses.

The basic requirement is to select some analytical transformation $R' = f(R)$ taking into account the relevant structural changes of the reference. All what we then need is an accurate interpolation procedure in order to linearize the sampling of the function $\tilde{\chi}(R')$.

From a compilation of the enormous amount of crystal structure data on metalloporphyrin stereochemistry, we have been led to retain three major types of perturbation :

(i) an axial shift δh of the metal with respect to the average plane of the porphyrin macrocycle :

$$R'^2 = R_{CT}^2 + (H+\delta h)^2 = R^2 + 2H\,\delta h + \delta h^2 \quad (1)$$

(ii) A variation of the macrocyclic cavity radius :

$$R' = R + \Delta_{cav} \quad (2)$$

is the indirect consequence of the "ruffled" conformations of the pyrrole rings. As pointed out by HOARD [15], small amplitude rotations ϕ of the pyrrole subunits around the Metal...N_{eq} axes are possible in accordance with S_4 symmetry, provided that the bond parameters are satisfying the exact relationship (figure 2a) :

$$a_N + b\cos\frac{\alpha}{2} - c\sin\psi = (b\sin\frac{\alpha}{2} + c\cos\psi)\cos\phi \quad (3)$$

More often, the adjustments in bond parameters are largely taken into account by the Metal...N_{eq} bonds so that an apparent contraction of the cavity radius is observed. In the first approximation, we usually assume Δ_{cav} = constant. This is quite reasonable for the N_{eq}, C_A, C_B shells but less satisfactory for the C_{meso} signatures. In specific cases where Δ_{cav} becomes too large, we have developed an additional perturbation which is aimed at correcting selectively the signal of the C_{meso} shell [16].

(iii) "Doming" effects can be easily generated by slightly tilting the pyrrole rings around axes which are now perpendicular to the Metal...N_{eq} bonds (figure 2b). Such distortions of the porphyrin core make it possible to keep short Metal...N_{eq} bond lengths while allowing a quite large displacement of the metal above the 4 N_{eq} plane. Useful parameters for describing this perturbation are the distance d separating the 4 N_{eq} plane from the mean plane of the porphine skeleton and the less commonly given tilt angle θ (figure 2b) :

$$R'^2 = R^2 + H'^2 - H^2 + 2R\left(1 - \frac{1}{2}\frac{H^2}{R^2}\right)(a'_N - a_N) + (a'_N - a_N)^2 \quad (4)$$

with : $a'_N = a_N\cos\theta + (H-d)\sin\theta$ and $H' = (H-d)\cos\theta - a_N\sin\theta$

"Doming", "antidoming" or "saddle shaped" conformations of the macrocycle simply differ by the sign of θ and d for each pyrrole subunit.

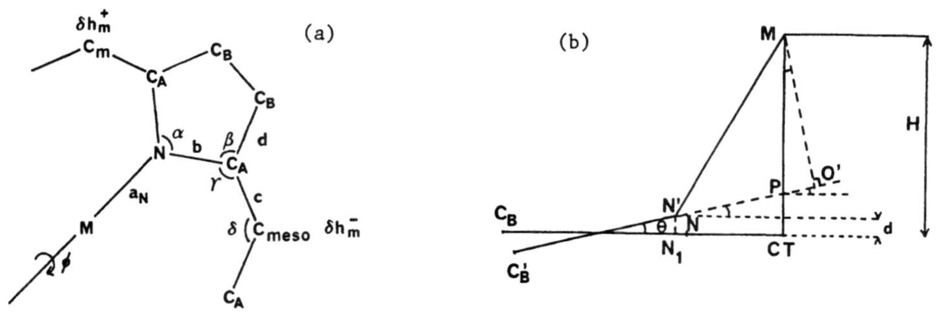

Figure 2 : Basic deformations of a porphyrin core :
a. "ruffling" conformation , b. "doming" conformation .

An interesting question is whether or not difference EXAFS analyses can discriminate between a change δh of the axial displacement H of the metal and a variation of the cavity radius Δ_{cav}. In principle the answer is affirmative because the two perturbations exhibit a quite different R dependence, as illustrated by equation (5) :

$$R' = R + \Delta_{cav}\left(1 - \frac{H^2}{2R^2}\right) + H\frac{\delta h}{R} + \frac{1}{2}\frac{\delta h^2}{R} \qquad (5)$$

where second order terms Δ_{cav}^2 , $\left(\frac{H^2}{R^2}\right)^2$ are neglected for the sake of simplicity : this means that it is not possible to achieve a good cancellation of all signatures of the porphyrin core, i.e. the N_{eq}, C_A, C_{meso}, C_B, C_ϕ, without taking into account both corrections. Accuracies of the order ± 0.05 Å for δh and ± 0.015 Å for Δ_{cav} have been obtained in favourable cases. Unfortunately, in practice "doming" and related effects often add correlations degrading the quality of the structural information available. There are however cases where "doming" like corrections are of definitive requirement in order to obtain an acceptable cancellation of all signatures of the porphyrin core : a typical example was provided by the comparison of the spectra reproduced in figure 3a for **1** = TMP:Mn(V)≡N and **2** = TMP:Mn(III)-Br. A paradox, not to be accounted for by equation (5) is that the shift between the two spectra increases with R. Figure 3b shows that the perturbed difference spectrum generated with "doming" like corrections ($\delta h = 0.15$ Å , $\Delta_{cav} = 0.025$, $\Delta d = 0.15$ Å) exhibits the positive Mn...N and negative Mn...Br signatures at the expected distances while there is obviously a nearly perfect extinction of all signals of the porphyrin core. As the crystal structure of **2** is not known yet, we have been unable to refine the determinations of $H_1 \simeq 0.45$ Å (?) and $d_1 \simeq 0.1$ Å (?), but the distances Mn...N (1.50 ± 0.01 Å) and Mn...Br (2.60 ± 0.02 Å) are fully consistent with other crystal structure data [17,18] or EXAFS investigations at the Br K-edge.

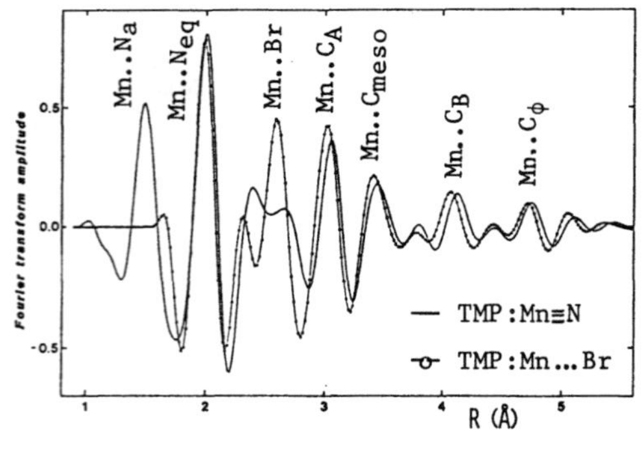

Figure 3a : Comparison of the FT spectra Im[χ̃(R)] of **1** = TMP:Mn≡N and **2** = TMP:Mn-Br. These spectra are implicitly corrected for the scattering amplitude and phase shifts of the dominant Mn*...N shell.

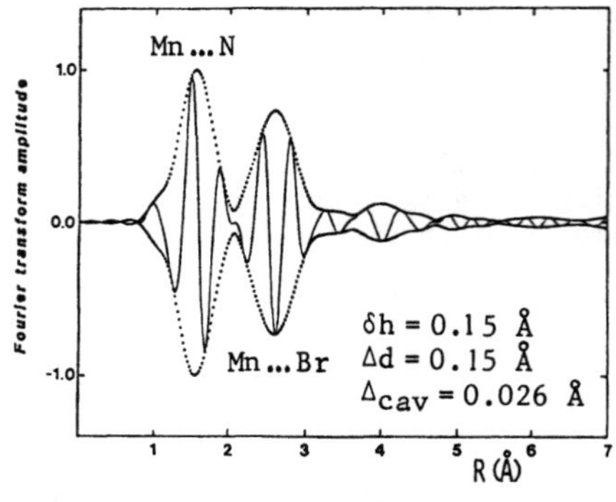

Figure 3b : Perturbed difference FT spectrum of (**1** - **2**).

Perhaps of more relevance to biostructural problems is the comparison of the spectra Im[χ̃(R)] reproduced in figure 4a and which concerns **4** = TPP:Fe(II)(N-MethylImidazole)$_2$ and the naked complex **3** = TPP:Fe(II). According to its known crystal structure [19], **3** has a ruffled core with rather short Fe...N_{eq} distances (1.97$_1$ Å) and thus the perturbed difference spectrum (**4** - **3**) shown in figure 4b was performed with a slightly expanded cavity radius ($\Delta_{cav} = 0.015$ Å). It is noteworthy that this spectrum reproduces all the expected features associated with the axial 2-methylimidazole ligands : the Fe...$N_{3,3'}$ and Fe...$C_{A,A'}$ signals peaking at R_1 = 2.01 ± 0.01 Å and R_2 = 3.02 ± 0.05 Å respectively, and even more characteristic the enhanced multiple scattering of the Fe...N_3...N_1 or Fe...N_3...C_B paths peaking with an inversed phase at distances $R_3 \simeq 3.8$ Å which are c.a. 0.3 Å short of the exact

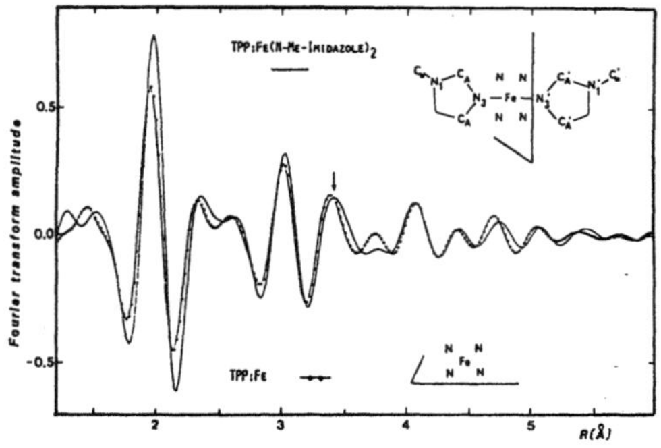

Figure 4a : Comparison of the FT spectra of **3** = TPP:Fe and **4** = TPP:Fe(N-Methyl-Imidazole)$_2$. The spectra are implicitly corrected for the Fe...N shell.

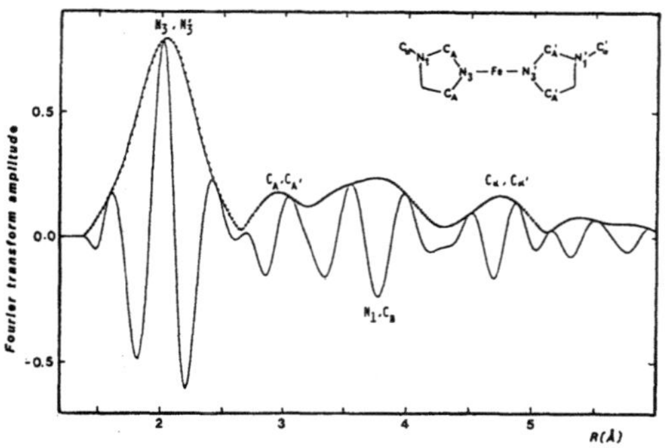

Figure 4b : Perturbed difference FT spectrum of (**4** - **3***).

values. Indeed this perturbed difference spectrum exhibits a remarkable analogy with the FT spectra of copper or zinc tetrakis(imidazole) complexes which have been used as reference compounds for multiple scattering calculations over these recent years [21,22]: calculations of that complexity would have been beyond the scope of current endeavour on the original data of **4** while they are becoming quite possible via our perturbed difference spectrum. These results also make us rather confident of the reliability of our analyses.

3. STRUCTURAL INVESTIGATION OF BIOMIMETIC COMPLEXES

3.1. High-valent metal oxo- or oxo-like complexes.

A good illustration of the potentiality of our perturbed difference analyses is to be found in a paper which has been published recently [1] and relates the structural characterization of high valent manganese porphyrin complexes. As full details are available elsewhere, we shall summarize shortly here the major results of this study which was initiated by the isolation of these reactive species at Toulouse [1]. Oxidation by NaOCl of **5** = TMP:Mn(III)Cl gives a high valent complex **6** = TMP: Mn(O)(solv.) which is capable of selective epoxidation of olefins. We produced convincing evidence that in **6** the metal is lying in or very close to the mean porphyrin plane and is hexacoordinated. Using **2** again as a reference, a perturbed difference analysis indicated that of the two axial bonds, only one is well defined ($R = 1.84 \pm 0.02$ Å) and is thus to be unambiguously assigned to a Mn...O single bond whereas the second is provisionally assigned to a more loosely bound water molecule at $R \simeq 2.3$ Å. Thus for this specific system, our EXAFS results ruled out the model of a real Mn^V=O compound. Further work is in progress in order to isolate and characterize other metal oxo- compounds.

3.2. Carbenes and σ-bonded aryl Fe(II) complexes.

It has been shown by MANSUY and his collaborators [4,22,23] that the reaction of iron(III) porphyrins with polyhalomethanes in the presence of a large excess of reductant produces a variety of porphyrin-iron bonded carbene species. The original impetus for these endeavours was the postulate by ULLRICH and MANSUY [3] that the anesthetic halothane reacted with cytochrome P-450 to give such a porphyrin iron carbene. Up to the present, however, only few structural investigations of iron carbene complexes have been reported and it was especially attractive to check if our perturbed difference analyses can yield accurate determinations of the iron...carbon double bonds.

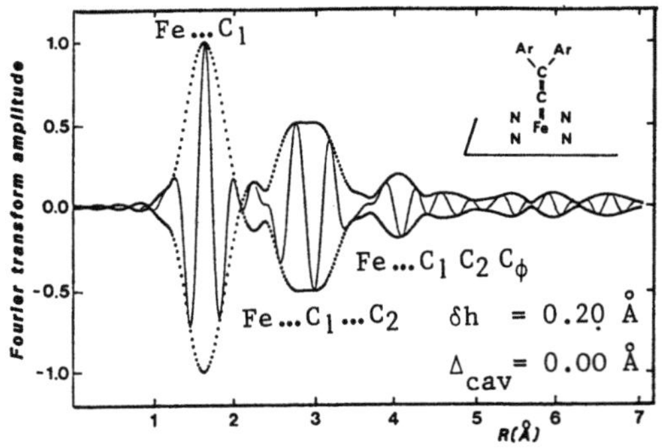

Figure 5 : Perturbed difference FT spectrum of (**7** - **3***) where **7** = TPP:Fe=C=C(Ar-Cl)$_2$, δh = 0.20 Å , Δ_{cav} = 0.00 Å

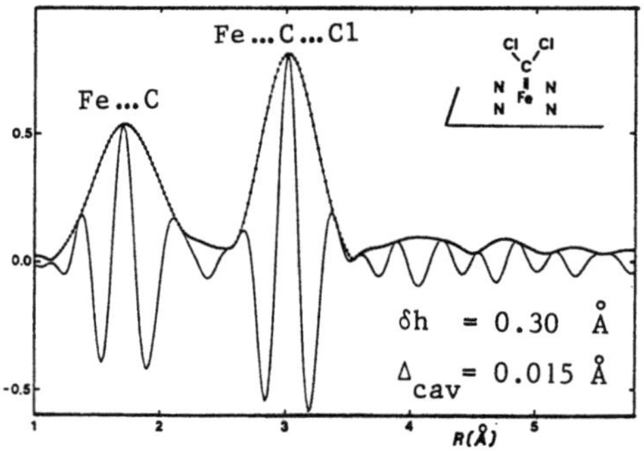

Figure 6 : Perturbed difference FT spectrum of (**8** - **3***) where **8** = TPP:Fe=CCl$_2$, δh = 0.30 Å , Δ_{cav} = 0.015 Å .

A typical example of this class of structure is the vinylidene carbene complex **7** = TPP:Fe=C=C(Ar-Cl)$_2$ which was prepared by reaction of the naked complex **3** with the insecticide DDT = Cl$_3$CCH(Ar-Cl)$_2$ and is the most stable of all prepared iron carbene entities [23]. Reference is also made to **7** in order to explain synergistic action of insecticides [4]. Another example to be also considered here is the dichlorocarbene **8** = TPP:Fe=CCl$_2$ prepared by reaction of **3** with carbon tetrachloride [22]. Perturbed difference FT spectra of **7** and **8** were calculated using **3** as a reference and are reproduced in figures 5 and 6. Both complexes are penta-coordinated and the metal is in both cases slightly shifted out from the mean porphyrin plane with H_7 = 0.20 ± 0.05 Å and H_8 = 0.30 ± 0.03 Å. The porphyrin core is certainly "ruffled" in the case of **7** (Δ_{cav} = 0.00 Å) but slightly less in the case of **8** (Δ_{cav} = 0.015 Å). The axial Fe...C bond lengths were thus obtained with an excelent accuracy : R_7(Fe...C$_1$) = 1.66 ± 0.02 Å , R_8(Fe...C$_1$) = 1.70 ± 0.02 Å . The former value is quite comparable to the Fe...C bond length of 1.67$_5$ Å found in the crystal structure of the μ-carbido dimer **9** = TPP:Fe=C=Fe:TPP [24] which is involving a formally dicarbenic carbon atom bridging two metal centers. A preliminary perturbed difference analysis of the EXAFS spectrum of **9** is reproduced in figure 7 and gave exactly the same bond length of 1.66 ± 0.02 Å. Thus our result does confirm the carbenic structure of **7** which was recently questioned [25]. As regards **8** , our estimation of the Fe...C bond length is definitely shorter than the value of 1.83 Å found in the crystal structure of the hexacoordinated species TPP:Fe(CCl$_2$)(H$_2$O) [26] but such a difference is quite consistent with the fact that **8** was prepared in anhydrous non-coordinating solvents.

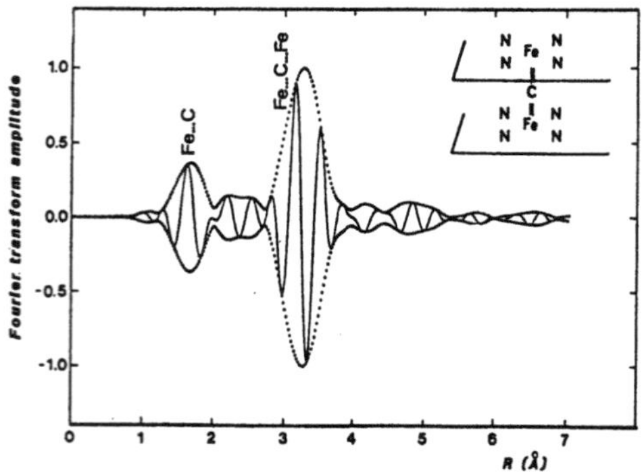

Figure 7 : Perturbed difference FT spectrum of (**9** - **3***) where **9** = μ-(TPP:Fe=)$_2$C , δh = 0.25 Å , Δ_{cav} = 0.00 Å .

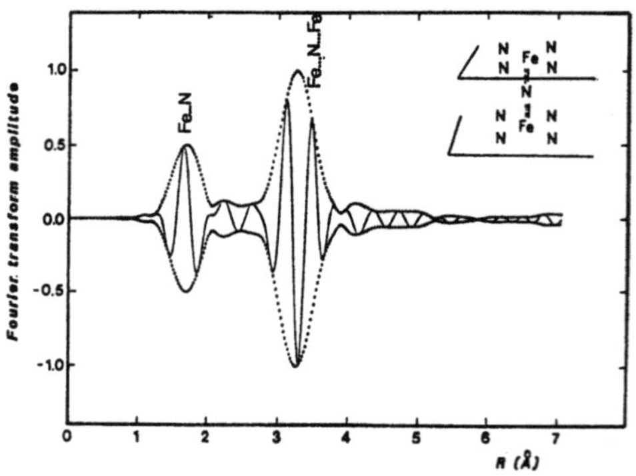

Figure 8 : Perturbed difference FT spectrum of (**10** - **3***) where **10** = μ-(TPP:Fe$^{\pm}$)$_2$N , δh = 0.30 Å , Δ_{cav} = 0.015 Å .

Spectacular multiple scattering effects have to be taken into account if one wishes to analyze the signatures observed beyond the first shell : their phase is typically inversed with respect to what we expect in the absence of multiple scattering whereas amplitude enhancements are quite evident. Indeed these effects are not at all surprising for **7** or **9** because the sequences Fe...C_1...C_2 (R = 2.39 ± 0.03 Å) or Fe...C...Fe are linear. It is our interpretation that in **8** the Fe...C...Cl multiple scattering contributions might become dominant because the low frequency bending modes of the bent Fe...C...Cl sequence should damp out selectively the direct Fe...Cl scattering contribution [27]. Similar anomalies were also currently detected in the cases of metal phosphates for bent sequences Metal...O...P [28,29].

An important question is whether difference EXAFS analyses can also provide us with accurate determinations of the Fe...C bond lengths in the case of σ-aryl or σ-alkyl bonded iron porphyrins. Thus we decided to record the EXAFS spectra of one of the most stable representatives of these short living compounds, i.e. **11** = TPP:Fe-C_6H_5 . A puzzling result was that apart from some sort of a different scaling, the two FT spectra of **11** and **3** looked very similar... Fortunately, the crystal structure of **11** was solved recently by DOPPELT [30] and it becomes immediately clear from the structural data quoted in Table I that this problem is rather ill-conditioned as regards difference EXAFS analyses :

. The metal is only slightly displaced from the porphyrin plane ($\delta h \simeq 0.17$ Å), this effect being partially compensated by a ruffled core contraction (Δ_{cav} = - 0.020 Å).

. Due to the nature and geometry of the axial ligand, the current tests used for evaluating the cancellation of the porphyrin signatures become extremely ambiguous.

SHELL	**11** = TPP:Fe-C_6H_5 [29]			**3** = TPP:Fe [19]		
j	SCATTERERS	$\bar{N}_j/\bar{N}_1$	$\bar{R}_j$ (Å)	$\bar{R}_j$ (Å)	$\bar{N}_j/\bar{N}_1$	SCATTERERS
1	4 N_{eq} + C_{Ph1}	1	1.96_1	1.97_1	1	4 N_{eq}
2	8 C_A + 2 C_{Ph2}	2	2.99_5	3.01_5	2	8 C_A
3	4 C_{meso}	0.8	3.41_8	3.40_7	1	4 C_{meso}
4	8 C_B + 2 C_{Ph3}	2	4.23_1	4.23_2	2	8 C_B
5	4 C_ϕ + C_{Ph4}	1	4.88	4.90_8	1	4 C_ϕ

Table I : Predicted radial distributions for **11** and **3** according to their crystal structures.

As predicted by Table I and nicely illustrated by figure 9a, the only discrimination test is the amplitude of the C_{meso} signal. Indeed EXAFS results and crystal structure data are fully consistent, the lack of an easy discrimination between the two spectra being in itself indicative of the fact that the iron...carbon bond length (1.95_5 Å) has to be close to the average distance $\bar{R}_1$ = 1.96 Å. Furthermore, the perturbed difference spectrum shown in figure 9b (Δ_{cav} = 0.015 Å , δh = 0.20 Å) is already quite encouraging and led to the exact Fe...C distance : R(Fe...C_{pl}) = 1.96 ± 0.02 Å. From our analysis, the axial shift of the metal should be of the order of 0.20 ± 0.05 Å. This result is comparable with the crystal structure determination but is associated with a slightly positive value for Δ_{cav} instead of the above mentioned core contraction (Δ_{cav}<0).

3.3. Nitrene and nitrosoalkane complexes

We shall retain here only two compounds for which short axial Fe...N bond lengths were detected by our perturbed difference EXAFS analyses using **3** = TPP:Fe as a

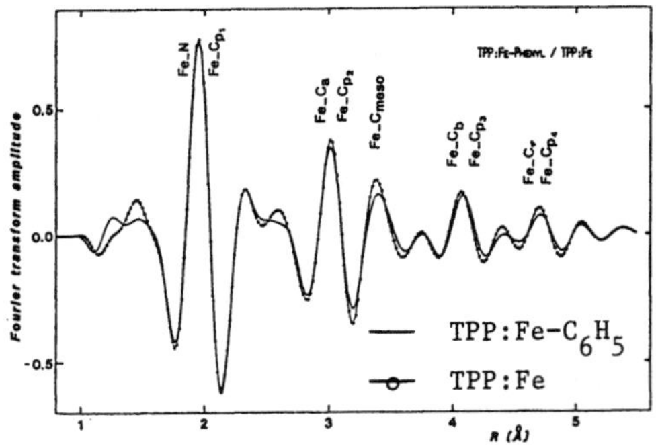

Figure 9a : Comparison of the FT spectra of **11** = TPP:Fe-C_6H_5 and **3** = TPP:Fe corrected implicitly for the Fe...N shell. For the sake of consistency with Table I, both spectra have been renormalized to unity with respect to the dominant shell.

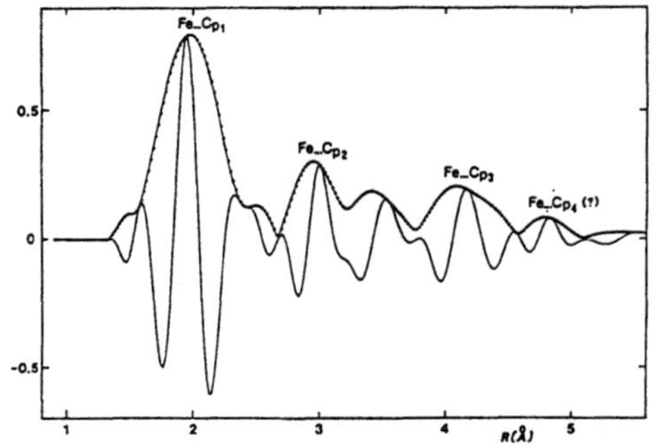

Figure 9b : Perturbed difference FT spectrum of (**11** - **3**) where **11** = TPP:Fe-C_6H_5 , δh = 0.20 Å , Δ_{cav} = 0.015 Å .

standard reference. Most convincing is the case of the μ-nitrido dimer **10** = TPP:Fe⋯N⋯Fe:TPP illustrated by figure 8. Again the short Fe...N bond length (1.66 ± 0.01 Å) and the axial shift $\delta h = 0.30 \pm 0.05$ Å found by our method are perfectly consistent with the known crystal structure [31]. The multiple scattering effects relative to the Fe...N...Fe linear sequence are also spectacular.

Far more complicated is the case of the hexacoordinated nitrosoalkane complex **12** = TPP:Fe(i-PrNO)(N-Methyl-Imidazole) because one may expect the two axial Fe...N bond lengths to be non equivalent but not well resolved [5]. This is confirmed by figure 10a in which we are comparing the FT spectrum of **12** to some perturbed spectrum of the reference species **3**. It is noteworthy that some kind of a partially destructive interference effect results in a strong distortion of the first peak of the spectrum of **12**. Although this analysis is still preliminary, the perturbed difference FT spectrum reproduced in figure 10b led to the following tentative estimation of the axial distances: Fe...N' = 1.78 ± 0.07 Å , Fe...N_3 = 2.10 ± 0.05 Å . Further work is still in progress in order to refine these estimations which are to be compared with the axial distances 1.86 Å and 2.10 Å found for the crystal structure of TPP:Fe(i-PrNO)(i-PrNH_2).

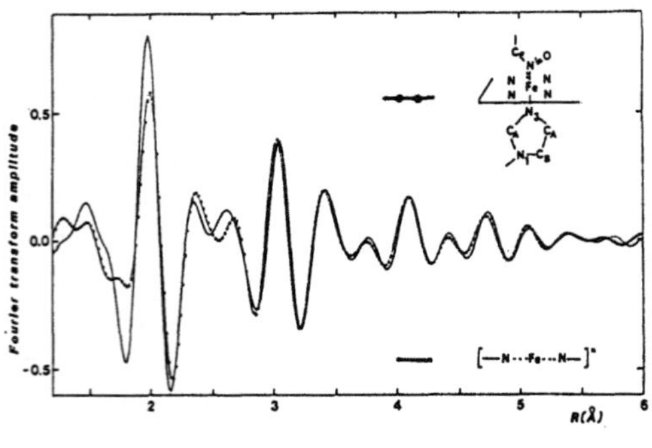

Figure 10a : Comparative display of the FT spectrum of 12 = TPP:Fe(i-PrNO)(N-Methyl-Imidazole and a perturbed spectrum of **3** obtained for δh = 0.00 Å and Δ_{cav} = 0.018 Å .

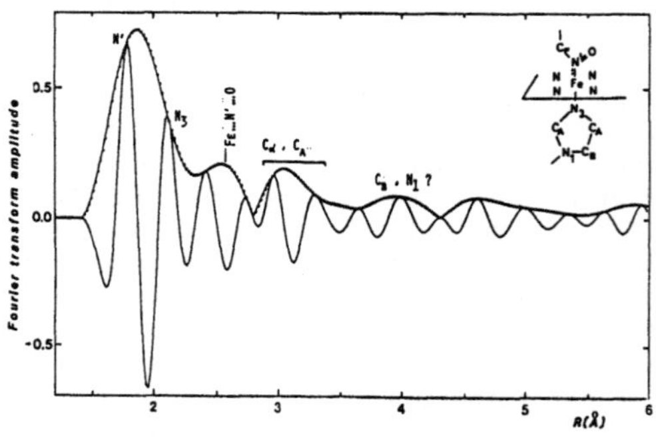

Figure 10b : Perturbed difference FT analysis of (**12** - **3**) for the very tentative parameters : δh = 0.00 Å , Δ_{cav} = 0.018 Å .

4. CONCLUSION

Perturbed difference FT analyses have been shown to be reliable and to open a new way to unravel the hidden structural informations of EXAFS spectra. Small effects such as an axial shift of the metal, ruffling or doming conformations of the porphyrin core can be detected and can even be evaluated with a reasonable accuracy under favourable conditions. A pre-requisite to using the method and benefiting from its sensitivity is to obtain good quality experimental data with an excellent statistics. This might be a serious limitation as regards applications to real hemoproteins which we are now investigating.

Abbreviations used : TPP = meso-5,10,15,20-tetraphenylporphyrinatodianion,
TMP = meso-5,10,15,20-tetramesitylporphyrinatodianion,
i-Pr = isopropyl- , Ar = aryl-

REFERENCES

1. O. Bortolini, M. Ricci, B. Meunier, P. Friant, I. Ascone and J. Goulon, Nouv. J. Chimie 10 (1986) 39.
2. J.E. Penner-Hahn, T.J. McMurry, M. Renner, L. Latos-Grazinzki, P.K. Eble, I.M. Davis, A.L. Balch, J.T. Groves, J.M. Dawson and K.O. Hodgson, J. Biol. Chem. 258 (1983) 12761.
3. D. Mansuy, W. Nastainczyk and V. Ullrich, Arch. Pharmacol. Exp. Path. 285 (1974) 315.
4. D. Mansuy, J.P. Battioni, J.C. Chottard and V. Ullrich, J. Am. Chem. Soc. 101 (1979) 3871.
5. D. Mansuy, P. Battioni, J.C. Chottard, C. Riche and A. Chiaroni, J. Am. Chem. Soc. 105 (1983) 455.
6. J.P. Mahy, P. Battioni, D. Mansuy, J. Fischer, R. Weiss, J. Mispelter, I. Morgenstern-Badarau and P. Gans, J. Am. Chem. Soc. 106 (1984) 1699.
7. P. Battioni, J.P. Mahy, G. Gillet and D. Mansuy, J. Am. Chem. Soc. 105 (1983) 1399.
8. J.L. Poncet, R. Guilard, P. Friant, C. Goulon-Ginet and J. Goulon, Nouv. J. Chimie 8 (1984) 583.
9. J. Goulon, P. Friant, J.L. Poncet, R. Guilard, J. Fischer and L. Ricard, "EXAFS and Near Edge Structures", Springer Series in Chem. Phys. 27 (1983) p. 100, Ed. by A. Bianconi, L. Incoccia and S. Stipcich.
10. P. Friant, J. Goulon, J. Fischer, L. Ricard, M. Schappacher, R. Weiss and M. Momentau, Nouv. J. Chimie 9 (1985) 33.
11. M.F. Perutz, S.S. Hasnain, P.J. Duke, J.L. Sessler and J.E. Hahn, Nature 295 (1982) 535.
12. J. Goulon, P. Friant, C. Goulon-Ginet, A. Coutsolelos and R. Guilard, Chem. Phys. 83 (1984) 367.
13. S.P. Cramer, T.K. Eccles, F. Kutzler, K.O. Hodgson and S. Doniach, J. Am. Chem. Soc. 98 (1976) 8059.
14. W.T. Elam, E.A. Stern, J.D. McCallum and J. Sanders-Loehr, J. Am. Chem. Soc. 105 (1983) 1919.
15. J.L. Hoard, Ann. N.Y. Acad. Sci. 206 (1973) 18.
16. M. Loos, P. Friant, I. Ascone, J. Goulon, J.M. Barbe, A. Coutsolelos and R. Guilard, Proc. of the EXAFS-IV International Conference at the Abbaye de Fontevraud (1986) to be published in J. Physique (1986).
17. C.L. Hill and F.J. Hollander, J. Am. Chem. Soc. 104 (1984) 7138.
18. J.W. Buchler, C. Dreher, K.L. Lay, Y.J.A. Lee and W.R. Scheidt, Inorg. Chem. 22 (1983) 888.

19. J.P. Collman, J.L. Hoard, N. Kim, G. Lang and C.A. Reed,
J. Am. Chem. Soc. 97 (1975) 2676.
20. R.F. Pettifer, D.L. Foulis and C. Hermes,
Proc. of the EXAFS-IV International Conference at the Abbaye de Fontevraud (1986) to be published in J. Physique (1986).
21. R.W. Strange, S.S. Hasnain, N.J. Blackburn and P.F. Knowles,
Proc. of the EXAFS-IV International Conference at the Abbaye de Fontevraud (1986) to be published in J. Physique (1986).
22. D. Mansuy, M. Lange, J.C. Chottard, P. Morliere and D. Brault,
J. Chem. Soc. Chem. Commun. (1977) 648.
23. D. Mansuy, M. Lange and J.C. Chottard,
J. Am. Chem. Soc. 100 (1978) 3213.
24. V.L. Goedken, M.R. Deakin and L.A. Bottomley,
J. Chem. Soc. Chem. Commun. (1982) 607.
25. C.E. Castro and R.S. Wade,
J. Org. Chem. 50 (1985) 5342.
26. D. Mansuy, M. Lange, J.C. Chottard, J.F. Bartoli, B. Chevrier and R. Weiss,
Angew. Chem. Int. Ed. 17 (1978) 781.
27. J.J. Boland and J. Baldeschwieler,
J. Chem. Phys. 80 (1984) 3005.
28. J.E. Harries, D.W.L. Hukins and S.S. Hasnain,
Proc. of the EXAFS-IV International Conference at the Abbaye de Fontevraud (1986) to be published in J. Physique (1986).
29. L. Rodehuser, P. Rubini, P. Friant and J. Goulon,
Unpublished results on gallium phosphates.
30. P. Doppelt,
Inorg. Chem. 23 (1984) 4009.
31. W.R. Scheidt, D.A. Summerville and I.A. Cohen,
J. Am. Chem. Soc. 98 (1976) 6623.

Investigation on Metal Binding Sites in DNA by Means of X-Ray High Resolution Spectroscopy

G.ONORI, Dipartimento di Fisica - Università di Perugia,
M.BELLI, A.SCAFATI, M.MATZEU, E.RONGONI* - Laboratorio di Fisica, Istituto Superiore di Sanità, and Sezione Sanità, INFN, Roma
S.MOBILIO, A.BALERNA - Laboratori Nazionali dell'INFN, Frascati
A.BIANCONI Dipartimento di Fisica, Università La Sapienza, Roma

1 - Introduction

Metal ions are required for virtually all biological processes in which nucleic acids are engaged and play an essential role in their structural and dynamic properties. Nevertheless the presence of the "wrong" metal ions, or even the essential ones in the "wrong" concentration may cause harmful biological effects (toxic, mutagenic, carcinogenic).

There are several lines of evidence showing that these effects can be described in terms of metal-nucleic acid, or metal-nucleotide, interactions. Thus the properties of DNA-metal ion complexes are of great interest and have been studied in the recent years by various authors using several biochemical and physical approaches[1,2].

Nucleic acids contain three classes of potential donor sites to which metal ions can bind: phosphate groups, ribose hydroxyls and donor atoms (N and O) of the bases. Experiments have shown that all potential sites can become actual interaction sites, depending upon the metal ion and the composition of the medium. Roughly, alcali metals (Ca^{2+}, Mg^{2+}) produce electrostatic or ionic binding,whereas metals of the first transition group with their incomplete electron structure produce covalent coordination binding.

These studies present a particular interest in the case of the first transition group metals , in view of their inhibiting or activating effects on enzymatic reactions[2]. In this group the Cu^{2+} ion has the strongest ability to bind with DNA and therefore is suitable for studying the binding mechanism.

In biophysical research, high resolution X-ray spectroscopy has been first applied to metalloproteins[3]; recently, we used a high-resolution X-ray absorption analysis for the study of metal ions binding to nucleotides [4] and nucleic acids[5,6].

The present study reports the results of our recent investigation of the binding of Cu atoms to DNA and the "in situ" reduction of Cu (II) to Cu(I) caused

---

* Present address: Dipartimento di Fisica - Università di Perugia.

by chemical reductants or UV radiation at 310 nm[6,7]. It has been suggested that copper reduction and oxydation is involved in the control of DNA as a primer[12]. Photochemical reduction of DNA-Cu complex may be important in connection with biological damage caused by solar radiation.

The results here reported have been obtained by XANES analysis on X-ray absorption spectra collected at the Wiggler beam line of the Adone storage ring, associated with other techniques as circular dichroism and UV differential absorption spectroscopy.

2. Materials and Methods

2.1. Materials

Calf thymus DNA was obtained by Boehringer Mannheim GmbH, dialysized against 10 mM NaCl with 2mM EDTA for removing divalent metal ions, and suspended in 10 mM NaCl. DNA-Cu(II) complex was obtained by mixing DNA solution with microliter quantities of $CuCl_2$ solution to have a ratio (r) of 0.25 for Cu/nucleotide concentration.

Nucleotide concentration was determined by absorption measurements at 260 nm using the value of 6600 mol^{-1} cm^{-1} for the absorption coefficient[8]. The sample concentration used for the spectrophotometric measurements corresponds to A = 1 and that used for X-ray spectroscopy to A = 40.

Chemical reduction of the DNA-Cu(II) complex was performed by addition of $NaBH_4$ at the same concentration as that used for $CuCl_2$.

All the experiments were made with unbuffered solution at pH about 7, to avoid chelation of Cu(II) by buffer components.

2.2. UV irradiation

The DNA sample was irradiated with monochromatic light at 310 $\pm$ 10 nm using a Hilger & Watts D96 quartz prism monochromator equipped with a 1000 W Oriel high pressure Xe-Hg, lamp. The samples were stirred during the irradiation. The average fluence was 750 KJ/m^2, which corresponds to the maximum observed effect. The fluence was measured with a thermopile-nanovoltmeter calibrated system.

2.3. Spectrophotometric measurement

The absorption spectra were measured at 20 °C by a double beam Cary 118 spectrophotometer equipped with an external thermostatting device .

The modifications induced in DNA absorption spectrum by Cu(II) binding, or reductants addition, or UV irradiation, were measured by a differential technique, using the untreated solution as a reference.

Circular dichroism spectra were recorded by means of a Jasco J41A spectropolarimeter.

2.4. High resolution X-ray spectroscopy

For the XANES analysis, the pre-edge background was subtracted, and the absorption coefficient, α, was normalized to the high energy absorption value, taken at 100 eV above the absorption edge. The energy E_0 of the edge was determined as the maximum of the absorption coefficient derivative.

3. Results and discussion

3.1 DNA-Cu(II) complex

The interaction of Cu(II) ions with DNA has been the subject of a great number of studies[(2)]. These studies indicate that Cu(II) ions can bind to the phosphate groups and/or to the nitrogens of the bases. Cu(II) ions have marked effect upon DNA conformation and, at certain concentrations, may produce a drastic decrease in the helix stability of DNA against thermal denaturation. These effects may be of biological interest since it has been shown that carefully isolated DNA samples contained specific amounts of copper ions.

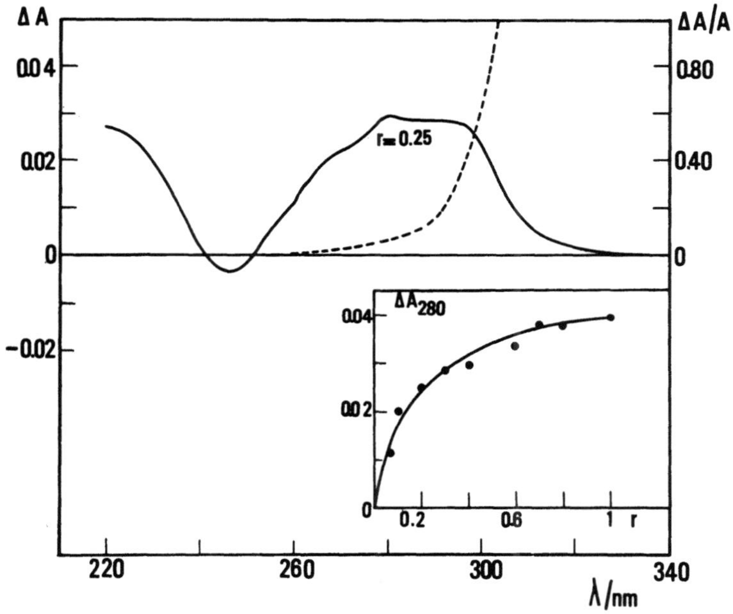

Fig.1 . Absorption change ΔA of calf thymus DNA after Cu(II) addition (solid line) and relative spectra change $\Delta A/A$ (dashed line). The inset shows ΔA as a function of r.

3.1.1. UV absorption measurements

The small modification in the nucleic acid absorption spectrum due to metal binding can be related to changes in the "stacking" interactions between bases and/or to localized interactions of ions with bases at more electron-dense sites (N and O atoms)[9].

Fig.1 shows the difference spectrum $\Delta A(\lambda)$ obtained when calf thymus DNA was mixed with the $CuCl_2$ (r = 0.25) solution and absorption was measured against a reference solution containing an equal concentration of DNA.

The difference spectrum shows a maximum between 276 and 280 nm and a negative minimum at 248 nm; it is very similar to that observed by other investigators in complexes of copper ions and other first group transition metals with DNA molecules [10,8]. It suggests a perturbation of the base electronic structure due to the direct binding of metal ions to the bases. The changes of absorption depend on the $Cu(II)/PO_4$ ratio (r) and are almost complete at r = 1. The relative effect $\Delta A/A$, is mainly evident in the mid-UV region.

3.1.2. CD measurements

Circular dichroism measurements are very sensitive even to minor conformational changes of DNA in solution. CD spectrum of DNA becomes strongly modified by DNA-Cu(II) complex formation (Fig.2): a marked decrease of the positive band

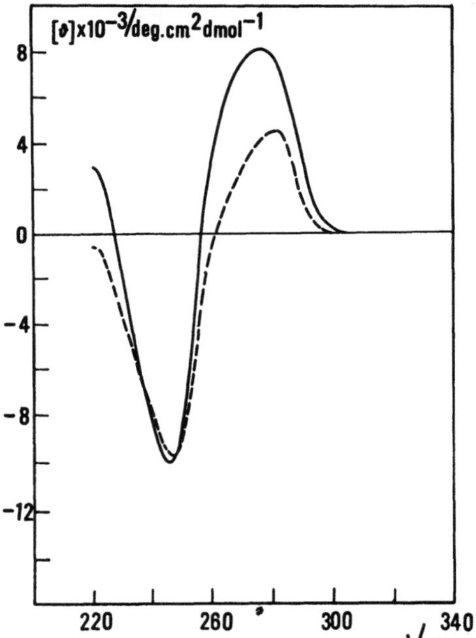

Fig.2. Circular dichroism spectra of DNA (solid line) and DNA-Cu(II) at r = 0.25 (dashed line).

around 275 nm is observed while the negative band is scarcely affected. CD spectra of the resulting complex may be interpreted in terms of conformation change of DNA from the B-form to a C-like form on ligand binding [11].

3.1.3. X-ray absorption

The XANES is determined by the multiple scattering processes of the excited photoelectron on the neighbour atoms and it is sensitive to the local electronic and geometric environment of the absorbing atom.

Fig.(3) reports the XANES spectrum of DNA-Cu(II) complex. In this spectrum we

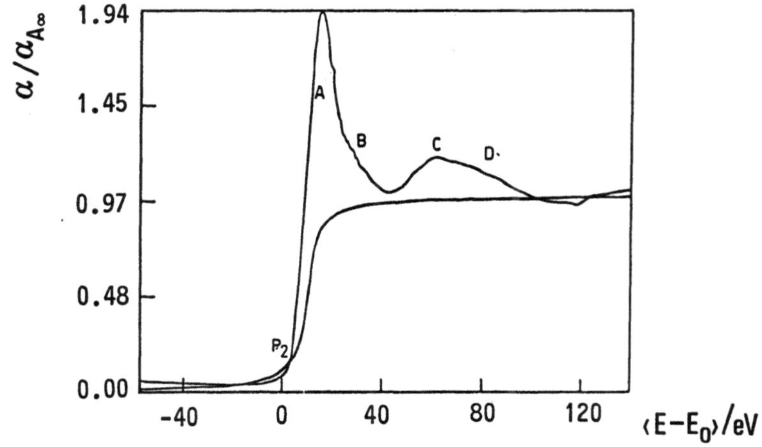

Fig.3. XANES spectrum of DNA-Cu(II) complex.

identify the characteristic pre-edge dipole-forbidden 1s 3d transition p_2 and the main multiple scattering resonance A for excitations of the photoelectron of p symmetry in the continuum. The A,B,C and D features at higher energies are due to multiple scattering and are similar to those found for the Cu(II) in solution. The absence of other features in the rising part of the edge indicates that there are no significant distortions from an octahedral symmetry (see also the paper by Onori et al. published in this volume).

3.2. DNA-Cu (I) complex

Divalent copper ions bound to DNA may be reduced "in situ" by addition of ascorbic acid or borohydride(12) and this affects the local stability of the double stranded DNA(7).

3.2.1. UV absorption measurements

Addition of reducing agents to a DNA-Cu(II) complex solution causes changes in DNA absorption spectrum. Fig.4 shows the observed change in DNA-Cu(II) spectrum ($r = 0.25$) in presence of an excess of sodium borohydride. This difference spectrum, in agreement with the finding of Minchenkova and Ivanov(12), corresponds to a decrease in extinction between 238 and 266 nm and an increase between 220 and 238 and at λ>266 nm. Isosbestic points are present at 238 and 266 nm.

The observed absorption changes in this system caused by the action of the

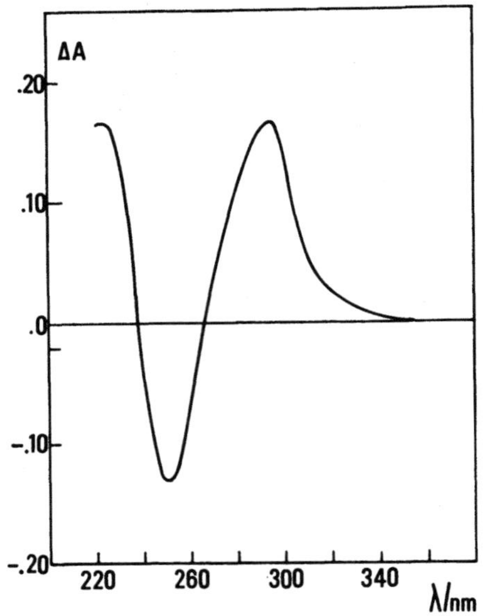

Fig.4. Changes in the DNA-Cu (r = 0.25) absorption spectrum on $NaBH_4$ addition.

reducing agent has been attributed to the formation of a DNA-Cu(I) complex with its binding site on a DNA base. Note that the changes in absorption spectrum due to chemical reduction of DNA-Cu(II) complex are just similar to that of cytosine protonation, suggesting the involvment of a G-C pair in the formation of DNA-Cu(I) complex.

3.2.2. CD measurements

The reduction of divalent copper ions bound to DNA by $NaBH_4$ gives marked variations in CD spectrum related to those observed in absorption spectrum. A positive band with maximum at 295 nm and a deep negative band with minimum of 268 nm are observed (see Fig.5). The Cu^{1+} ion is in a d^{10} state and therefore has an outer electron shell similar to that of Ag^{1+}. The difference of the optical properties of DNA-Cu(I) complex with respect to DNA is indeed very similar to the change caused by DNA complexing with Ag^{1+} (for $r < 0.2$). This suggests similar binding sites. As Ag^{1+} has a high tendency to make linear chelates and it has been located between two base pairs, one of them being a G-C pair, this could also be the case for Cu^{1+}, the localization of the positive ion charge on the G-C pair explains the similarity between the difference absorption spectrum after Cu reduction and that after cytosine protonation[15].

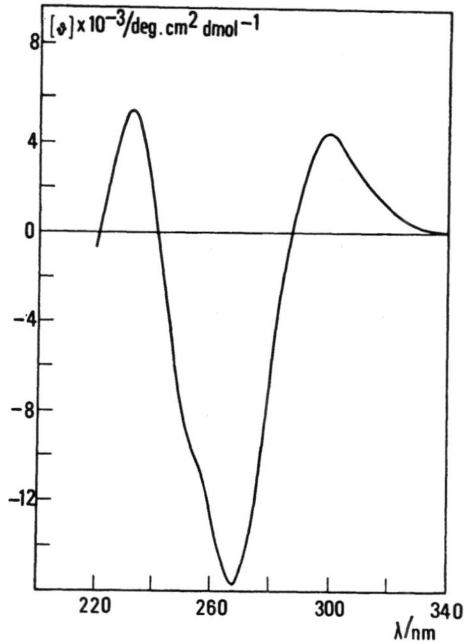

Fig.5. CD spectrum of: DNA-Cu(I) complex obtained by Cu chemical reduction.

3.2.3. X-ray absorption measurements

The XANES spectrum of the chemically reduced DNA-Cu(I) complex is shown in Fig.6. The intensity decrease of the main multiple scattering resonance A and the overall variation of the XANES spectrum (in respect to DNA-Cu(II) complex indicates a change of the Cu site structure.

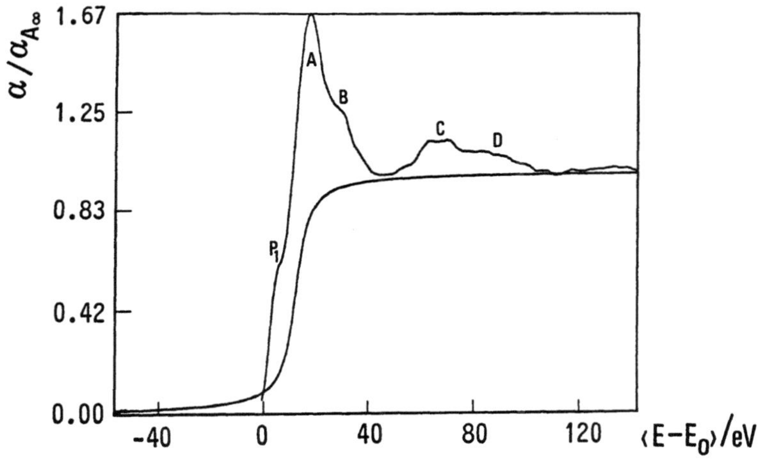

Fig.6. XANES spectrum of DNA-Cu(I) complex obtained by Cu chemical reduction.

A new feature P_1 appears at the threshold after addition of $NaBH_4$ while the

P_2 peak is suppressed because of the $3d^{10}$ configuration of Cu^{1+} ions. We have fixedthe zero of the energy scale E_o at the energy of the 1s $\rightarrow$3d weak pre-peak P_2 of the DNA-Cu(II) complex. In this energy scale the feature P_1 is at 3 eV. This transition appears in the polarized XANES spectra of single cristals only in the directions when there are no neighbor atoms(16).

Multiple scattered wave calculations for a linear coordination have been successful at qualitatively reproducing the prominent features of this spectrum (results not shown).

3.3. Effect of mid-UV radiation on DNA-Cu(II) complex

The formation of DNA-Cu(II) complex is accompanied by changes in optical absorption spectrum that are mainly evident in the 290-320 nm range (Fig.1).

This spectral range plays an important role in connection with the biological damage caused by solar radiation. One would expect that, because of the far higher absorbance, the DNA-Cu(II) complex would suffer far more damage when irradiated with mid-UV light, than would untreated DNA. Actually, we have shown that the photochemistry of DNA in the mid-UV is greatly altered when Cu(II) is bound.

3.3.1. UV absorption measurements

Irradiation at 310 nm of DNA-Cu(II) complex causes (Fig.7) a decrease of extinction between 240 and 280 nm (minimum at 257.5 nm) and an increase between

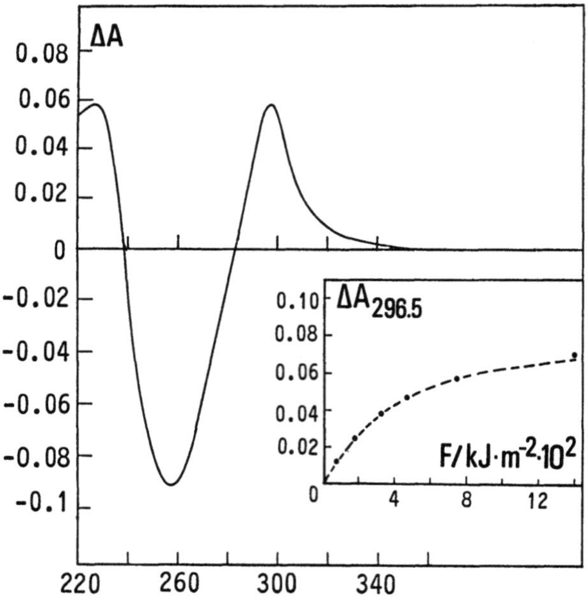

Fig.7. Changes in the DNA-Cu(II) (r = 0.25) absorption spectrum after UV irradiation at 310 nm and a fluence of 744 kJ/m^2. The inset shows the variation at 296.5 nm as a function of the fluence.

220 and 240 nm and at 280 nm (maximum at 296.5 nm).

This difference spectrum is similar to that obtained by adding reducing agents, such as sodium borohydride, to a DNA solution containing Cu(II) ions (Fig.4) thus suggesting that DNA - Cu(II) complex is transformed into a DNA-Cu(I) complex under UV irradiation.

Minor differences between the optical spectra of the chemically reduced and the UV irradiated DNA, in the range around 260 nm can be assigned to local denaturation and/or other structural alterations of irradiated DNA.

The maximum absorption change at 296.5 nm in the UV irradiated sample, even after very large fluences, is only about 35-40% of the maximum change observed by adding sodium borohydride, thus indicating that UV irradiation results in a photoreduction of only some of the copper sites.

A DNA solution irradiated under the same conditions as for DNA-Cu(II) complex showed no variations in absorption spectra.

3.3.2. CD measurements

After UV irradiation the CD spectrum of the DNA-Cu(II) complex (Fig.8) shows

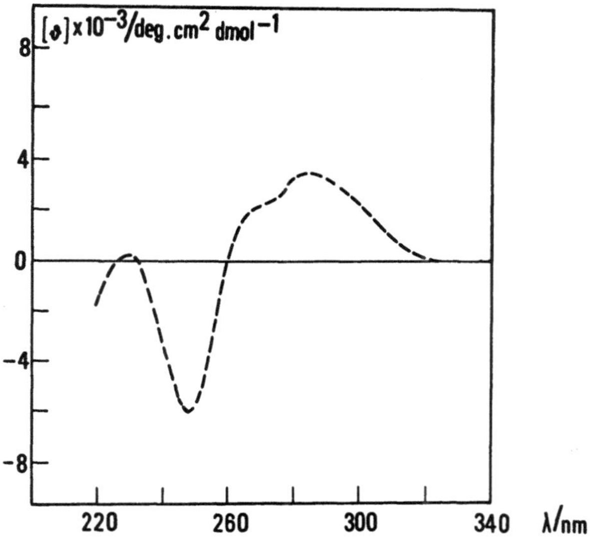

Fig.8. CD spectra of: DNA-Cu(II) complex after UV irradiation at the same conditions as for Fig.7.

a strong modification of the positive band that broadens up to 320 nm, corresponding to the absorption increase observed in this spectral range. Further a marked decrease in the negative CD band around 245 nm is found.

These variations are not similar to those observed in the chemically reduced sample. This result indicates that besides the partial reduction of copper ions, the UV radiation also damages the primary structure of DNA[7].

3.3.3. X-ray absorption measurements

The XANES spectrum of irradiated sample is shown in Fig.9. The feature P_1 like that seen in the chemically reduced sample (but less evident), appears at the threshold after irradiation.

The intensity of the P_1 feature in the UV irradiated sample has been used to calculate the percentage of Cu(I) sites formed by irradiation. The experimental spectrum has been fitted with a linear combination of the XANES spectra of DNA-Cu(II) untreated and $NaBH_4$ (fully) reduced complex. We have shown in this way that, in agreement with the UV absorption results, only 35% of the Cu sites are reduced, suggesting the presence of a specific class of sites affected by UV radiation.

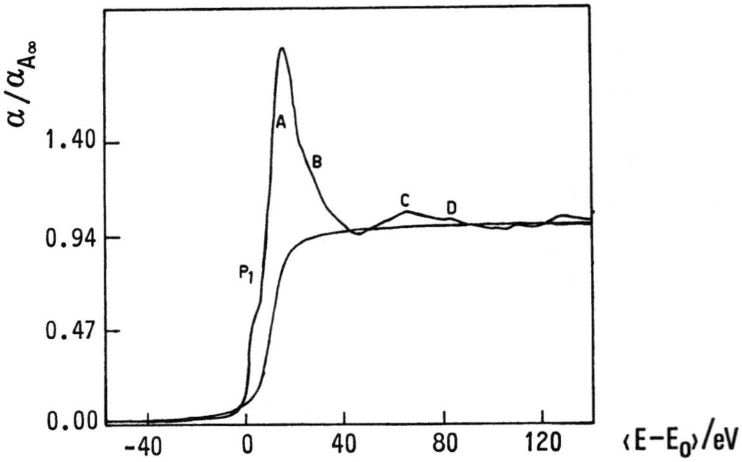

Fig.9. XANES spectrum of DNA-Cu complex after UV-irradiation at the same conditions as for Fig.7.

4. Conclusions

Optical and X-ray spectroscopy are complementary techniques which can be usefully combined to study the metal-nucleic acid interaction. In this way we have shown that DNA can be sensitized to mid-UV light by copper binding.

Interest in local structure determination of metal complexes in aqueous solution has been renewed by the possibility offered by the near-edge structure analysis. Indeed, a quantitative interpretation of XANES spectra can be obtained by the multiple scattering approach recently proposed(17,18). Furthermore, the use of the recent synchrotron radiation facilities makes it possible experiments on dilute systems with a metals concentration of few mM, such as that used in this work.

References

1) M.Daune: in Metal Ions in Biological Systems, edited by H.Sigel,V.3 (1974),p.1.
2) H.Pezzano and F.Podo: Chem.Rev., 80, 365 (1980)
3) S.I.Chan,et al.: J.Mol.Struc., 45, 239 (1978).
P.Eisemberger and S.Owaga: Nature,274, 30 (1978).
4) M.Belli et al: Nuovo Cimento, 2 D (1983).
5) M.Belli et al.: in Physics in Environmental and Biomedical Research, edited by S.Onori and E.Tabet, (1986),p.33.
6) E.Rongoni et al.: Biopolymers, 25, 217 (1986)
7) M.Matzeu and G.Onori: Photochem. Photobiol., 44, 59 (1986).
8) G.Luck and C.Zimmer: Eur J.Biochem., 29, 528 (1972).
9) G.Onori and F.Mazzei: Nuovo Cimento, 70, 265 (1986).
10) J.P.Schreiber and M.Daune: Biopolymers, 8, 139 (1969).
11) W.Forster, et al. Biopolymers, 18, 625 (1979).
12) L.E.Minchenkova and V.I.Ivanov: Biopolymers, 5, 615 (1967).
13) R.H.Jansen and Davidson: Biopolymers, 4, 17 (1966).
14) V.I.Ivanov, L.E.Minchenkova and V.P.Timofeev: Mol.Biol., 1, 572 (1967).
15) M.Daune, C.A.Dekker and H.K.Schechman: Biopolymers,4,51 (1966).
16) T.A.Smith et al. Springer Proc.Phys., 2, 58 (1984).
17) A.Bianconi, et al.: Phys.Rev.B, 26, 6502 (1982).
18) J.Garcia et al.: Solid State Comm., 58, 595 (1986).

Study of Cu-ATP Complexes

G.ONORI - Dipartimento di Fisica, Università di Perugia.
M.BELLI, A.SCAFATI - Laboratorio di Fisica, Istituto Superiore di Sanità and Sezione Sanità-INFN, Roma.
S.MOBILIO, E.BERNIERI - INFN-Laboratori Nazionali di Frascati
A.BIANCONI, J.GARCIA* - Dipartimento di Fisica, Università La Sapienza, Roma.

INTRODUCTION

High-resolution X-ray spectroscopy with synchrotron radiation has been used since several years by our group at the Frascati Synchroton Radiation Facility as a tool for studying the interaction of metal ions with nucleotides (1) and nucleic acids (2,3).

The biological relevance of this interaction is well established. Metal ions have a fundamental effect on the structure and activity of polynucleotides (4,5) and, since they may affect the fidelity of the DNA duplication and of the protein synthesis (6), alterations in their amount or in their nature may have toxic, mutagenic, and carcinogenic consequences.

High-resolution X-ray spectroscopy presents unique features for investigating the binding sites of metals in biological molecules, particularly for samples in aqueous solution, which cannot be studied with the usual X-ray diffraction techniques.

In this work we report some results on the Cu(II) binding to the 5'-adenosine triphosphate (ATP) as a function of pH. The ATP molecule, which acts as an activator in several enzymatic reactions, requires the presence of divalent metal ions. The solution structure of metal-ATP complexes is of great importance for understanding their biochemical function as well as the basic features of the metal-nucleic acid binding. Many metal-ATP complexes have been extensively investigated by various authors to evaluate the binding constants and to identify the sites for metal coordination (7-9). These studies are complicated by the metal ability to bind at a number of different sites (e.g.,phosphate, base, ribose hydroxyl

* Present address: Dept. of technology, Zaragoza University, Spain.

groups), and there is considerable disagreement about the sites where a given metal ion is coordinated (9).

Most of these studies pointed to the effect produced by the metal binding on the chemical and physical properties of the nucleotide. A different approach may be useful, based on the effect that the ligand field has on the electronic properties of the metal itself. We used this kind of approach, by means of X-ray spectroscopy with synchrotron radiation to probe the 1s electronic transitions and of optical absorption spectroscopy to probe the d-d transitions of the cupric ion (10,11) in the Cu-ATP complexes.

EXPERIMENTAL

Copper was used as $CuCl_2 \cdot 6H_2O$ reagent grade. ATP was purchased from Boehringer-Mannheim GmbH. No buffers were used, in order to avoid the interaction of Cu(II) with other ligands.

Solutions were prepared by slowly adding concentrate (2N) NaOH to aqueous solution initially at pH 3. The change in volume was negligible. The concentrations for the solutions used were as a rule 15 mM for the Cu(II) salt and 30 mM for the ATP, tripolyphosphate, and D-Ribose.

Optical absorption spectra were measured with a Mod 360 Shimadzu spectrophotometer, 3-ml, 1-cm long quarz cells being used.

X-ray absorption spectra were collected at the Wiggler beam line of the Adone storage ring, and both XANES and EXAFS analyses were performed with the transmission technique. Special teflon cells with kapton windows, 2-mm thick were employed. High resolution was achieved by using a Si (111) monochromator with 1-mm entrance slit.

All spectra were recorded immediately after the preparation of solutions. All measurements were performed at room temperature.

X-RAY ANALYSIS

For the XANES analysis, the pre-edge background was subtracted, and the absorption coefficient α was normalized to the high energy absorption value, taken at 100 eV above the absorption edge. The energy E_o of the edge was determined as the maximum of the derivative of the absorption coefficient.

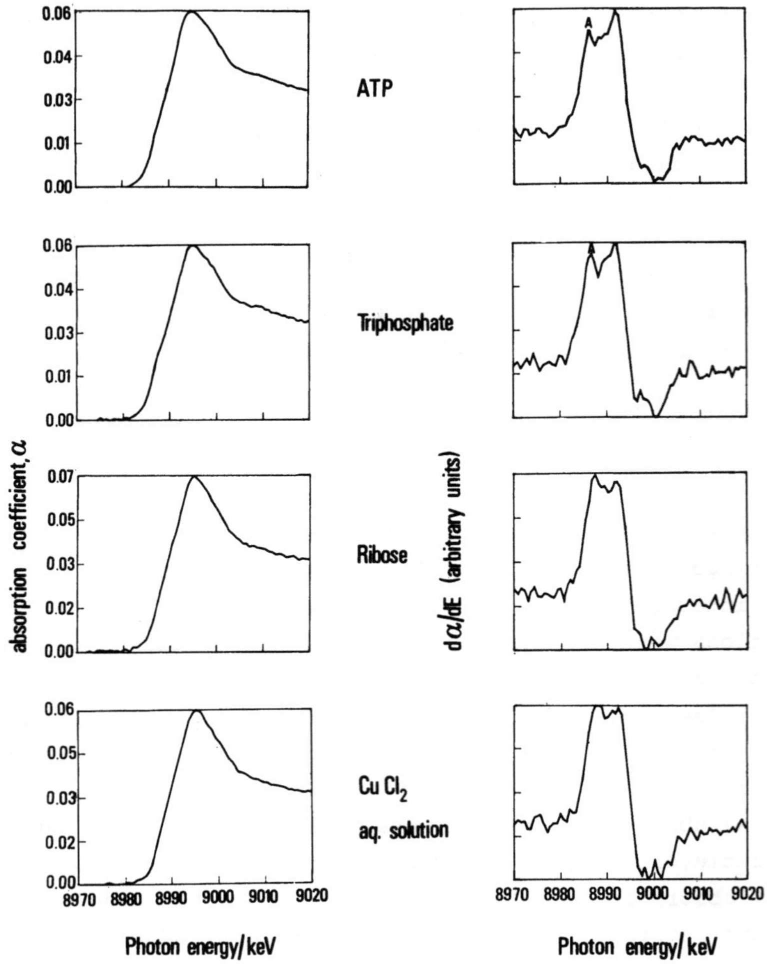

Fig.1 XANES spectra (left) and their derivatives (right) of the indicated samples at pH 3.

This finding suggests that the Cu ion in the Cu-ATP complex is bound to the ribose moiety of ATP at high pH, and this conclusion is confirmed by the Fourier analysis.

The interpretation of the XANES spectra is possible by using a multiple scatterig calculation (12). Some calculated spectra are reported in Fig.4 for a) a perfectly octahedral cluster of 6 oxygen atoms at 2.0 Å distance around the Cu ion, b) for a small 0.3 Å elongation of the two axial oxygens with respect to the 4 planar oxygens at 2.0 Å distance, and c) for a higher axial distortion of 0.6 Å.

This calculation shows that the axial distortion produces a shoulder in the rising absorption edge which corresponds to the feature A observed in the experimental spectra. This feature can be resolved only if the distortion is larger than 0.3 Å. According to this calculation, our XANES of Cu-ATP at pH 3 indicates that the Cu binding site consists of 6 oxygens and presents a small axial distortion, between 0.3 and 0.6 Å.

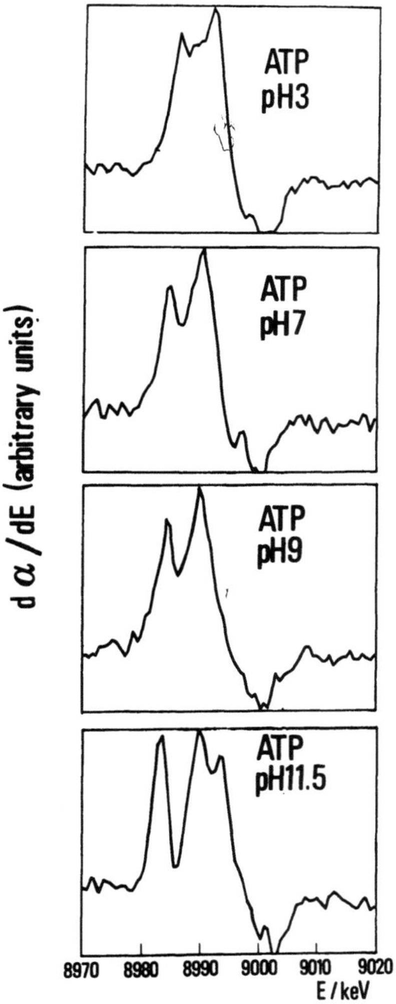

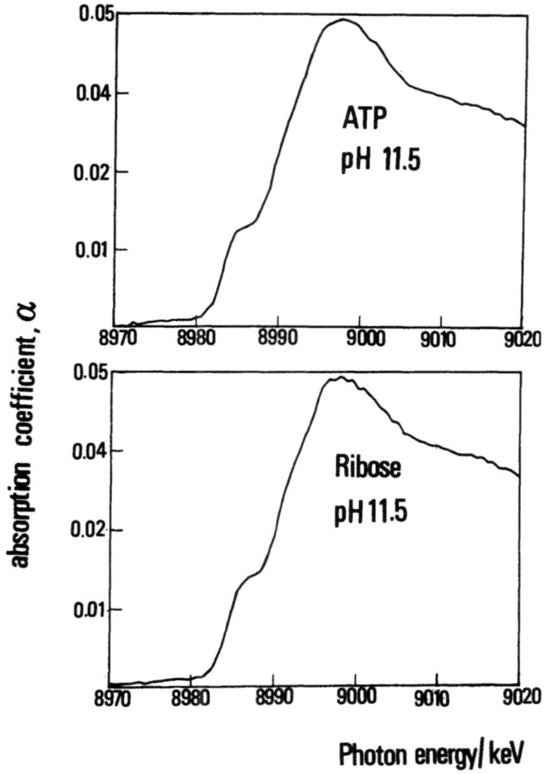

Fig.3 Comparison of the XANES of Cu-ATP and Cu-D-Ribose complexes at pH 11.5.

Fig.2 XANES derivatives of Cu-ATP complex at the indicated pH's.

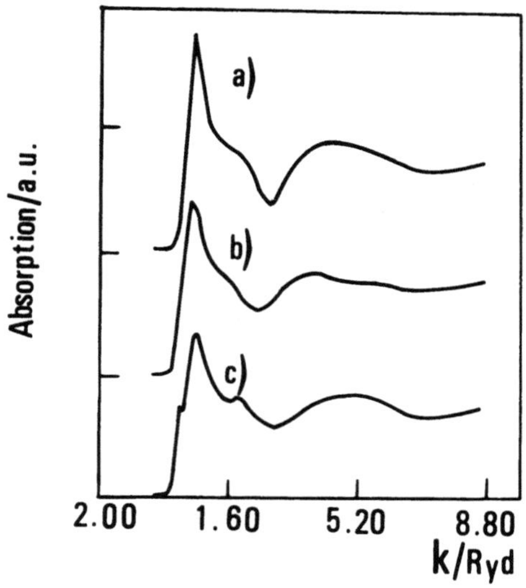

Fig.4 Calculated XANES spectra for CuO_6 clusters:(a) perfectly octahedral; (b) 0.3 Å axial elongation; (c) 0.6 Å axial elongation.

The fact that the spectra of Fig.2 show an increasing peak A with the pH, indicates an increase of the axial distortion of the copper site. The spectrum at pH 11.5 (Fig.3), shows a different lineshape, which is common to the ribose spectrum at the same pH.

Further multiple scattering calculations (13), indicate that feature A changes its lineshape, when the copper sites become square-planar. We tentatively assign the spectra at pH 11.5 to copper sites with this kind of configuration. In any case the axial distance should be larger than 2.6 Å.

b) EXAFS

The Fourier analysis of the EXAFS spectra at pH 3 gives similar main peaks for the first coordination shell (Figs.5 and 6). The different copper sites can be identified from the weak contributions from other shells. On these bases we can distinguish three types of copper sites: the first one for ribose and $CuCl_2$ at pH 3, the second one for Cu-ATP and Cu-Tripolyphosphate at pH 3 and the third one for

Cu-ATP and Cu-D-Ribose at pH 11.5. In all the spectra at pH 3 the best fit analysis of the first peak was not able to resolve the axial distortion.

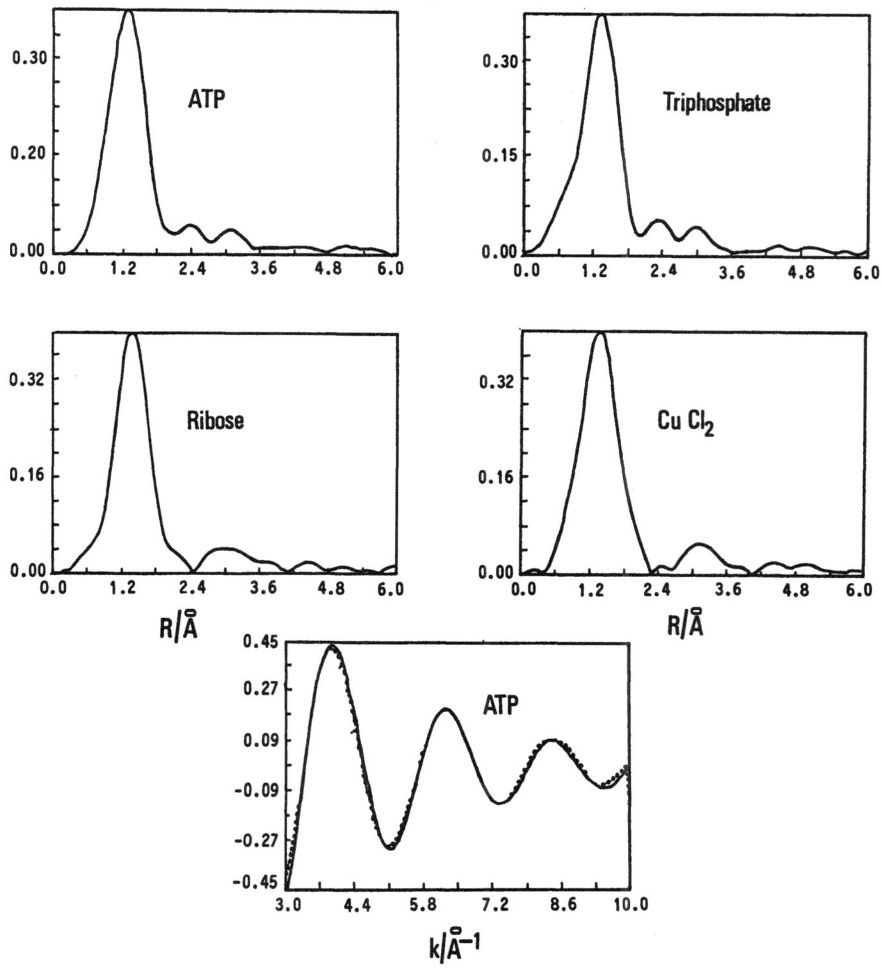

Fig.5 Fourier transforms of the EXAFS spectra at pH 3. Lower panel shows the best fit (dotted line) of the back-Fourier between 0.85 and 1.85 Å for Cu-ATP. The other samples give similar results.

For the EXAFS analysis, the oscillating part $\chi(k)$ of the absorption coefficient was extracted by standard computer analytical methods, and its Fourier transform was determined. The various components in the Fourier space were separately analyzed and a best fit by plane wave approximation of the back-Fourier transform was used to obtain the distance R_i and coordination number N_i of the i-th shell of atoms around the Cu ion according to the fitting function:

$$\chi(k)=(1/k)\Sigma_i\left\{\left[F_i(k)/R_i{}^2\right]N_i\ e^{-\sigma^2k^2}e^{-R_i/\Lambda_i k}\ \sin\left[2kR_i+\Phi(k)\right]\right\}$$

The exact E_o value of the energy scale was here adjusted as a best fit parameter. The back-Fourier analysis was performed on the first peak between 0.85-0.90 and 1.85-1.95 Å.

RESULTS

a) XANES

The XANES and the EXAFS spectra of Cu-ATP, were compared with those of Cu-Tripolyphosphate, Cu-D-Ribose complexes and of the $Cu(H_2O)_6$ ion in a 15 mM $CuCl_2$ solution.

Fig.1 reports the four XANES spectra and their derivatives at pH 3. The similarity of the Cu-ATP and Cu-Tripolyphosphate spectra indicates that the Cu ions bind to the phosphate chain of ATP at this pH, and the similarity between the Cu-D-Ribose spectrum and that of the aqueous spectrum of the cupric ion, indicates that ribose does not bind Cu ions at all. The derivatives of the XANES spectra show a small peak in the rising part of the edge (peak A) for the Cu-ATP and Cu-Tripolyphosphate complexes. A corresponding structure is also present in Cu-D-Ribose and in $CuCl_2$ solution although of the same order of the experimental noise, and suggests for the free aqueous ion a small Jahn-Teller distortion.

Fig.2 reports a sequence of the derivatives of the XANES spectra of Cu-ATP complex for pH 3, 7, 9, and 11.5. A significant increase with the pH of the small peak A already found in Fig.1 is observed.

Fig.3 reports the XANES spectra at pH 11.5 showing the similarity between the Cu-ATP and the Cu-D-Ribose complexes. We do not report the spectra for the Cu-Tripolyphosphate and the aqueous ion because of the formation of insoluble Cu hydroxide.

The parameters for the samples at pH 3, obtained with a one-shell (6 oxygens) model are reported in Table 1, where r is a parameter representing the overall accuracy of the fit.

Table 1. Best fit results of EXAFS spectra at pH 3.

	N_1	R_1/Å	σ_1/Å²	Λ_1/Å²	r
Cu-ATP	6	1.90	0.003	1.25	0.013
Cu-Tripolyph.	6	1.88	0.004	1.03	0.017
$CuCl_2$	6	1.95	0.007	1.03	0.005

At pH 11.5 the best fit analysis for Cu-ATP and Cu-D-Ribose gives N_1=4, R_1=1.89 Å and these results are in agreement with the

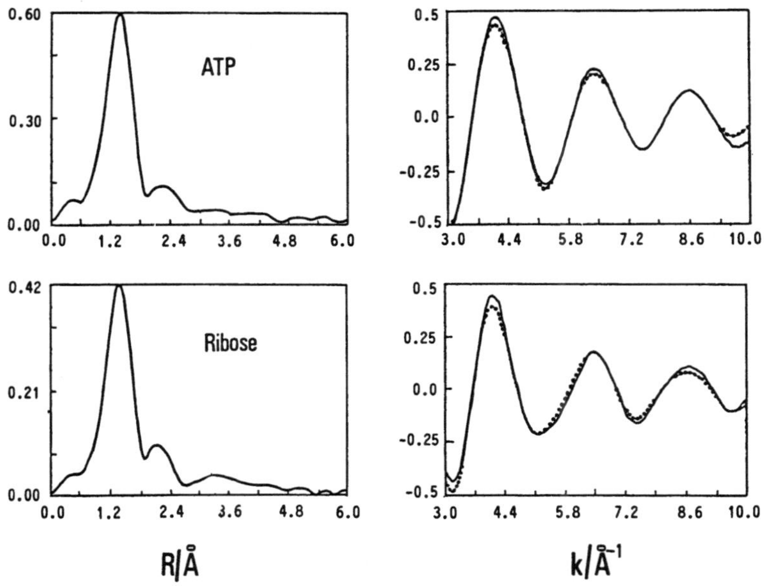

Fig.6 Fourier transforms of the EXAFS spectra (left) of Cu-ATP and Cu-D-Ribose at pH 11.5. Right panels show the best fits (dotted lines) of the back-Fourier transform between 0.9 and 1.95 Å.

indications coming from XANES, of an axial distortion higher than 0.6 Å. Indeed, the amplitude of the EXAFS oscillations due to atoms at distances higher than 2.6 Å are very small and affect only very slightly the total signal.

CONCLUSIONS

The pH of the solutions is one important factor in determining the nature of interaction sites in metal-ATP complexes. In fact, a competition is expected between metal ions and protons for the ionizable groups which act as ligands.

The metal-ATP interaction may be expected to depend on the state of ionization of the adenine ring (pK$\sim$4), of the terminal phosphate group (pK$\sim$7) and of the ribose hydroxyl group (pK$\sim$12.5).

Our results show that Cu(II) binds to phosphates in the Cu-ATP complex at pH 3, when the N(1) position of the adenine ring is protonated, while at pH 11.5 the metal is bound to the ribose moiety. This is in agreement with optical absorption measurements (10,11).

The XANES and EXAFS give additional information about the axial distortion of the copper site. At pH 3 the copper is coordinated by six oxygens with a small axial distortion.

The increase of peak A with pH can be interpreted, taking into consideration the optical findings, as an increase of the fraction of copper sites with larger distortion.

The data obtained at pH 11.5 are consistent with a square-planar configuration, according to the model reported in Fig. 7.

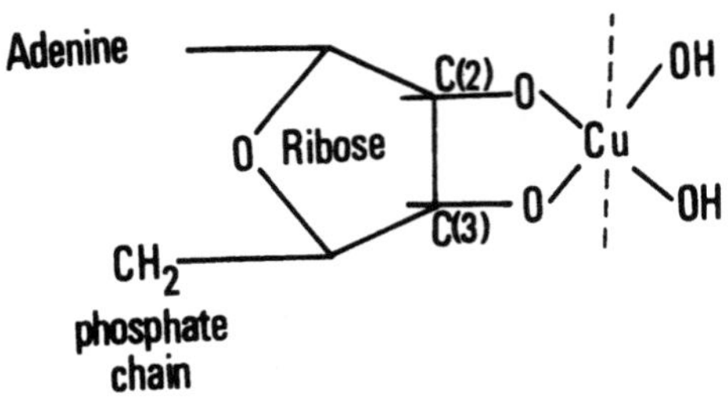

Fig.7. Possible model for metal coordination in Cu-ATP complex at pH 11.5.

The change with pH observed both in X-ray and optical spectra does not occur in the case of adenine-deoxynucleotides, where the copper coordination with the phosphate groups predominates in the whole pH range considered (11). This result further supports similar conclusions reported in the literature (14-16) and is of some significance in biology since metal ions appear to be able to distinguish between ribose and deoxyribose nucleotides in enzymatic reactions.

Acknowledgements. We thank Mr. Maurizio Vischetti for his techinical assistance in X-ray experiments and Mr.Giacomo Monteleone for his photographic work.

REFERENCE

1. M.Belli, A.Scafati, A.Bianconi, E.Burattini, S.Mobilio, C.R.Natoli, L.Palladino. Nuovo Cimento, 2D, 1281, 1983.
2. E.Rongoni, A.Scafati, M.Matzeu, M.Belli, G.Onori, A.Reale, A.Balerna, A.Bianconi,E.Bernieri. Biopolymers, 25, 217, 1986.
3. M.Belli, A.Balerna, E.Bernieri, A.Bianconi, E.Burattini, M.Matzeu, F.Mazzei, S.Mobilio, C.R.Natoli, G.Onori, L.Palladino, A.Reale, E.Rongoni, A.Scafati. In " Physics in Environmental and Biomedical Research", World Scientific,(1986),p.33.
4. I.Sissoef,I.Grisvard, E.Guillé. Progr.Biophys.Mol.Biol. 31, 165, 1976.
5. M.M. Teeter, et al. In" Metal Ions in Genetic Information Transfer" Elsevier Publ.(1981),p.233.
6. L.A. Loeb, A.S.Mildran. In " Metal ions in Genetic Information Transfer".Elsevier Publ.(1981), p.233.
7. R.B. Martin and Y.H.Mariam. In "Metal ions in biological systems".H.Siegel Ed., M.Dekker, New York, vol.8,p.57,1979.
8. A.T. Tu and M.J. Heller:"In Metal Ions in biologycal systems" H.Siegel Ed., M.Dekker, New York, vol.1 p.1 1974.
9. I.G. Marzilli. Prog. Inorg. Chem., 23, 255, 1977.
10. G.Onori and D.Blidaru. Nuovo Cimento, 5D, 339, 1985.
11. G.Onori and D.Blidaru. Nuovo Cimento, 5D, 348, 1985.
12. J.Garcia et.al. Solid State Commun.,58, 595, 1986.

13. J.Garcia et al.Proceeding of EXAFS and Near Edge Structure IV, edited by J.Petiau. Jour. de Physique (to be published).
14. N.A. Berger, E.Tarien and G.L. Eichhorn. Nature New Biol., 239, 237, 1972.
15. Y.U. Chao and D.R. Kearns. J.Am. Chem.Soc., 99, 6425, 1977.
16. M.Gabriel, D.Larcher, C.Thirion, J.Torreilles and A.Crastes de Paulet. Inorg. Chim. Acta, 24, 187, 1977.

X-Ray Absorption Spectroscopy of Manganese and Iron in the Photosynthetic Apparatus

Ann E. McDermott, R. D. Guiles, Vittal K. Yachandra
James Cole, R. David Britt, S. L. Dexheimer
Kenneth Sauer and Melvin P. Klein

Laboratory of Chemical Biodynamics, Lawrence Berkeley Laboratory
University of California, Berkeley, CA. 94720 USA

The process of photosynthesis in green plants may be considered as a photoinduced chlorophyll mediated transfer of electrons from water to CO_2 resulting in the formation of carbohydrates and O_2. Strong oxidants produced by photosystem II (PS II) remove electrons from the O_2-evolving complex resulting in water oxidation. Photosystem I (PS I) produces powerful reductants that donate electrons through a series of membrane bound proteins, one of which has been identified as an iron-sulfur protein containing at least two Fe-S clusters, to soluble ferredoxin and NADP, which are ultimately responsible for CO_2 reduction.

The primary focus of our work is directed toward the Mn-containing O_2-evolving complex (Mn-OEC) in PS II and the Fe-S acceptor centers A, B and X in PS I.

Iron X-ray Absorption Spectra of Acceptors in PS I

The stable electron acceptors in PS I, centers A, B and X, are thought to be [2Fe-2S] or [4Fe-4S] ferredoxins [1]. Their EPR spectra resemble the spectra of the [2Fe-2S] and [4Fe-4S] ferredoxins, having anisotropic signals with g_{ave} <2.0, and their EPR spectra are quite different from the spectra of the [1Fe], [3Fe-4S] or [3Fe-3S] ferredoxins. PS I contains 10-14 Fe and approximately 12 acid labile sulfide per P700 [2,3], and contains one center A, one center B and one to two X per P700 [4-7]. On the basis of these observations two models of PS I seem likely: 1) PS I contains three [4Fe-4S] ferredoxins or 2) it contains two [4Fe-4S] ferredoxins and two [2Fe-2S] ferredoxins. Evans [8,9] addressed the structure of A, B and X using Mössbauer spectroscopy and concluded that the spectra of all the reducible iron in PS I resembles spectra of [4Fe-4S] ferredoxins and not of [2Fe-2S] ferredoxins. Only approximately 65% of the iron present was reduced in that study.

In this study, we address the structures of centers A, B and X using X-ray absorption spectroscopy. We have used iron K-edge spectra to assess how much of the iron is bound as ferredoxins. K-edge spectra probe bound state transitions from the 1s level and are sensitive to oxidation state and site symmetry [10]. We have also used Extended X-ray Absorption Fine Structure (EXAFS) to address whether the ferredoxins are [2Fe-2S] or [4Fe-4S] clusters. EXAFS provides information about types and numbers of ligands, and provides sensitive bond length information. In the case of ferredoxins it is diagnostic for [2Fe-2S] and [4Fe-4S] clusters [11].

Materials and Methods

PS I preparations from *Synechococcus* and from market spinach were prepared using Triton X-100 extraction and sucrose gradient centrifugation as detailed elsewhwere [12,13]. The preparations had 11-14 Fe and 9-13 acid labile sulfide per P700 and showed EPR signals due to A, B and X but showed no signals due to heme, soluble [2Fe-2S] spinach ferredoxin, or g=4.3 iron. The preparations contained peptides of molecular weight 60, 22, 19 and 18 kDa by SDS PAGE.

X-ray absorption spectra were collected as described elsewhere [12-14]. The PS I preparation was assayed for integrity both before and after X-ray exposure by EPR and optical methods. EXAFS data were analyzed by curve fitting according to the Teo-Lee method [15].

Results and Discussion

Figure 1 shows the iron K-edge spectrum of PS I from spinach and *Synechococcus* along with the spectra of soluble spinach [2Fe-2S] ferredoxin, $(Et_4N)_2$ Fe_4S_4(S-benzyl)$_4$ and $(Et_4N)_2$ $Fe_2S_2(S_2$-o-xyl$)_2$. The iron K-edges of the PS I preparations are nearly identical to those of the soluble [2Fe-2S] ferredoxin including the size of the 1s to 3d transition, an indicator of the amount of non-centrosymmetric iron. Roe *et. al.* [10] have shown that this feature is systematically larger in tetrahedral complexes than in octahedral complexes. We have found that 15% adventitiously bound iron in our soluble ferredoxin model proteins changes the edge shape dramatically, so the edge is a sensitive reporter of ferredoxin content. We estimate that over 90% of the iron in PS I is bound in ferredoxins.

Figure 2 shows the Fourier filtered spinach PS I EXAFS spectrum plotted over spectra from soluble spinach [2Fe-2S] ferredoxin and over data from soluble *Clostridium pasteurianum* [4Fe-4S] ferredoxin. In the region of k=7.5Å^{-1} the [4Fe-4S] and [2Fe-2S] ferredoxin data are out of phase owing to the much greater amount of iron backscattering in the [4Fe-4S] cluster from three equivalent Fe neighbors. The PS I spectrum, while not identical to either spectrum, more nearly resembles the [2Fe-2S] spectrum. Curve fitting analysis suggests that the PS I spectrum is better explained by a sum of [2Fe-2S] and [4Fe-4S] ferredoxins than by all [4Fe-4S] ferredoxins. Figure 3 shows the simulations of the PS I iron EXAFS spectra from spinach and *Synechococcus*. In each case the simulation was restricted to using four sulfur neighbors and either 2.3 or 3.0 iron neighbors to

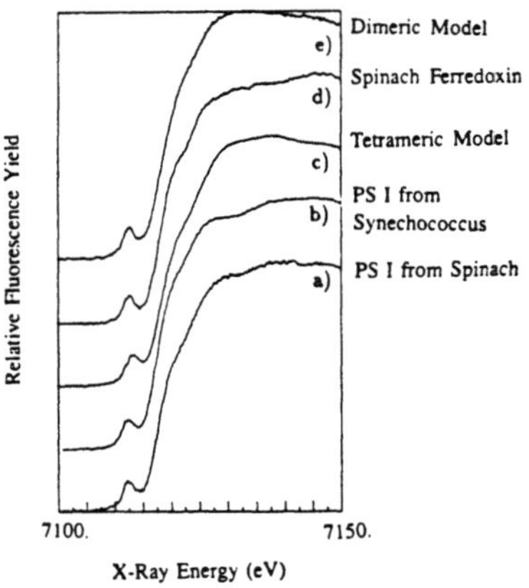

Figure 1. Iron K-edge spectra of of oxidized PS I with iron-sulfur models. From the bottom: a) PS I from spinach, b) PS I from *Synechococcus*, c) $(Et_4N)_2Fe_4S_4$(S-benzyl)$_4$, d) spinach $[2Fe\text{-}2S]^{2+(2+,1+)}$ ferredoxin and e) $(Et_4N)_2Fe_2S_2(S_2\text{-o-xyl})_2$. The pre-edge transition has been assigned as a 1s-3d transition, and is a good indicator of non-centrosymmetric environments. The PS I from spinach and *Synechococcus* are quite similar to each other and to the [2Fe-2S] soluble spinach ferredoxin.

simulate either a system of two [2Fe-2S] clusters and two [4Fe-4S] clusters or simulate a mixture of different [4Fe-4S] clusters. All other parameters used in the simulations were chosen by comparison with our data on soluble ferredoxins or published data [11]. The central portions of these fits around $k=7.5Å^{-1}$ are most informative, because this 'beat' region is sensitive to the admixture of iron and sulfur backscattering. The lower k regions are more prone to artifacts due to contributions from other processes, and the higher k regions have poorer signal to noise ratios. For both our spinach and *Synechococcus* data the fits assuming a mixture of [2Fe-2S] and [4Fe-4S] ferredoxins had least squares errors lower by a factor of more than two. These fits agree much better with the data around $k=7.5Å^{-1}$.

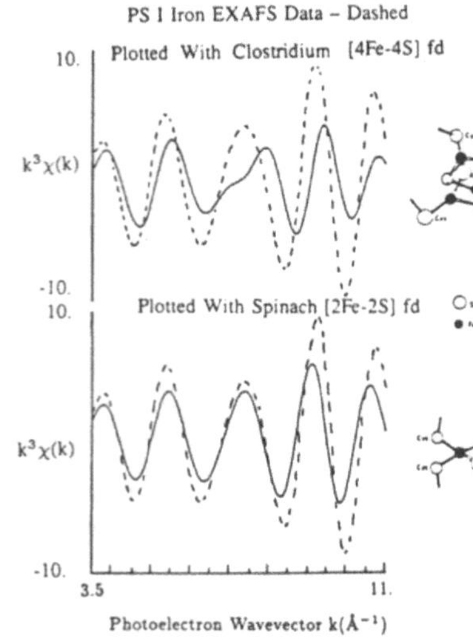

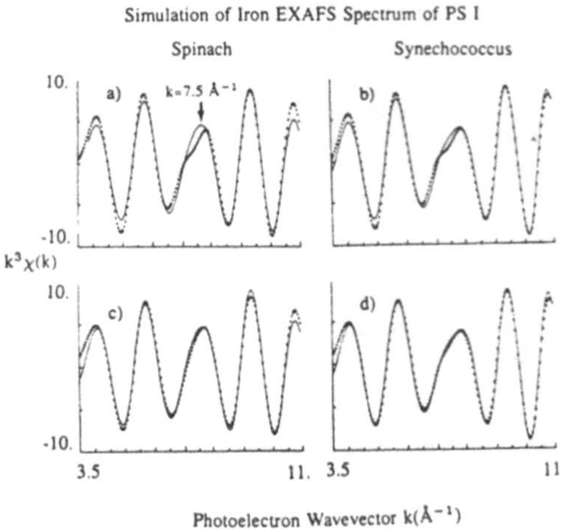

Figure 2. (on the left) The Fourier filtered k^3-weighted EXAFS spectrum of PS I from spinach (dashed line) plotted with: (a) the spectrum of soluble *Clostridium pasteurianum* [4Fe-4S] ferredoxin (solid line) and (b) the spectrum of soluble spinach [2Fe-2S] ferredoxin (solid line). The beat region around $k=7.5Å^{-1}$ is diagnostic of [2Fe-2S] and [4Fe-4S] ferredoxins. The beat region indicates that PS I is likely to contain some [2Fe-2S] ferredoxins.

Figure 3. (on the right) The Fourier filtered k^3-weighted EXAFS spectrum of PS I (solid lines) is plotted with calculated simulations using the Teo-Lee method [15]. The fits (dotted lines) are as follows: a) spinach PS I simulated assuming all [4Fe-4S] ferredoxins, b) PS I from *Synechococcus* simulated assuming all [4Fe-4S] ferredoxins, c) PS I from spinach simulated assuming two [4Fe-4S] and two [2Fe-2S] ferredoxins and d) PS I from *Synechococcus* simulated assuming two [2Fe-2S] and two [4Fe-4S] ferredoxins. The quality of the fits in the beat region indicates that PS I is likely to contain [2Fe-2S] as well as [4Fe-4S] ferredoxins.

The Mössbauer results of Evans *et al* indicate that A, B and X are all [4Fe-4S] ferredoxins [8,9]. We are considering the possibility that the [2Fe-2S] ferredoxins reflected in the EXAFS spectrum may be contained in the 35% of iron which was not reduced in the Mössbauer experiment. This iron is likely to be associated with X [7], since X is more difficult to reduce quantitatively than is center A or B. Recently we have observed a near infrared absorption band near 1050 nm in PS I which is typical of [4Fe-4S] ferredoxins. This band can be titrated reversibly at -500mV. This confirms the assignment that centers A and B are the [4Fe-4S] ferredoxins in PS I, and so X is likely to be the [2Fe-2S] ferredoxin(s).

While the EPR spectrum of X does not saturate below 100mW, it has an optimum temperature of observation of approximately 8 K, below which the signal diminishes rapidly (data not shown). This indicates that the state giving rise to the EPR signal is above, but close to the ground state. This is unusual for [2Fe-2S] or [4Fe-4S] ferredoxins, and we are considering whether this be explained by an exchange coupled pair of [2Fe-2S] clusters, separated by over 3.5 Å, so that intercluster distances are not seen in the EXAFS spectrum. This exchange coupling may also explain why the power saturation and EPR linewidth of X are different from those of other ferredoxins.

The State of Manganese in the Photosynthetic Apparatus

The Mn-OEC is the least understood of the light-driven electron transfer protein complexes in chloroplasts, even though it has been the subject of extensive studies [16,17]. Studies of O_2 evolution using a train of saturating light flashes have given rise to a model for the accumulation of oxidizing equivalents in which some intermediates, labeled S_0-S_4 operate in a cyclic fashion [18].

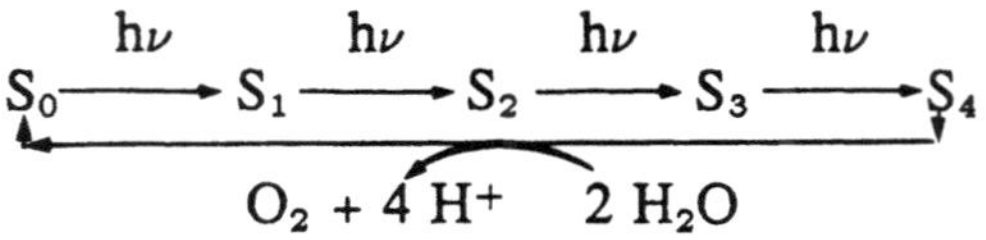

Recently established procedures for isolating subchloroplast PS II preparations [19,20] and the discovery of a light-induced multiline EPR signal [21-23] attributable to the S_2 state of the Mn-OEC have facilitated the preparation of well characterized samples stabilized in the S_1 and S_2 states. We have used such detergent extracted but functional photosystem II particles to study the light-induced changes in the Mn oxidation state and ligand environment [24].

Recently, we have been able to prepare and stabilize samples in the S_0 and S_3 states for study. In this report we present the edge spectra of samples poised in the S_0 and S_3 states, the EXAFS results of S_1 and S_2 samples from spinach, and the effects of peptide release on the structure of Mn in the OEC. Also we have prepared highly active O_2-evolving PS II particles from the thermophilic cyanobacterium Synechococcus and we present the Mn K-edge spectra of the S_1 and S_2 states and EXAFS results of the S_1 state.

Materials and Methods

Preparation of oxygen-evolving PS II sub-chloroplast membranes from spinach was accomplished by a modification of two different Triton X-100 fractionation procedures and is described in detail in Ref 14. Highly active O_2-evolving PS II particles from the thermophilic cyanobacterium Synechococcus were obtained by β-octyl glucoside extraction [13]. The samples from spinach had rates of oxygen evolution of 300-400 μmoles of O_2 (mg of Chl)$^{-1}$h^{-1} and contained ~4 Mn atoms and 260 Chl per reaction center. The cyanobacterium preparation yielded O_2 rates of about 1000-1500 μmoles of O_2 (mg of Chl)$^{-1}$h^{-1} and contained about ~3 Mn atoms and 70-90 Chl per reaction center. Reaction center sizes were determined by integration of EPR signal II_S.

X-ray absorption samples were suspended in 50 mM MES buffer, pH 6.0, 10 mM NaCl, 5 mM $MgCl_2$ and ~30% glycerol. The samples of 20-30 mg Chl ml^{-1} concentration were mounted in lucite sample holders, and the procedure of dark adaptation, illumination, EPR and X-ray measurements was carried out directly in these sample holders. X-ray absorption edge spectra and EXAFS spectra were collected in the fluorescence mode at the Stanford Synchrotron Radiation Laboratory, Stanford, CA. Energy calibration was maintained by simultaneous measurement of the strong and narrow pre-edge feature of $KMnO_4$ at 6543.3 eV. The uncertainty in our edge energy measurements is 0.1 eV. During X-ray measurements, samples were suspended in a double walled kapton cryostat maintained at 170 K by a liquid N_2 boil-off jet. Details of the X-ray experimental procedure and data analysis are described in Refs [14,24-27]. EPR spectra were recorded on a Varian E-109 spectrometer. Samples were run at 8 K at a microwave frequency of 9.21 GHz and with 100 kHz field modulation.

The S_1 samples were prepared by dark adaptation of the PS II particles for ~2 h and then freezing the sample in liquid N_2. The S_2 samples were initially dark adapted and then were equilibrated at 190 K in a Varian V 6040 NMR temperature controller and illuminated with a 400 W tungsten lamp through a 5 cm water filter. The samples were then frozen in liquid N_2. The production of the S_2 state was monitored by observing the appearance of the characteristic multiline EPR signal at 8 K.

Samples were poised in a S_0-like state by treatment with NH_2OH followed by illumination at 190 K or at 273 K in the presence of DCMU. NH_2OH is known to set the S-state cycle back two steps [28,29]. Upon illumination at 273 K in the presence of DCMU the system advances to a state resembling S_0 (S_0^*). Both these illumination protocols ensure that the PS II acceptor side is blocked between the primary and secondary acceptor site hence allowing the S-state cycle to advance only by one step. S_3 samples were prepared in a manner analogous to the S_2 samples. Subsequent to the generation of the S_2 state, the samples were warmed to ~273 K for 30 sec allowing electron transport to occur from the primary acceptor to the secondary acceptor freeing the primary acceptor to receive another electron from the donor side of PS II [30,31]. The sample was then equilibrated at 235 K and illuminated for 10 min. The reduction of multiline signal and the regeneration of the primary acceptor quinone EPR signal at g=1.9 were used as the criteria for the generation of the S_3 state.

Results and Discussion

Mn X-ray Absorption Edge Studies

The PS II preparations have permitted us to demonstrate that a significant change occurs in the Mn K-edge energy of PS II particles upon advancing from the S_1 to the S_2 state [24,32]. These observations established for the first time that Mn is directly involved in the storage of oxidizing equivalents. A 1s-3d pre-edge feature is also seen in the edge spectra indicating the non-centrosymmetric environment of Mn in the complex.

We have studied the Mn K-edges of samples poised in the S_0^* and S_3 states. The Mn K-edge spectra for spinach PS II samples in the S_0^*, S_1 and S_2 states are shown in Fig 4a and in the S_2 and predominantly in the S_3 states are shown in Fig 4b. The relative fraction of S_2 was determined by monitoring the amplitude of the multiline EPR signal. The fraction in S_3 was monitored by measuring the diminution of the multiline signal on illumination of a S_2 sample at 235 K.

The Mn K-edge inflection of the S_1 sample occurs at about 6551.3 eV, which is close to the K-edge of Mn(III) complexes [33]. The S_2 edge inflection exihibits a shift to higher energy by about 1 eV and occurs at ~6552.5 eV which is between that observed for Mn(III) and Mn(IV) complexes. The S_0^* edge inflection is observed at 6550.3 eV about 1 eV less than that of the S_1 edge. The lower K-edge energy suggests that S_0^* is reduced relative to S_1. Though we have been unable to generate a pure S_3 state due to competing kinetic processes we estimate that the fraction of the S_3 state in our samples is ~50%. The K-edge inflection of the S_3 state is similar to that observed for the S_2 state as seen in Fig 4b. We conclude that the oxidation state of Mn increases in the transition from S_0 to S_1 and in the transition from S_1 to S_2 but is invariant on advancement from S_2 to S_3.

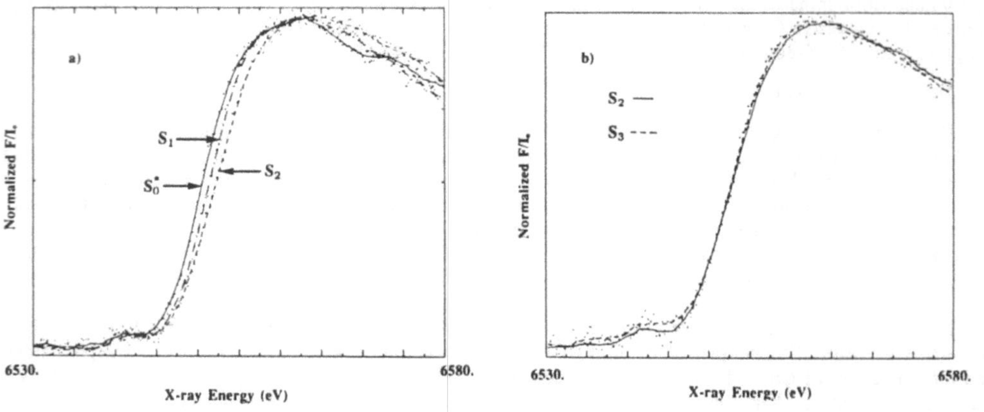

Figure 4. The X-ray absorption K-edge spectra of Mn in the S_0^*, S_1, S_2 and S_3 states of spinach PS II particles. A smoothed curve is drawn through the data points. The small pre-edge feature at ~6543 eV is due to the 1s-3d bound state transition. a) (on the left) The K-edge inflection energy for the S_0^* state is at 6550.3 eV, the S_1 state at 6551.3 eV and the S_2 state is at 6552.5 eV. The shifts from S_0^* to S_1 to S_2 indicate a progressive oxidation of Mn as the Mn-OEC steps through the S-State cycle. b) (on the right) The K-edge inflection of the S_3 state is approximately at the same position as the inflection for the S_2 state, indicating that the oxidation state has not changed during the S_2 to S_3 transition.

Recently we have prepared PS II particles from a thermophilic cyanobacterium Synechococcus. The particles are free of PS I and retain full O_2-evolving capacity. Using these particles we have generated a low temperature multiline EPR signal similiar to that observed in the S_2 state of spinach [21]. In Fig 5 are shown the Mn K-edge spectra for Synechococcus PS II particles in the S_1 and S_2 states. The K-edge inflection for the S_1 state occurs at ~6551.1 eV. On advancement to the S_2 state the edge energy shifts to higher energy of about 6552.1 eV. The edge position and shape are similar to those observed for spinach, including the pre-edge 1s-3d bound state transition. This indicates the similarity of the Mn complex in spinach and Synechococcus.

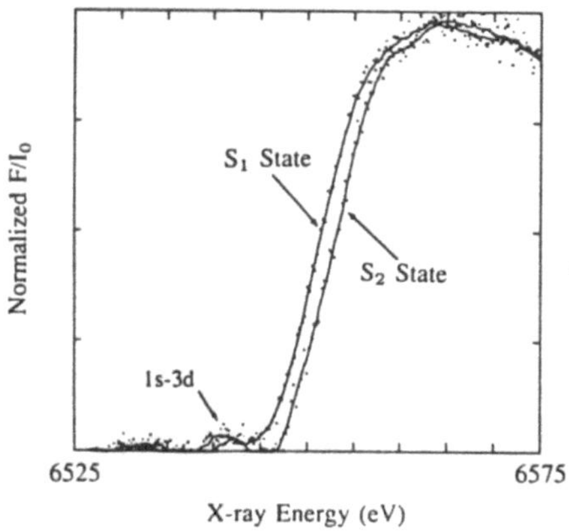

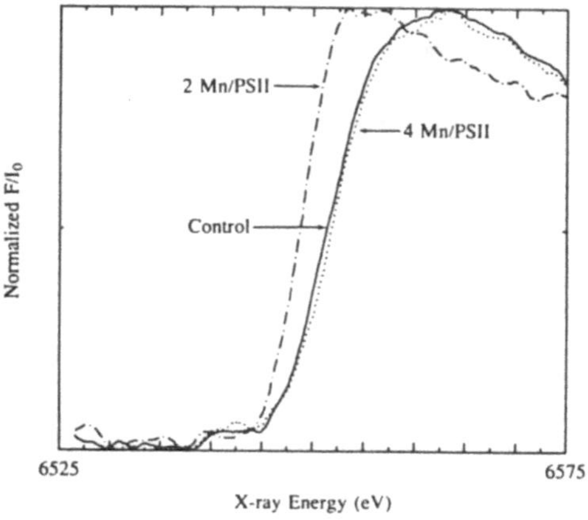

Figure 5. (on the left) The X-ray absorption K-edge spectra of Mn in PS II particles from the thermophilic cyanobacterium Synechococcus. The Mn K-edge inflection for the S_1 state is at 6551.1 eV and at 6552.1 eV for the S_2 state. Note the pre-edge transition at ~6543 eV. Smoothed curves are drawn through the data points.

Figure 6. (on the right) The Mn K-edge spectra of $CaCl_2$ washed PS II particles. Smoothed curves are drawn. The control is a S_1 sample. The 4 Mn/PS II sample is from PS II particles from spinach incubated in 800mM $CaCl_2$ buffers for 2 h and which are depleted of the 16, 24 and 33 kDa peptides, while the 2 Mn/PS II sample was subsequently incubated for 12 h in low ionic strength buffer and contains only 2 Mn/PS II. The inflection energy for the 4 Mn/PS II sample is at 6551.9 eV and at 6548.9 eV for the 2 Mn/PS II sample.

$CaCl_2$ washing of PS II particles is known to release three peptides of molecular weight 16, 24 and 33 kDa, which have been shown to be involved in O_2 evolution [34]. Recently, we have extended our X-ray K-edge studies to $CaCl_2$ washed PS II particles in order to understand the structural changes produced in the Mn complex as a result of the release of the three peptides. Incubation of PS II particles in 800 mM $CaCl_2$ buffers for 2 h releases the three peptides. Subsequent 12 h incubation at low ionic strength additionally releases 2 Mn atoms/PS II reaction center from the 4 Mn/PS II that are normally present in active preparations. The Mn K-edge spectra of samples depleted of the three peptides and containing 4 Mn/PS II and 2 Mn/PS II are shown in Fig 6 The K-edge spectrum of the 4Mn/PS II sample is similar to the control S_1 sample, while the 2 Mn/PS II sample is distinctly different in shape and position with an inflection point around 6548.9 eV, suggestive of a +2 oxidation state for Mn. We conclude that the Mn complex is still intact in the 4 Mn/PS II samples in spite of the loss of the three peptides and inactivation of O_2 evolution. However, when 2 Mn/PS II are also released a major change occurs in the structure of the Mn complex as evidenced by the dramatic change in the edge shape, the disappearance of the pre-edge feature, and the shift in position of the inflection.

Mn EXAFS Studies

The EXAFS results for the spinach PS II particles show that the salient features of the Mn structure are essentially identical for samples in the S_1 and S_2 states. (The greater amplitude of the first Fourier peak for the S_2 sample is due to a background removal artifact.) These features are a Mn neighbor at 2.70 Å, N or O ligands at 1.75 Å and additional N or O ligands at 2.00 Å. Such coordination is consistent with a di-μ-oxo binuclear complex [35]. The great similiarity of the EXAFS for samples prepared in S_1 and S_2 indicates that the light induced edge shift results from a change in Mn oxidation state with no change in the coordination of Mn in the O_2-evolving complex.

In Fig 7 are shown EXAFS Fourier transforms of the Synechococcus PS II particles and from spinach PS II particles in the S_1 and S_2 states. The three R-space features observed in spinach are also present in Synechococcus. The parameters obtained from simulations to the fourier filtered data from the fourier peaks labelled I, II and III in Fig 7 are shown in Table 1. These data demonstrate that the structure of the Mn complex in the O_2-evolving system is largely conserved over the two billion year evolutionary period that separates the two species. We therefore suggest that the bridged Mn complex is an essential feature of the O_2-evolving complex.

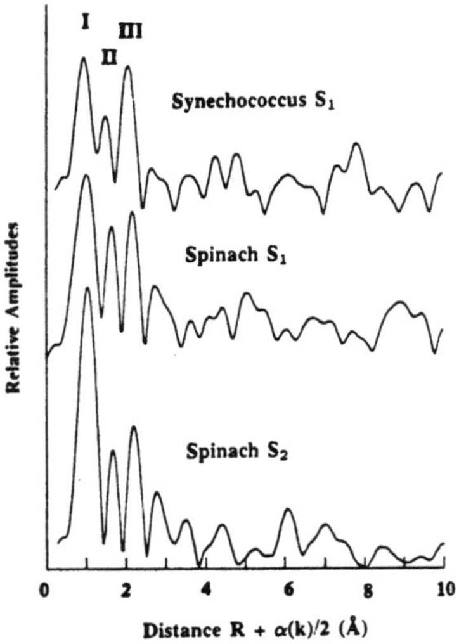

Figure 7. Fourier transforms of the $k^1\chi(k)$ Mn EXAFS data from spinach S_1, spinach S_2 and Synechococcus. The peaks labelled I and II are characteristic of bridging and terminal N or O ligands and peak III is due to a neighboring Mn atom and the distances are typical for μ-oxo bridged binuclear Mn clusters.

Parameters Extracted from the Mn EXAFS Data of PS II Particles in the S_1 and S_2 State from Spinach and from the Thermophilic Cyanobacterium Synechococcus

	Distance (Å)		
	Spinach		Synechococcus
	S1	S2	S1
I Mn-O or N Bridging Ligand	1.75	1.76	1.85
II Mn-O or N Terminal Ligand	2.00	1.98	2.08
III Mn-Mn	2.70	2.72	2.71

Table I We estimate the uncertainty in the distances to be 0.03 Å except for the bridging ligand distances where it is 0.05 Å. The fits were performed using theoretical phase and amplitude functions by the Teo-Lee method [15].

There is growing concensus that there are about four Mn atoms per PS II reaction center. Our third R-space Fourier peak fits best to 1-2 Mn atoms at the distance of 2.7 Å. A careful reexamination of the data shown in Fig 7 suggests that there may be an additional R-space peak beyond that now assigned to Mn. As discussed above and shown in Table I, there is virtual identity of the EXAFS for spinach PS II particles in the S_1 and S_2 states. We thus believed it reasonable to add the k-space spectra to improve the S/N. The Fourier transform of the k^1 weighted sum is shown in Fig 8, which shows a fourth R-space peak significantly above the noise level. The improvement in S/N is apparent. Also shown in Fig 8 is the Fourier transform of the k^2-weighted data set.

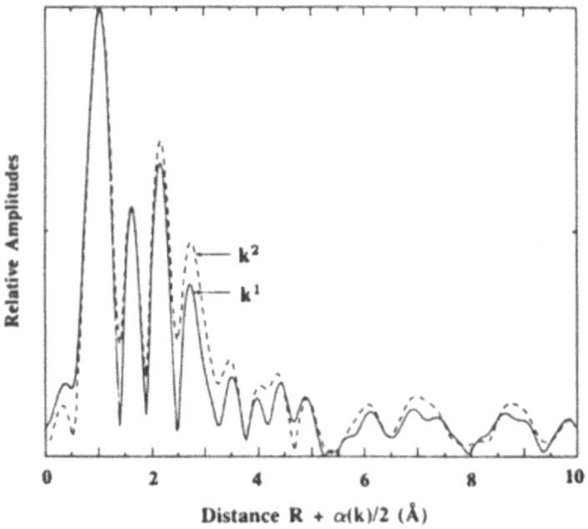

Figure 8. Fourier transform of the sum of the S_1 and S_2 EXAFS data weighted by k^1 and k^2. The solid line is the k^1-weighted data set and the dashed line is the k^2 data set. Note that the third and fourth peaks grow with increasing k-weighting indicating a heavy scatterer like manganese.

One of the procedures used to discriminate between lower-Z and higher-Z backscatterers is to compare the FT of the data at different k-weightings. The higher Z scatterers are emphasized by higher k-weighting because the peak in their amplitude function occurs at higher k values. The proportional growth of both the third and fourth R-space peaks shown in Fig 8 with increased k-weighting suggests that the fourth peak may also be assigned to Mn. If this assignment is correct, this would imply that the four Mn atoms in the Mn-OEC are present in the form of a tetrameric cluster. A reasonable fit is obtained to the back-transformed data from the fourth peak to one or more Mn atoms at ~3.3 Å. This peak can also be fit to two low-Z shells, such as C, N or O at 3.0 and 3.2 Å, that would occur from binding to a histidine ring, a common metal coordinating group in proteins. Although the k-weighting behaviour of the wave is indicative of a heavy atom, one cannot unequivocally rule out the possibilty of contributions from second shell imidazole atoms.

Chloride ion is essential for oxygen evolution and it has been suggested that this ion might be a ligand of Mn [36,37]. We have addressed this question by both EXAFS and EPR. We have attempted to force a Cl backscattering shell into the EXAFS simulations for samples in both S_1 and S_2. In all cases, either the quality of fit is reduced or the number of Cl atoms included approaches zero. This result also addresses the question of sulfur coordination to Mn as the backscattering parameters of Cl and S are quite similar. We have prepared PS II particles in which Br^- has replaced Cl^- with complete restoration of oxygen evolving capacity. The high resolution EPR spectra of corresponding samples again indicates that there is no statistically significant difference between them suggesting that halide does not directly coordinate Mn [38].

Acknowledgements

We thank Dr. S. Katoh for providing the strain of *Synechococcus* and Dr. Akhihiko Yamagishi for advice concerning the preparation of PS I from *Synechococcus*. We are grateful to Dr. David Goodin for many helpful discussions concerning X-ray fluorescence detection. We thank Rick Storrs for help with data collection. This work was supported by a grant from the National Science Foundation (PCM 82-16127 and PCM 84-16676) and by the Director, Office of Energy Research, Office of Basic Energy Sciences, Division of Biological Energy Conversion and Conservation of the Department of Energy under contract DE-AC03-76 SF00098. Synchrotron radiation facilities were provided by the Stanford Synchrotron Radiation Laboratory which is supported by the U.S. Department of Energy, Office of Basic Energy Sciences, and by the NIH Biotechnology Program, Division of Research Resources.

References

1. Evans, M.C.W. (1982) in "Iron Sulfur Proteins", Spiro, T., ed.: pp 249-284 John Wiley and Sons, New York.
2. Golbeck, J. (1980) Methods Enzymol. 69, 129-141
3. Lundell, D., Glazer, A., Melis, A. and Malkin, R. (1985) J. Biol. Chem. 260, 646-654
4. Bearden, A. and Malkin, R. (1972) Biochim. Biophys. Acta. 283, 456-468
5. Williams-Smith, D., Heathcote, P., Charanjit, K. and Evans, M.C.W (1978) Biochem J. 170, 365-371
6. Heathcote, D., Williams-Smith, D. and Evans, M.C.W. (1978) Biochem. J. 170, 373-378
7. Bonnerjea, J. and Evans, M. (1984) Biochim. Biophys. Acta. 767, 153-159

8. Evans, H., Rush, J., Johnson, C. and Evans, M.C.W. (1979) Biochem. J. 182, 861-865
9. Evans, H., Dickson, D., Johnson, C., Rush, J. and Evans, M.C.W. (1981) Eur. J. Biochem. 118, 81-84
10. Roe, A., Schneider, R., Mayer, R., Pryz, J., Widom, J. and Que, L., Jr. (1984) J. Am. Chem. Soc. 106, 1676-1681
11. Teo, B., Shulman, R., Brown, G. and Meixner, A. (1979) J. Am. Chem. Soc. 101, 5624-5631
12. McDermott, A., Yachandra, V., Guiles, R., Britt, R., Dexheimer, S., Sauer, K. and Klein, M., *submitted*
13. McDermott, A. Ph. D. Dissertation, University of California, Berkeley *in preparation.*
14. Yachandra, V., Guiles, R., McDermott, A., Britt, R., Dexheimer, S., Sauer, K. and Klein, M.P. (1986) Biochim. Biophys. Acta 850, 324-332
15. Teo, B. and Lee, P. (1979) J. Am. Chem. Soc. 101, 2814-2831
16. Sauer, K. (1980) Acc. Chem. Res. 13, 249-256
17. Amesz, J. (1983) Biochim. Biophys. Acta 726, 1-12
18. Kok, B., Forbush, B. and McGloin, M. (1971) Photochem. Photobiol. 14, 307-321
19. Kuwabara, T. and Murata, N. (1982) Plant Cell Physiol. 23, 533-539
20. Berthold, D.A., Babcock, G.T. and Yocum, C.F. (1981) FEBS Lett. 134, 231-234
21. Dismukes, G.C. and Siderer, Y (1981) Proc. Natl. Acad. Sci. USA, 78, 274-278
22. Hansson, Ö. and Andrèasson, L.-E. (1982) Biochim. Biophys. Acta 679, 261-268
23. Dismukes, G.C., Ferris, K. and Watnick, P. (1982) Photobiochem. Photobiophys. 3, 243-256
24. Goodin, D.B., Yachandra, V.K., Britt, R.D., Sauer, K. and Klein, M.P. (1984) Biochim. Biophys. Acta 767, 209-216
25. Jaklevic, J., Kirby, J.A, Klein, M.P., Robertson, A.S., Brown, G.S. and Eisenberger, P. (1977) Solid State Commun. 23, 679-682
26. Powers, L., Chance, B., Ching, Y. and Angiolillo, P. (1981) Biophys. J. 34, 465-498
27. Stern E.A. and Heald, S.M. (1979) Rev. Sci. Instrum. 50, 1579-1582
28. Bouges-Boucquet, B. (1973) Biochim. Biophys. Acta 292, 772-785
29. Radmer, R. and Ollinger, O. (1982) FEBS Lett. 144, 162-166
30. Joliot, A. (1974) Biochim. Biophys. Acta 357, 439-448
31. de Paula, J.C., Innes, J.B. and Brudvig, G. (1985) Biochem. 24, 8114-8120
32. Goodin, D.B., Yachandra, V.K., Guiles, R.D., Britt, R.D., McDermott, A., Sauer, K. and Klein, M.P. (1984) in EXAFS and Near Edge Structure III (eds., Hodgson, K.O., Hedman, B. and Penner-Hahn, J.E.), Springer-Verlag, New York, 130-135
33. Kirby, J.A., Goodin, D.B., Wydrzynski, T. Robertson, A.S. and Klein, M.P. (1981) J. Am. Chem. Soc. 103, 5537-5542
34. Ghanotakis, D.F. and Yocum, C.F. (1985) Photosynth. Res. 7, 97-114
35. Cooper, S.R., Dismukes, G.C., Klein, M.P. and Calvin, M. (1978) J. Am. Chem. Soc. 100, 7248-7252
36. Izawa, S., Heath, R.L. and Hind, G. (1969) Biochim. Biophys. Acta 180, 388-398
37. Sandusky, P.O. and Yocum, C.F. (1983) FEBS Lett. 162, 339-343
38. Yachandra, V.K., Guiles, R.D., Sauer, K. and Klein, M.P. (1986) Biochim. Biophys. Acta 850, 333-342

EXAFS and XANES Studies of Calcium Polysaccharides

L. ALAGNA, T. PROSPERI, A.A.G. TOMLINSON
I.T.S.E., C.N.R. Research Area, Monterotondo Stazione ITALY
R. RIZZO, Chemistry Department, University of Naples, Via Mezzocannone 18 Naples, Italy.

Abstract

EXAFS studies of the calcium environments in polysaccharides have provided important insights into chain conformation and gel formation. Work on Ca^{2+} poly(α-D-galacturonate) in solid and gel forms and on solid Ca^{2+} oligo(α-L-guluronate) is reviewed, with particular emphasis on the need to choose appropriate models mimicking well defined Ca-O bond distributions. The difference in environment for the solids provides the first structural evidence that Ca^{2+}-polysaccharide interactions are sufficiently stereospecific as to participate in the control of gelation and other biologically important processes.

Introduction

Over the past decade, EXAFS and XANES studies have provided detailed structural information for many metallo-proteins and other biological systems. However, it is undeniable that the most definitive results have been obtained for metal ions such as Cu^{2+} and Fe^{2+}, Fe^{3+}, which are "active" in other electronic spectroscopies (d-d absorbance, e.p.r., m.c.d., etc.). In these cases, an extensive background of suggestions for active sites exists on which correlations with the EXAFS and XANES can be based (1). Where the probing of "silent" ions is concerned, background information is much sparser and is often involved with metal-ion substitution with "active" metal ions, in which case site -changes may be involved. The classic example is Ca^{2+}, which is often sustituted by lanthanides (2) for studying the sites involved in coordination. With calcium there is the added complication that this metal ion can give a variety of coordination geometries showing a wide scatter of Ca-O bond lengths, so that EXAFS studies have involved mainly average Ca-O distances (3).

However, the importance of Ca^{2+} binding to biopolymers, especially polysaccharides (4), makes it imperative to apply EXAFS and XANES studies to achieve structural information.

It has long been known from X-ray fibre diffraction that polysaccharides have a double helix structure (5). Pectic substances in particular adopt a 3 chain conformation (Fig. 1) and Walkinshaw and Arnott (6) have shown that in the solid state Ca^{2+} pectate maintains a 3_1 symmetry of antiparallel chains. These authors also put forward a structural model based on energy minimization of a geometrical calculation, and concluded that 5 Ca-O (sugar) + 1-4H O molecules may be involved in the junction sites. Pectin consists of a linear 1,4-linked α-D-galacturonic backbone, interrupted by 1,2-linked L-rhamnose and a proportion of branch chains composed mainly of neutral sugars (7). On de-esterification, pectins form gels in the presence of Ca and other divalent cations. In a thorough study involving circular dichroism spectroscopy, Morris et al. (8) concluded that the Ca^{2+} site binding was within dimers of 2_1 helical chain symmetry (shown schematically in Fig. 1). They concluded that a $2_1 \rightarrow 3_1$ polymorphous phase change occurs in Ca^{2+} poly-α-D-galacturonate on drying.

This is of particular interest because Ca^{2+}poly(α-L-guluronate) is also known to adopt an extended helical structure, but with two-fold symmetry axis, i.e. 2_1, in both solid state (5,9) and in solution (n.m.r. studies) (10). (α-L-guluronic acid is a polysaccharide forming the main component of alginic acid, which is extracted from brown algae -Phaeophyta-). There has been considerable interest in the molecular origin of its selectivity to gel formation with Ca^{2+} (11).

Mackie et al. (10) also reported computer modelling studies of host cavity dimension generated by contiguous (1-4) linked α-L-guluronic acid residues. They concluded that a variety of cations can be accommodated whilst maintaining the glycosidic torsion angles almost constant. The geometrical requirements of the cation binding sites favoured O(2), O(3), O(61'), O(62'), and O(5') as possible donor atoms (see Fig.1), the junction zones hence involving six O (sugar) atoms from two neighbouring chains plus 1-3 oxygen atoms from water molecules .

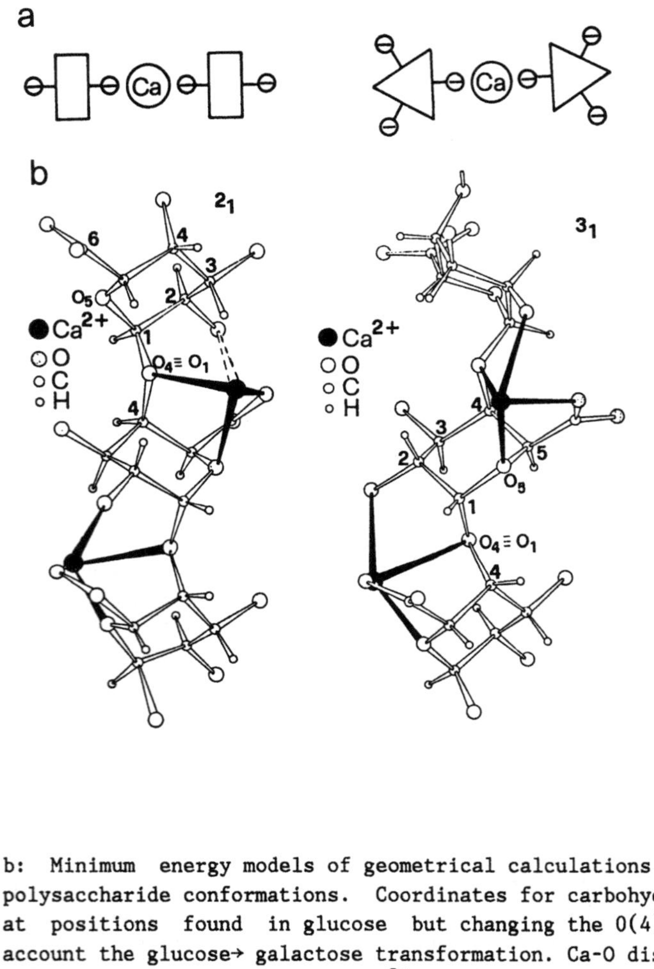

Fig. 1 a: Views along chain axis for a 2_1 and 3_1 conformation of poly-α-D-galacturonate and poly-α-L-guluronate.
b: Minimum energy models of geometrical calculations (see ref. 10) for 3_1 and 2_1 polysaccharide conformations. Coordinates for carbohydrate rings were kept fixed at positions found in glucose but changing the O(4) coordinate to take into account the glucose→ galactose transformation. Ca-O distances <2.6 Å from the Table were kept fixed; excursion of Ca^{2+} within the cavity of oxygen atoms formed was no greater than ca. 0.01 Å before going out of the energy minimum: Note the long Ca-O distance to a bridging glycosidic oxygen in the 3_1 model and the steric hindrance to closer approach of the calcium to O(2) in the 2_1 model.

Poly-α-L-guluronic acid and poly-α-D-galacturonic acids are related stereochemically, being exact mirror images, except for position C-3, as shown in Fig.2.

This structural analogy facilitates the EXAFS and XANES study by allowing comparison with materials having the same backbone. To further simplify the study an oligo (α-L-guluronic acid) of polymerization degree = 20 was investigated ; details are given in ref. 1.

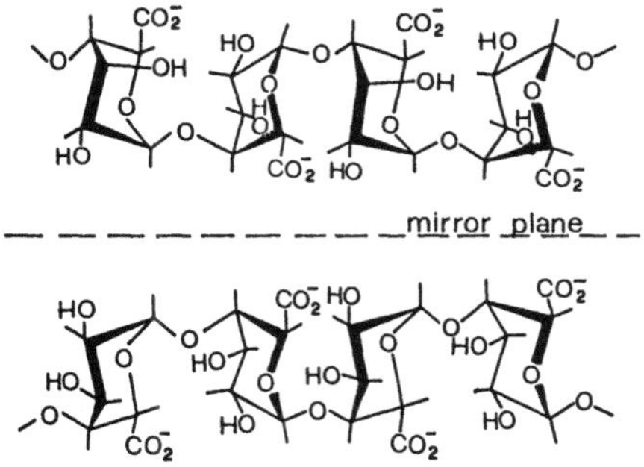

Fig.2 Relationship between poly α -L-guluronate (A), and poly α -D-galacturonate (B)

Model Compounds

Since the major aim was to assess the capability of the plane wave EXAFS equation to successfully partition Ca-O distances, choice of model compounds was found to be crucial. Dheu-Andries and Perez (13) have reviewed the crystallographic data available on calcium carbohydrates and rationalized the partitioning found. Calcium L-ascorbate dihydrate (coordination number 8) was a first choice because it has the narrowest Ca-O bond distribution (14). Calcium 2-keto-D-gluconate trihydrate (coordination number 9) provided a more stringent test because of the presence of a single, long Ca-O bond at 2.73 A (15): Sodium calcium α -D-galacturonate hexahydrate provided a further $Ca-O_9$ model (16) and calcium copper (II) tetraacetate tetrahydrate (17) was an octahedral $Ca-O_6$ model. Calcite was also measured as a finger print model of a Ca-O pair distribution strongly perturbed by second-shell interactions, deriving from multiple scattering due to the C=O bonds (18). The compounds were prepared as in the references. An air-tight O-ringed cell, with kapton windows, has been designed to accommodate variable thickness gels, whilst avoiding sample spillage under vacuum or gas pressure.

Ca^{2+}-Oxygen Environment

The parameters extracted from Ca^{2+}L-ascorbate dihydrate were used to fit the EXAFS of Ca^{2+}2-keto-D-gluconate trihydrate. The extensive fits, detailed in ref. 12, led to the conclusion that a satisfactory fit of this model was possible only for a 6 : 2 : 1 partitioning of the Ca-O distances. Assessments were based on a fit index defined by :

$$\text{Fit Index (F.I.)} = \Sigma\ (\ \chi\ (k)_{expt} - \chi\ (k)_{calcd})^2 k^3\ /\ (P_{expt} - P_{par})$$

where P_{expt} = number of points in the k-range from the Fourier back-transform and P_{par} = number of parameters. Particularly encouraging is the fact that a good fit was possible only if a long Ca-O bond at 2.73 Å , as in the crystal structure, was included.

When applied to the polysaccharides, the results detailed in the Table were obtained. A good fit to the Ca-O shell of solid Ca^{2+}poly(α- D-galacturonate) was possible only by including a very long Ca-O bond at 2.95 Å . However, two fits were equally satisfactory, each giving a 10 coordinate Ca^{2+} but with a different partitioning of the Ca-O bonds (6). The oligoguluronate instead provides a CaO_8 environment -the most common for Ca^{2+} simple carbohydrate compounds (13). The finding that poly-α -D-galacturonic acid and oligo-α-L-guluronic acid bind

TABLE. Structural Parameters Derived from EXAFS Analysis (a)

Material	Ca-O (Å)	N	σ (Å)	F.I.
Ca L- ascorbate·$2H_2O$	2.46	8	0.12	0.09
	2.43	6	0.09	
Ca 2 -keto -D- gluconate·$3H_2O$	2.60	2	0.06	0.08
	2.73	1	0.08	
Ca poly(α -D -galacturonate)	2.40	2	0.06	
(solid state form) (b)	2.55	6	0.10	0.06
	2.81	1	0.02	
	2.95	1	0.07	
	2.39	2	0.05	
	2.54	5	0.09	0.06
	2.78	2	0.10	
	2.95	1	0.07	
Ca oligo(α L- guluronate)	2.33	3	0.26	
(solid state)	2.45	3	0.11	0.06
	2.76	2	0.13	

(a) Fits for first shell Ca-O performed over k range of 9 $Å^{-1}$ for FT intervals of typically 1.30 Å.
(b) Both fits equally satisfactory.

differently to the Ca^{2+} provides the first definitive evidence that Ca^{2+} polysaccharide interactions are indeed specific, rather than being governed by simple columbic binding of the Ca^{2+} to anionic carboxylic groups (19). Furthermore, the Ca-O bond distribution in the poly-α -D-galacturonate is wider (even if the long Ca-O = 2.95 Å is discounted) than that, 2.69-2.79 Å, put forward by Walkinshaw and Arnott from X-ray fibre diffraction (6). The inference is that the sugar rings provide more oxygen atoms than in Walkinshaw and Arnott's model (which contemplates 5 Ca-O (sugar) bonds) hence increasing steric hindrance at the Ca^{2+} site, leading to a wider spread in Ca-O bond lengths. Unfortunately, specific assignment of Ca-O (sugar) and Ca-O (water) from the EXAFS alone is not possible : Ca-O (water) distances in the tabulation of ref. 13 lie in the range 2.30-2.60 Å . In addition, the long Ca-O = 2.95 Å bond lies outside the range found for simple Ca^{2+} carbohydrates (13).

However, further insight into the way that Ca^{2+}coordination directs the cooperative binding in the polysaccharide chains has been obtained by combining the EXAFS results with minimum-energy geometry calculations, performed as described in ref. 10. For the 3_1 conformation, keeping the shorter Ca-O distances (<2.60 Å) as fixed parameters generates a broad distribution for the remaining Ca-O bonds, with a total of 8 Ca-O(sugar) bonds and a long Ca-O(glycosidic) bond at 2.95 Å (not contemplated in the previous model (6)). Again using the EXAFS-derived Ca-O bonds <2.60 Å as constraints for the 2_1 chain conformation instead generates only 6 Ca-O(sugar) bonds with a distribution of the remaining Ca-O distances so different as to give a well-separated peak in the FT, as shown in Figure 3 (and despite the fact that the longest Ca-O distance is now only 2.76 Å). The lowest-energy chain conformations for the 2_1 and 3_1 cases are shown in Figure

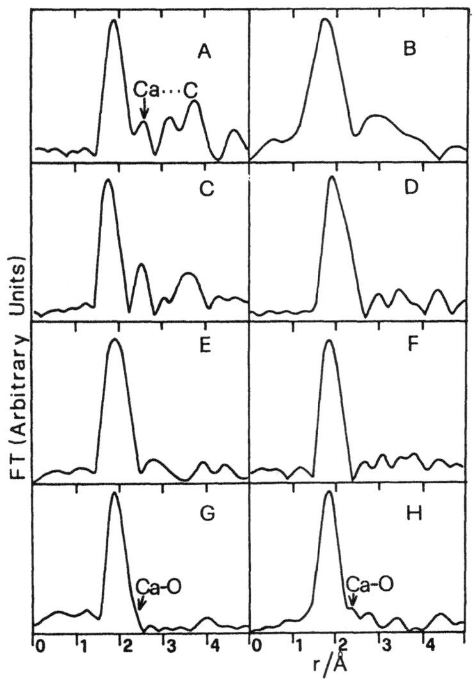

Fig.3 FT.s of A:calcite; B:$CaCu(CH_3CO_2)\cdot 6H_2O$ C:Calcium 2-keto-D-gluconate trihydrate; D:Ca,Na-α-D-galacturonate · $6H_2O$ E:Calcium L-ascorbate dihydrate; F:Ca^{2+} poly(α-D-galacturonate) solid; G:Ca^{2+} oligo (α -L-guluronate) solid; H:Ca^{2+}poly(α -D-galacturonate) gel. (Conditions : k^3 gaussian window).

The geometrical calculations also are in reasonable agreement with further peaks ascribable to Ca...C distances between 3.03 and 3.25 Å in the FT. Particularly satisfying is the fact that, although not quantifiable using the plane- wave approximation, the minimum-energy computer modelling shows that Ca...C distances do not coincide. In the Ca^{2+} 2-keto-D-gluconate trihydrate, very close-lying Ca...C distances (from X-ray diffraction) lead to a prominent peak between 2.5 and 3.1 Å in the FT.s of Fig.3.

Two water molecules presumably complete the coordination sphere of the Ca^{2+} in both solid Ca poly(α -D-galacturonate) and Ca^{2+}oligo(α -L-guluronate). This result is particularly important as regards the EXAFS of Ca^{2+}poly(α -D-galacturonate) gel, which is virtually superposable on that of the oligosaccharide, as illustrated in Fig.4. The FT of the gel clearly indicates that - surprisingly- the Ca^{2+} binding site available is more ordered than that in the solid (12).

Geometry Diagnostics For Ca^{2+} Sites From XANES

The fine structure in the XANES of Ca^{2+} compounds is known to be determined by both first (i.e. Ca-O) and second (i.e. Ca...C) neighbours (20). Furthermore, in several Ca^{2+} -carboxyl containing compounds, the main peak is reported to shift, by ca 1 eV, to higher energy going from 6- to 8-coordination (21). In addition, for such complexes when octahedral 6-coordination is present, in which the - CO group is usually monodentate and the C-O-Ca angle is 120-150°, peak B is more pronounced than when 7- or 8- cordination occurs (21). The higher-energy regions for $CaCu(CH_3CO_2)\ .6H_2O$ and Ca 2-keto-D-gluconate trihydrate - shown in Fig.5, together with XANES spectra of the other model compounds - are in line with these correlations. However when second and third shell effects are negligible, as in solid Ca^{2+} poly(α-D-galacturonate) and Ca^{2+}oligo(α -L-guluronate) only the underlying spectrum of p states is observable (peak A). The symmetrical potential now present around the Ca^{2+} generated by such a wide range of oxygen atoms has

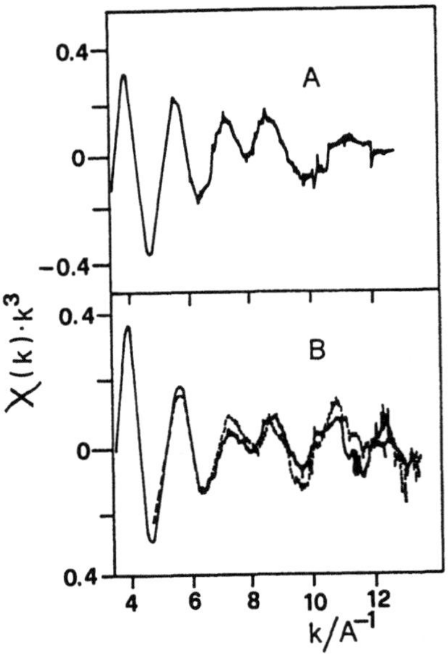

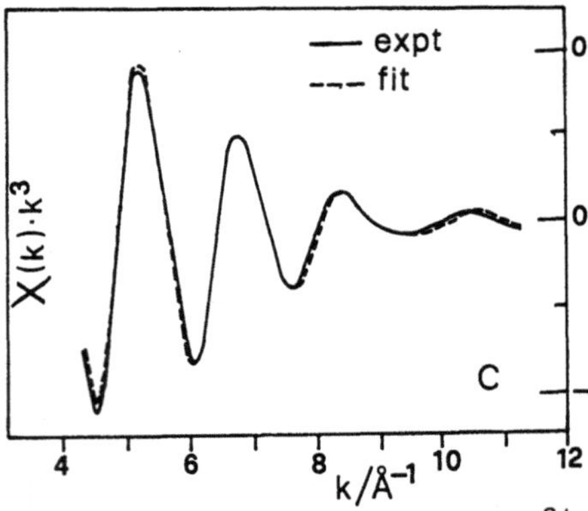

Fig. 4 A: EXAFS spectrum of Ca^{2+} poly(α -D-galacturonate) solid. B: Comparison between experimental EXAFS spectrum of Ca^{2+} oligo(α-L-guluronate) (___) and Ca^{2+} poly(α -D-galacturonate in gel. C: Best fit of Ca-O shell for solid Ca^{2+} poly(α-D-galacturonate) with the parameters of the Table.

little influence on the empty d orbitals. This contrasts with the highly directional character of partially filled (22) -or even entirely filled d-orbitals and is responsible for the featureless first-shell contribution. We conclude that a distinction between higher coordination numbers for Ca^{2+} is not possible without recourse to multiple scattering calculations.

Fig.5 XANES spectra of A:calcite; B:$CaCu(CH_3CO_2)$ $6H_2O$; C:Calcium 2-keto-D-gluconate trihydrate; D:Ca,Na-α-D-galacturonate $6H_2O$; E:$Calcium^{2+}$ L-ascorbate dihydrate; F:Ca^{2+}poly(α -D-galacturonate) solid; G:Ca^{2+}oligo (α -L-guluronate) solid; H:Ca^{2+}poly(α -D-galacturonate) gel.

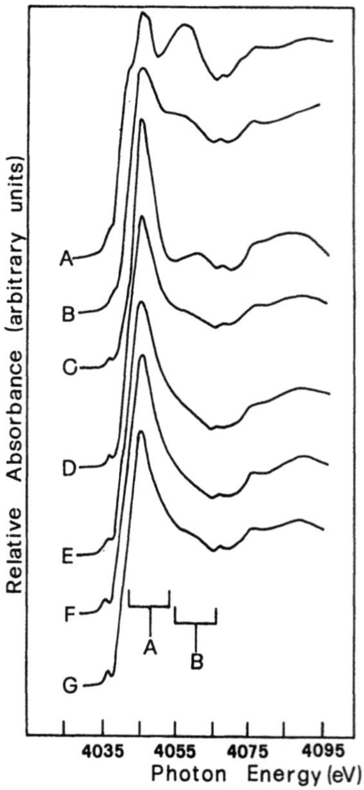

Conclusions

Despite the difficulties inherent in applying the plane-wave EXAFS equation to Ca^{2+} -containing materials, careful choice of model compounds can provide considerable insight into even such complex biopolymers as polysaccharides. Combination of EXAFS results with minimum-energy geometry calculations has demonstrated that Ca^{2+} binding to polysaccharides occurs preferentially at sites possessing sets of hydroxyl and carboxyl groups.

This supports the view that calcium carbohydrate interactions are sufficiently stereospecific as to participate in the control of gelation and other biologically important processes.

References

1) N.J. Blackburn and S.S. Hasnain, in "Biological and Inorganic Copper Chemistry" K.D.Karlin and J. Zubieta, Eds., Adenine Press, Guilderland, N.Y., 1986.
2) see for example C.H. Evans and B.J. Dronven, Biochem. J., 1983,213, 751, and and refs. therein.
3) R.G. Geere, P.H.Gaskell, G.N. Greaves, J. Greengrass and N. Binstead, in "EXAFS and Near Edge Structure", Springer-Verlag, Berlin, 1983, p. 256.
4) G.O. Aspinall, Polysaccharides, Pergamon, Oxford, U.K., 1970.
5) E.D.T. Atkins, I.A. Nieduzinski, W. Mackie, K.D. Parker and E.E. Smolko, Biopolymers, 1973,12, 1879; S. Arnott, A. Fulmer, W.E. Scott, I.C.M. Dea, R. Moorhouse and D.A. Rees, J. Mol. Biol., 1974, 90, 269.
6) M.D. Walkinshaw and S. Arnott, J. Mol.Biol., 1981, 153, 1075.
7) D.A. Rees and P. Wight, J. Chem. Soc. B, 1971, 1366.
8) E.R. Morris, D.A. Powell, M.J. Gidley and D.A. Rees, J. Mol.Biol., 1982, 155, 507.
9) E.R. Morris, D.A. Rees, D.Thom and J. Boyd, Carbohydr. Res., 1978, 66, 145.
10) W.Mackie, S. Perez, R. Rizzo, F. Taravel and M. Vignon, Int. J. Biol. Macromol., 1983, 5, 329.
11) R. Kohn and B. Larsen, Acta Chem. Scand., 1972, 26, 2455; A. Hang and O. Smidsrod, ibid. 1970, 24, 843; G.T.Grant, E.R. Morris, D.A. Rees, P.J.C. Smith and D. Thom, FEBS Lett., 1973, 32, 195.
12) L. Alagna, T. Prosperi, A.A.G. Tomlinson and R.Rizzo, J.Phys. Chem., 1986, in press.
13) M.L. Dheu-Andries and S. Perez, Carbohydr. Res., 1983, 124, 324.
14) R.A. Heam and C.E. Bugg, Acta Cryst., 1974, B30, 2705.
15) M.A. Mazid, R.A. Palmer and A.A. Balchin, Acta Cryst., 1976, B32, 885.
16) S.E.B. Gould, R.O. Gould, D.A.Rees and W.J. Scott, J. Chem. Soc. Perkin Trans. II, 1975, 237.
17) D.A. Langs and C.R.Hare, Chem. Commun., 1967, 890.
18) A. Bianconi, S. Doniach and D. Lublin, Chem. Phys. Lett., 1978, 59, 121.
19) I.G.F. Gilbert and N.A. Myes, Biochem. Biophys. Acta, 1960, 42, 469.
20) P.J. Durham, J.B. Pendry and C.A. Hodges, Solid State Commun., 1981, 38, 159.
21) A. Bianconi, A. Giovanelli, L. Castellani, S. Alemà, P. Fasella, B. Oesch and S. Mobilio, J. Mol. Biol., 1983, 165, 125.
22) L. Alagna, R.W. Strange, P.J. Durham, S.S. Hasnain, Jour. Am. Chem. Soc., 1986.

Application of X-Ray Absorption Spectroscopy to the Investigation of Biological Calcification

D.W.L. Hukins and J.E. Harries

Department of Medical Biophysics, University of Manchester, Manchester M13 9PT, England.

Introduction

X-Ray absorption spectroscopy (XAS) has been widely used to characterise calcified deposits from biological systems [1,2] since the original experiments were performed [3,4]. Here we present a progress report with emphasis on selected aspects of our recent research. After the introduction to biological calcification, which forms this section and indicates why XAS is so useful for characterising the deposits, we summarise the principles of the technique so that its qualitative and, finally, its quantitative interpretation can be discussed in more detail.

In some biological systems calcification is a normal process whereas in others it is pathological. Calcification is essential for the mechanical properties of bones and teeth whose mineral phase can best be described as a poorly crystalline form of hydroxyapatite (abbreviated to HAP), $Ca_5(PO_4)_3OH$ [5]. Embryonic bone may contain regions whose structure resembles that of the mineral brushite, $CaHPO_4.2H_2O$ [6]. Poorly crystalline HAP can form pathological deposits in the soft tissues of the body which are implicated in some rheumatic diseases; a preliminary XAS study of these arthropathic deposits has been published [7]. Similar deposits can form on synthetic materials in contact with body fluids *eg.* on the surface of in-dwelling urinary catheters [8]. Calcium pyrophosphate can also be deposited in the body, especially as the dihydrate, $Ca_2P_2O_7.2H_2O$; XAS studies have been performed on an amorphous form of calcium pyrophosphate in intracellular granules [9,10]. Other calcium-containing salts may also be deposited [11].

Calcified deposits are often poorly crystalline and may even lack any long-range order; furthermore, amorphous model compounds need to be investigated in order to understand biological calcification. XAS is equally sensitive to structure in amorphous, poorly crystalline and highly crystalline solids [12,13]. Thus it has been used to investigate the non-crystalline calcium phosphates in the casein micelles of milk [14,15] and which can be formed during *in vitro* mineralisation of casein and other phosphotproteins [16]. Furthermore, calcium occurs in a wide range of environments [17], so that XAS, is very sensitive to the phase of the deposit.

The two calcified systems considered here, bone and encrusting deposits on urinary catheters, both contain a mineral which resembles poorly crystalline HAP. However, in order to understand these systems, we need also to consider an amorphous calcium phosphate (ACP) which gradually matures into a poorly crystalline form of HAP unless it is stabilised chemically (*eg.* by addition of a magnesium salt) or by drying [18]. Until recently all the XAS spectra from these systems were recorded above the calcium K edge (E_0 = 4.838 keV; λ = 0.307 nm) and provided information on the local environment of their calcium ions [1-4]. Now complementary information on the local environment of the phosphorus atoms (present in the phosphate groups) has been obtained from spectra recorded above the phosphorus K edge (E_0 = 2.144 keV; λ = 0.578 nm) [19].

Principles

The X-ray absorbance of a sample is a function of photon energy, E. Above the energy, E_0, of an absorption edge, this absorbance consists of an oscillatory function, $\chi(E)$, superimposed on a smoothly varying function. The XAS consists of the oscillatory term $\chi(E)$, which may also be presented as a function of the modulus of a scattering vector

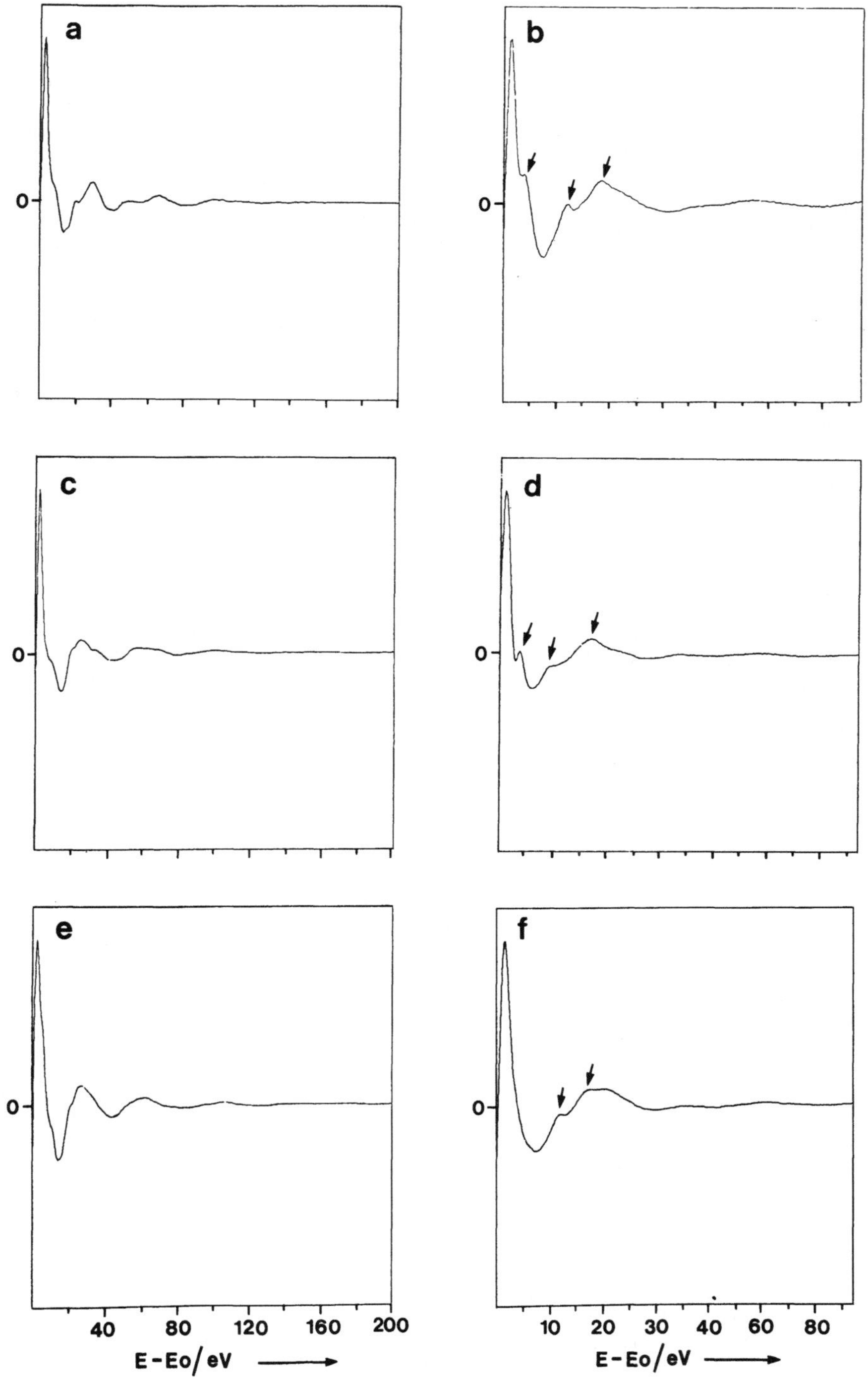

Fig. 1 XAS spectra recorded above the calcium and phosphorus K edges from crystalline synthetic calcium phosphates: (a) hydroxyapatite (HAP) (Ca edge), (b) HAP (P edge), (c) brushite (Ca edge), (d) brushite (P edge), (e) monetite (Ca edge), and (f) monetite (P edge). Spectra are unweighted and plotted against photon energy above the absorption edge.

$$k = [(2m/\hbar^2)(E-E_0)]^{\frac{1}{2}}$$

where m is the mass of an electron [12,13].

Synchrotron radiation is ideally suited to the XAS technique. Its intensity is much greater than from a conventional X-ray source and does not fluctuate rapidly with photon energy. Different techniques are required to record spectra above the calcium and phosphorus K edges because the low energy photons required to record an XAS above the phosphorus edge are ready absorbed by air. We have used apparatus available at the Synchrotron Radiation Source of the SERC Daresbury Laboratory [20,21] as described elsewhere [1,19].

When an X-ray photon whose energy is slightly greater than E_0 is absorbed, a photoelectron is emitted; the energy of this photoelectron is modified by scattering from any atom which it encounters in its path (mean free path < 1 nm). Thus the features of the XAS provide information on the immediate surroundings of the emitting atom; the type of atom is selected by recording at the energy of the appropriate absorption edge.

Conventionally a spectrum may be divided into the X-ray absorption near edge spectrum (XANES) and the extended X-ray absorption fine structure (EXAFS) region. There is a well developed theory of EXAFS which allows a theoretical spectrum to be calculated from the radial distribution of atoms around the emitter, and their Debye-Waller factors which attempt to account for thermal and static disorder [22,23]. Often only a single scattering event has to be considered for each photoelectron [22,24] but sometimes multiple scattering events may have to be taken into account [25]. Multiple scattering greatly complicates the analysis of XANES spectra [26]. Phase shifts which occur on emission and scattering, can be calculated from the structure [27] and need to be taken into account in the theoretical calculation of $\chi(k)$ [22,23]. The range of coordination distances which needs to be considered in the theoretical calculation of $\chi(k)$ can be estimated by Fourier transformation of the experimental EXAFS spectrum, taking due account of phase shifts [2], since the modulus of the transform then yields the radial distribution of scattering matter around the emitting atom [28].

Qualitative interpretation

Fig. 1 shows that the absorption spectra recorded above the calcium and phosphorus K edges of HAP, brushite and monetite ($CaHPO_4$) are clearly different. Consequently the technique is sufficiently sensitive to differences in local structure in these compounds that it can distinguish between them. This example shows that it may be used to identify unknown calcium phosphate structures by comparison of their XAS spectra with those from known compounds. Until recently only spectra recorded above the calcium edge had been used for this purpose [1-4,7-10,14-16] but now spectra recorded above the phosphorus edge, which are sensitive to the local environment of the phosphorus atoms in the phosphate groups, are showing promise for providing complementary information [19]. These spectra, shown in Figs. 1(b), (d) and (f), have three distinct peaks within 30 eV above the phosphorus K edge - marked with arrows in the appropriate figures. The first of these peaks is absent for monetite, can be seen in the spectrum of HAP and is clearly distinct for brushite. In contrast,the second peak is sharpest for HAP where it occurs at a higher energy (12 eV above the edge) than the brushite peak (at 10 eV). The third peak has a quite different appearance in each of the spectra. Overall the spectra are distinctly different, reflecting the different radial distribution of atoms surrounding phosphorus in the three crystal structures.

Fig. 2 shows how the EXAFS region, above the calcium K edge, is sensitive to the extent of the order in related calcium phosphate phases. The moduli of the Fourier transforms of these spectra demonstrate how the order in the structure becomes apparent at increasing distances, from the emitting atom, as the crystallinity increases. Thus the spectrum of ACP shows less fine structure than the spectrum of a poorly crystalline HAP, into which it converts, as a result of ACP showing no evidence for regular structure beyond 0.3 nm from the emitting atom, which acts as an origin for the radial distribution function. Fully crystalline HAP yields a

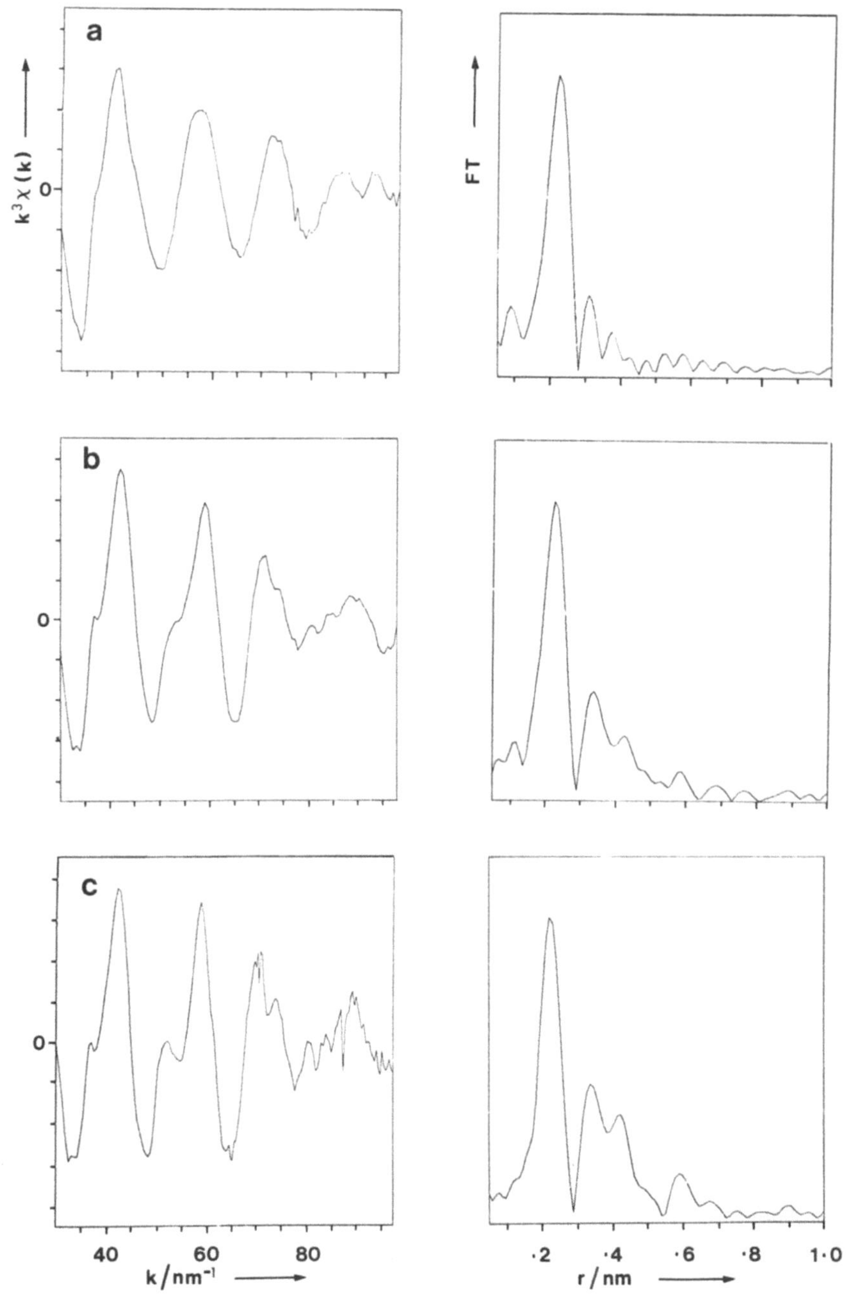

Fig. 2 EXAFS spectra, recorded above the calcium K edge, with the moduli of their Fourier transforms, for (a) amorphous calcium phosphate (ACP), (b) a poorly crystalline form of hydroxyapatite (HAP) prepared by maturation of ACP, and (c) a highly crystalline form of HAP. Spectra are presented as $k^3\chi(k)$ plotted against k.

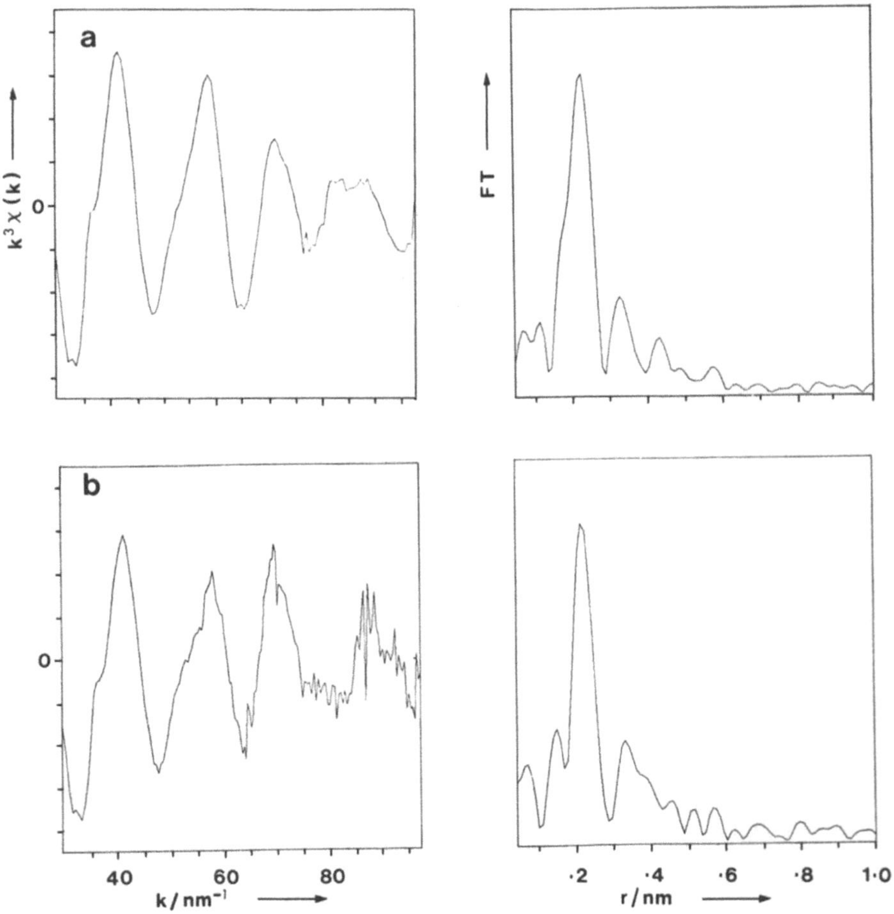

Fig. 3 EXAFS spectra, recorded above the calcium K edge, with the moduli of their Fourier transforms, for (a) bone mineral and (b) an encrusting deposit from a urinary catheter. Spectra are presented as $k^3\chi(k)$, for comparison with Fig. 2.

spectrum with well defined structure; the modulus of the Fourier transform, in Fig. 2(c), indicates that scattering of photoelectrons by atoms up to 0.6 nm contributes to the feature of this spectrum.

Information on the calcium and phosphorus environments in biological calcium phosphates may be obtained by comparison of their spectra with those from the model compounds. For example, the EXAFS spectrum, recorded above the calcium K edge, from bone in Fig. 3(a) more nearly resembles that of the poorly crystalline HAP, in Fig. 2(b), than those from the other model compounds in Fig.2. Thus the calcium environment in bone mineral is similar to that in poorly crystalline HAP but is different from that in highly crystalline HAP. Similarly, Fig. 4 shows that the phosphorus environment in bone more nearly resembles that in the poorly crystalline HAP than in any of the other model compounds. thus the XAS spectra recorded above the calcium and phosphorus K edges are complementary in the information they can provide about bone mineral.

EXAFS spectra, recorded above the calcium K edge, have been used to characterise the calcium phosphate deposited in urinary catheters because it is mixed with highly crystalline struvite ($NH_4MgPO_4.6H_2O$). Since the calcium phosphate appears to be amorphous, it cannot be detected by X-ray powder diffraction [8]; XAS spectra recorded above the phosphorus edge would also be difficult to interpret because of the presence of phosphate groups in struvite. The EXAFS spectrum of

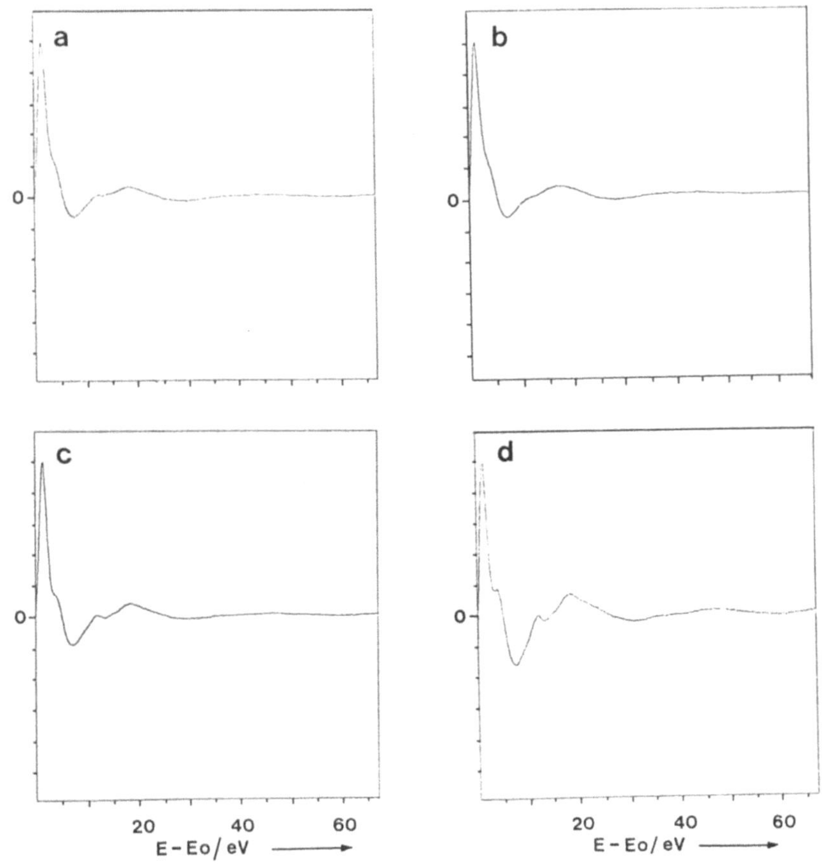

Fig. 4 XAS spectra recorded above the phosphorus K edge from (a) bone mineral, (b) amorphous calcium phosphate (ACP), (c) a poorly crystalline form of hydroxyapatite (HAP) prepared by maturation of ACP, and (d) a highly crystalline form of HAP (identical to Fig. 1(a) but repeated to aid comparison). Spectra are unweighted and plotted against photon energy above the absorption edge.

Fig. 3(a), together with the modulus of its Fourier transform, clearly shows that the calcium environment in this deposit is intermediate between that in ACP and poorly crystalline HAP - by comparison with spectra in Figs. 2(a) and (b). However, in order to characterise the biological calcium phosphates more precisely it sometimes proves necessary to analyse their XAS as described in the next section.

Quantitative interpretation

Fig. 5 compares the theoretical EXAFS spectrum, calculated from the coordinates of the known crystal structure of HAP, with that obtained experimentally, above the calcium K edge, from a highly crystalline sample; a detailed account of the calculation is provided elsewhere [29] and a summary of the main conclusions has been included in a previous publication [2]. Atoms up to 0.6 nm from the emitting calcium had to be included in this calculation which used the exact 'curved wave' theory appropriate for a powdered sample [23] and *ab initio* phase shifts - also calculated from the crystal structure [27]. Because the crystal structure of HAP is so complicated [30], atoms of the same kind with only slightly different radii had to be grouped into the same shells; the radii and Debye-Waller factors of the shells were then systematically varied to obtain the best agreement between the theoretical and experimental spectra [29]. However, atomic radii had never to be changed by more

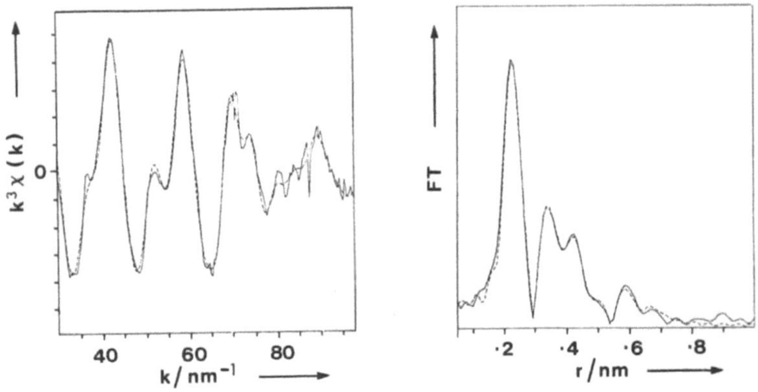

Fig. 5 Experimental (continuous line) and theoretical (dashed line) EXAFS spectra above the calcium K edge, with the moduli of their Fourier transforms, from highly crystalline hydroxyapatite. Spectra are presented as $k^3\chi(k)$ plotted against k; the continuous curve is identical to Fig. 2(c).

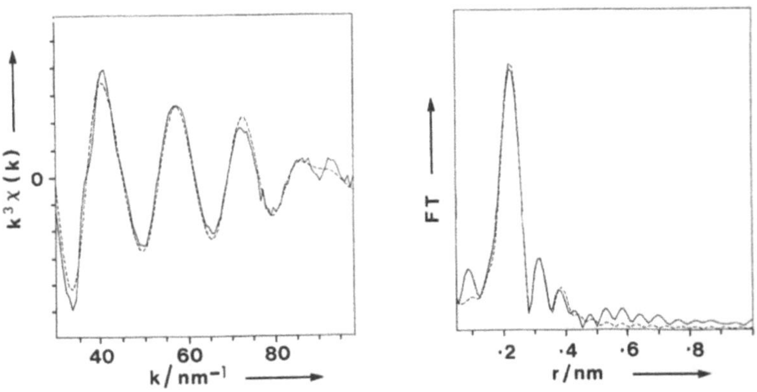

Fig. 6 Experimental (continuous line) and theoretical (dashed line) EXAFS spectra above the calcium K edge, with the moduli of their Fourier transforms, from amorphous calcium phosphate. Spectra are presented as $k^3\chi(k)$ plotted against k; the continuous curve is identical to Fig. 2(a).

than 0.01 nm (usually very much less) from their values in the accepted crystal structure. Some shells of atoms had to be omitted from the calculation if Debye-Waller factors were to have reasonable values ($< 6 \times 10^{-4}$ nm^2). Omission of these shells could be justified because photoelectrons scattered by these shells would interface destructively with photoelectrons scattered by other shells; furthermore, some of these interference effects arose because multiple scattering of the photoelectrons was not always negligible [29].

Now that a model has been developed which explains the EXAFS spectrum of HAP, it can be modified to explain the spectrum of related compounds. For example, Fig. 6 compares the experimental spectrum of ACP with a theoretical spectrum based on a modification of the HAP model. The modulus of the Fourier transform indicates that only shells up to 0.3 nm from the emitting calcium need be considered in this calculation. Thus the four inner shells of the HAP model were used; their radii and Debye-Waller factors were varied as little as possible to achieve the best fit between theoretical and experimental ACP spectra, with the result shown in Fig. 6. Indeed the final radii of the oxygen and phosphorus shells differed by no more than 0.007 nm from their values in the HAP model, although the

Debye-Waller factors increased somewhat (by about 2 x 10^{-4} nm^2), presumably in an attempt to describe the greater static disorder in ACP. It is anticipated that the same approach can be applied to analysis of EXAFS spectra, recorded above the calcium K edge, from bone, catheter deposits and the various other biological calcium phosphates which we are investigating.

Acknowledgements

We are especially grateful to Dr S.S. Hasnain for collaboration in the research described here and to Dr C. Holt who has collaborated in several aspects of it. We also thank Pat Mellor and Karen Davies for producing the typescript and illustrations. Facilities were provided by the director and staff of SERC Daresbury Laboratory; financial support was provided by SERC.

References

1. S.S. Hasnain, Springer Proc. Phys. 2, 145-150 (1984).
2. D.W.L. Hukins, J.E. Harries and S.S. Hasnain, Biochem. Soc. Trans. 14, 545-549 (1986).
3. R.M. Miller, D.W.L. Hukins, S.S. Hasnain and P. Lagarde, Biochem. Biophys. Res. Commun. 99, 102-106 (1981).
4. E.D. Eanes, L. Powers and J.L. Costa, Cell Calcium, 2, 251-261 (1981).
5. A.L. Boskey, Clin. Orthop. 157, 225-257 (1981).
6. L.C. Bonar, M.D. Grynpas and M.J. Glimcher, J. Ultrastruct. Res. 86, 93-99 (1984).
7. J.E. Harries, J.S. Shah and S.S. Hasnain, Springer Proc. Phys. 2, 142-144 (1984).
8. A.J. Cox, J.E. Harries, D.W.L. Hukins, A.P. Kennedy and T.M. Sutton, Brit. J. Urol., in press.
9. G.N. Greaves, K. Simpkiss, M. Taylor and N. Binsted, Biochem. J. 221, 855-868 (1984).
10. M. Taylor, K. Simpkiss and G.N. Greaves, Biochem. Soc. Trans. 14, 549-552 (1986).
11. R.I. Gibson, Amer. Mineral. 59, 1177-1182 (1974).
12. P. Einsenberger and B.M. Kincaid, Science, 200, 1441-1447 (1978).
13. S.J. Gurman, J. Mater. Sci. 17 1541-1570 (1982).
14. C. Holt, S.S. Hasnain and D.W.L. Hukins, Biochim. Biophys. Acta, 719, 299-303 (1982).
15. J.C. Irlam, C. Holt, S.S. Hasnain and D.W.L. Hukins, J. Dairy Res. 52, 267-273 (1985).
16. J.C. Irlam, D.W.L. Hukins, C. Holt and S.S. Hasnain, Springer Proc. Phys. 2, 139-1141 (1984).
17. R.J.P. Williams, Symp. Soc. Exp. Biol. 30, 1-17 (1976).
18. G.H. Nancollas, in 'Biological Mineralisation and Demineralisation', ed. G.H. Nancollas, pp. 79-99, Springer Verlag, Heidelberg, (1982).
19. J.E. Harries, W.J. Myring, C. Holt, D.W.L. Hukins, A.A. MacDowell and S.S. Hasnain, Biochem. Biophys. Res. Commun., submitted.
20. G.P. Diakun, G.N. Greaves, S.S. Hasnain and P.D. Quinn, SERC Daresbury Lab. Report DL/SCI/P507E (1986).
21. A.A. MacDowell, D. Norman and J.B. West, SERC Daresbury Lab. Report DL/SCI/P507E (1986)
22. P.A. Lee and J.B. Pendry, Phys. Rev. B11, 2795-2811 (1975).
23. S.J. Gurman, N. Binsted and I. Ross, J. Phys. C, 17, 143-151 (1984).
24. P.A. Lee, P.H. Citrin, P. Eisenberger and B.M. Kincaid, Rev. Mod. Phys. 53, 796-806 (1981).
25. S.J. Gurman, N. Binsted and J. Ross, J. Phys. C, 19, 1845-1861 (1986).
26. P.J. Durham, J.B. Pendry and C.H. Hodges, Comput. Phys. Commun. 25, 193-205 (1982).
27. J.B. Pendry, 'Low Energy Electron Diffraction', Academic Press, New York, (1974).
28. D.E. Sayers, E.A. Stern and F.W. Lytle, Phys. Rev. Lett. 27, 1205-1207 (1971).
29. J.E. Harries, D.W.L. Hukins and S.S. Hasnain, J. de Phys. C8-1181 (1986).
30. M.J. Kay, R.A. Young and A.S. Posner, Nature, 204, 1050-1052 (1964).

Ca Site Structure in Bungarotoxin: Possible Generalization to Ca Low Affinity and High Specificity Sites in Ca Binding Proteins

A. Giovannelli

Dipartimento di Fisica ,Universita` degli Studi di Roma "La Sapienza", 00185 Rome, Italy

INTRODUCTION

The problem of estimating the structural factors which control the affinity and specificity of Ca uptake and release in Ca binding proteins is central to understanding the molecular machinery of many Ca dependent processes occurring in excitable and somatic cells[1].

Evidence of conformational changes upon Ca binding is reported from many different technics, that is UV difference spectroscopy, ^{1}H-NMR, circular dichroism in the UV range, aromatic aminoacid fluorescence, and electrophoresis; in most cases the ion concentration necessary for the conformational transition is in agreement with that required for the biological activity[1,2]. However the limit of all these spectroscopies, except for NMR, is that they are unable to obtain direct information on the local structure of the Ca binding site.

High resolution crystal structure has been determined only for a few Ca binding proteins and, on the basis of the avalaible structural and sequence data, the characteristic EF hand found in carp parvalbumin seems to be shared, with subtle modifications, by most of Ca modulated proteins[3,4,5,6].

In spite of these concentrated efforts, the mechanism of binding selectivity and affinity has still to be clarified in terms of structural interactions capable of producing particular binding properties. This means that although sequence studies show a high conservation of the Ca binding loop and small modifications can produce substantial changes in the binding behaviour, a general correlation between these two factors has still not been found[1].

In this work EXAFS and XANES[7] spectroscopy have been used to investigate the local structural features of the low and high affinity binding sites comparing the spectra obtained on bungarotoxin with those of four Ca modulated proteins: calmodulin, troponin C, carp parvalbumin and S100 in the dimeric form, whose binding properties are reported in table 1.

TABLE 1 (Kd values. H high affinity, L low affinity)

	I	II	III	IV	Reference
CALMODULIN		≈ 10E-5÷10E-6		H: K_d<10E-6	8
PAVALBUMIN	10E-6H	10E-6H			1
S-100 PROTEIN	10E-6H				1
TROPONIN C	10E-4H	10E-4H	10E-7L	10E-7L	9
BUNGAROTOXIN	10E-4H				10

The CaBP bungarotoxin[10], the active peptide present in the venom of the snake Bungarus Careuleus, is a phospholypase A_2 like the protein first characterized as an high affinity acetilcholine receptor binding protein. It comprises a two chain dimer, α and β and its active form requires the presence of one Ca^{2+} bound on the β chain in a low affinity and high specificity site. Besides the great interest about the molecular action in the excitable ion channel biology, it offers an almost unique example of an isolated high specificity low affinity Ca^{2+} binding site suitable for X ray absorption analysis.

The analysis of the spectra obtained reveals a higher mean Ca-O (first shell) distance as the affinity value of the saturated binding site becomes lower, as found in bungarotoxin.This suggest that the mean Ca-O distance can be a parameter of control of the affinity behaviour of a Ca specific site. Further, differences in the geometry extending at least up to the third shell have been observed on the basis of XANES features, and this last effect has been found less in more closely evolutionary related proteins. Finally, for the Ca binding proteins having four Ca sites the spectra suggest a cooperativity in Ca binding limited to the couple of Ca bound at the same affinity values, that is the high and the low affinity sites.

EXPERIMENTAL

Bungarotoxin has been purified from the whole snake venom as described in literature[10]. The purification procedure of TnC, CaM, CPa and S100 has been carried out as reported elsewhere[9], except for S100 and CaM in which an affinity procedure has been used. The final step of each protein purification has been a desalting procedure by gel filtration on a G25 FINE (Pharmacia) column eqilibrated with calcium free water (bidistilled water passed through a CHELEX 100 resin). Sample quality has been controlled by SDS gel electrophoresis[11] and UV spectra. Calcium content in each protein has been verified by atomic absorption using protein solutions of 2 mg/ml in calcium free water, which has been used as reference. In spite of the desalting procedure, atomic absorption data on calmodulin and troponin C revealed the presence of one Ca bound per protein molecule. The samples for X ray spectroscopy have been prepared from pure apoprotein by adding $CaCl_2$ to reach the desired stoichiometric ratio on the basis provided by the atomic absorption determination of Ca content of each purified protein, that is 1:1 for BTX and S100 dimer; 2:1 for CPa; 1:1, 2:1, 3:1, for CaM and TnC. Precisely, Ca has been added in little defect to be sure to keep unsaturated the lower affinity sites. This way in these last two samples the possibility to compare the spectra from the putative high affinity sites (III and IV) with those from the all saturated configuration is achieved. Titration of Ca binding has been verified by circular dichroism and fluorescence spectra. All the experiment has been performed using lyophilized samples.

The spectra have been measured at the synchrotron radiation facility PULS with the storage ring ADONE working between 60 and 20 mA and at 1.5 GeV. A Si(220) has been used as monochromator, providing the resolution of $\Delta E/E=10^{-4}$ and a high signal to noise ratio. The XANES spectra were registred between 4000 and 4150 eV by steps of .3 eV, while the EXAFS between 3900 and 4600 eV with steps of 1. eV. EXAFS spectra have not been measured on BTX.

Data analysis has been carried out on a VAX 11/760 computer by plotting the relative absorption α/α_0 where α is the measured absorption coefficient and α_0 is the value of the absorption jump edge from the pre edge value to the high energy Ca atomic X ray absorption obtained by fitting the EXAFS oscillations. This method allows, by the normalization of the spectra of different proteins to the same physical parameter α_0 , to get information from the variation of the intensities of XANES peaks.

RESULTS

The XANES spectra of S100, CaM, and whiting[1] CPa (having only one Ca high affinity site) at a 1:1 molar ratio Ca/protein are compared in fig. 1.

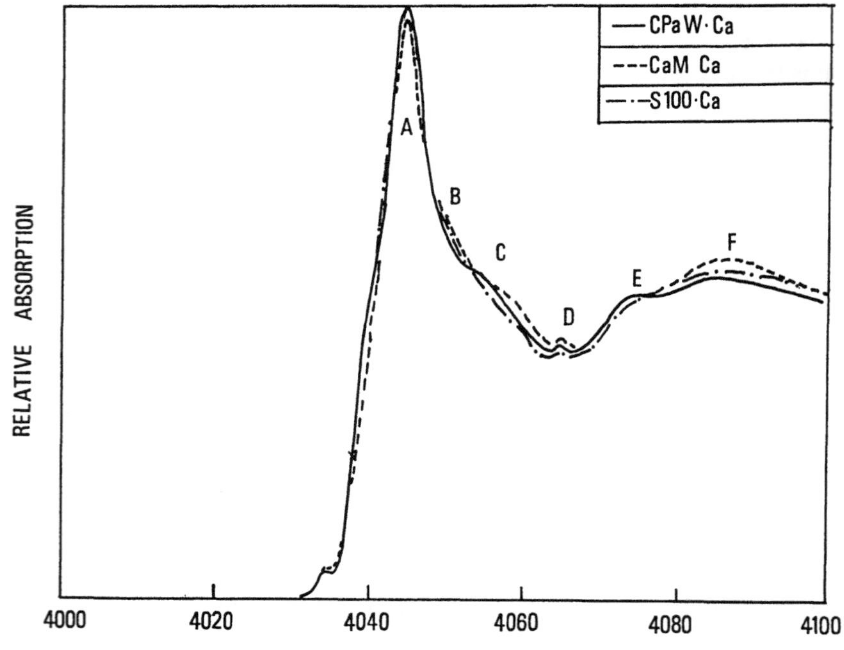

Fig. 1 Comparison between the normalized spectra of CaM, CPa whiting and S100; binding stoichiometry one Ca ion per protein

In all these samples stoichiometry suggests that Ca saturates the high affinity and specificity sites only. Similar results are found in the comparison between $CPaCa_2$ and $CaMCa_2$ (fig. 2).

The spectra show the same value of the the energy separation between the edge (peak E_1) and the first multiple scattering resonance (peak A) ($\Delta E=E_A-E_1$) which can be interpreted as a similar mean Ca-O distance according to the multiple scattering derived expression $\Delta ER^2=\cos(t)$[12].The different energy position and shape of the peaks from B to F can be assigned to small differences in the site geometry up to the third shell detected by the ejected photoelectron in the multiple scattering regime interacting with the carboxyl and carbonyl groups of both the residues and the protein chain.

The effect of the saturation of the remaining two low affinity sites on the XANES has been measured on $CaMCa_4$ (fig. 3), where a greater ΔE value has been found respect to that obtained in the case of the partially saturated CaMCa (prepared from the calmodulin solution without adding Ca) and $CaMCa_2$ in which the high affinity sites were occupied only.

The strength of these results is however limited by the impossibility to observe the spectrum of an isolated low affinity site. The data obtained on Ca BTX (fig. 4 and 5) represent a case in which this limitation vanishes, having BTX an only Ca low affinity site.

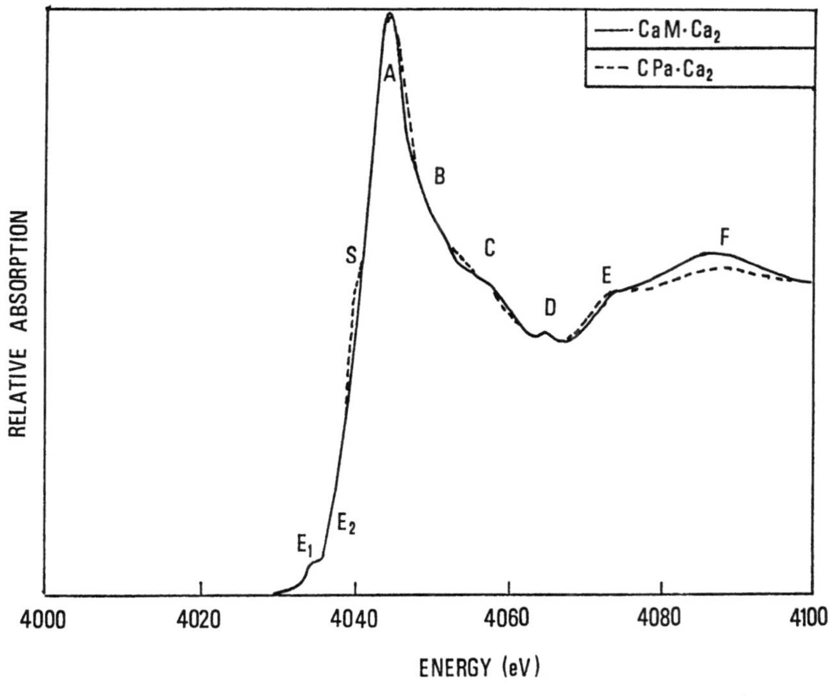

Fig. 2 The high similarity between the XANES spectra of $CPaCa_2$ and $CaMCa_2$ (high affinity) is evidence of common features in Ca geometry

The comparison of the spectra of CaMCa and $CaMCa_4$ with that obtained on BTX-Ca shows that the value of ΔE is closer to the values found in calmodulin only in the case in which the low affinity sites are occupied by Ca. In this case the spectroscopic effect of the low affinity sites "disguises" that produced by the high affinity ones in calmodulin. The strong departure of the XANES peaks (mainly C and F) in BTX from the features present in CaM can be expected, since the two proteins seem to be not closely related from the evolutionary point of view. In spite of these geometrical differences, the value of ΔE found in BTX is very close to that found in all the investigated calcium binding proteins in which the low affinity and high specificity sites are saturated, indicating a shorter mean Ca-O distance for this kind of sites.

A singular result has instead been obtained in TnC (data not shown), where a lower ΔE value occurs both with the high (low specificity, since they bind Mg too) and the low affinity sites saturated, indicating greater first shell distance. These results are confirmed by EXAFS analysis (not reported), which directly obtain the Ca-O mean distance from the pair distribution function.

Taking as a reference the mean distance found in CaM, the following values have been found respectively in TnC, CPa, and S100 sites with uncertainty of $\pm$ 0.02: CaM, high affinity sites (2.41), CPa (2.40), S100 (2.38).

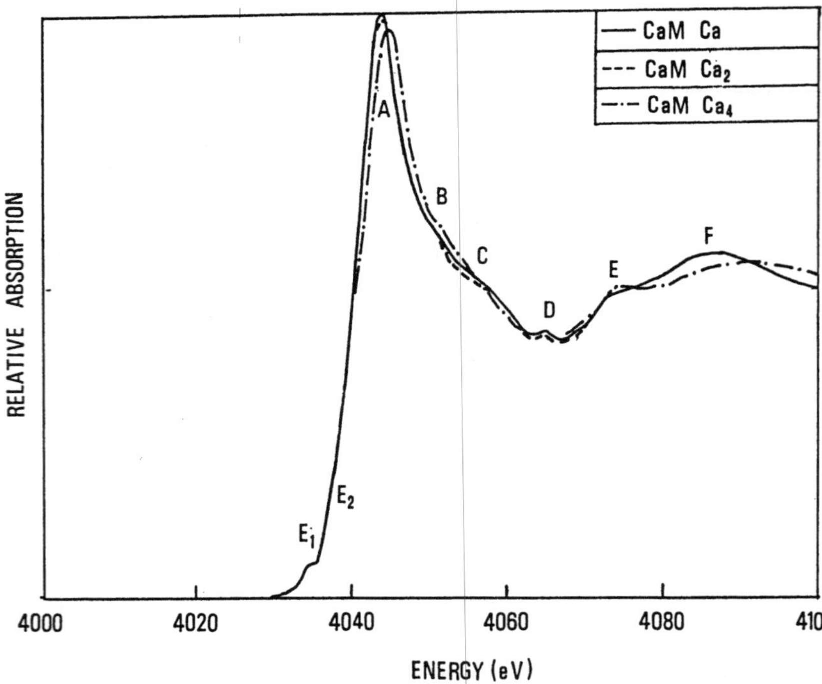

Fig.3 Spectra of Calmodulin with 1,2 and 4 Ca_2+ bonded. The saturation of the low affinity sites produces subtle effects on the XANES peaks stronger in the high energy portion of the spectra

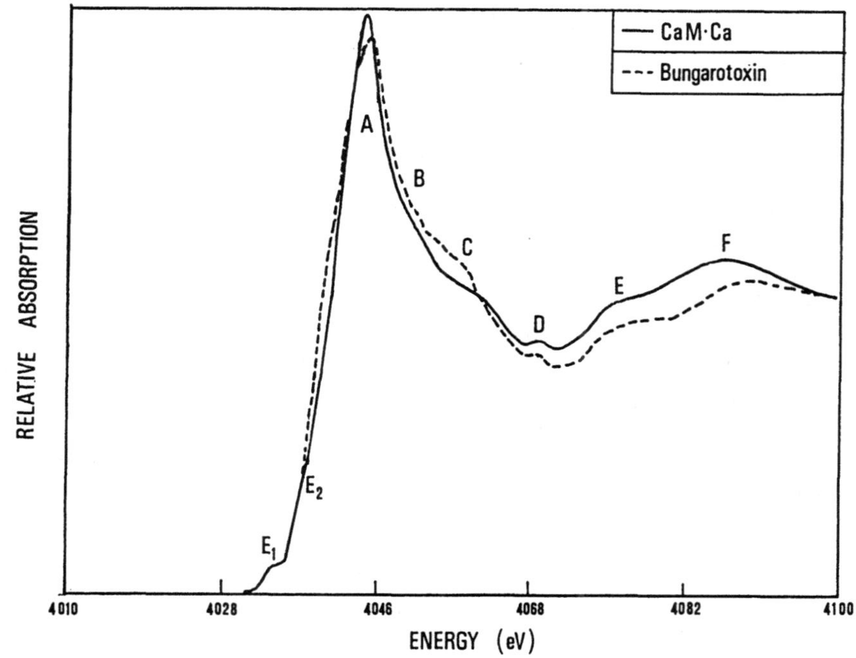

Fig. 4 The spectrum of Ca-Bungarotoxin is plotted normalized to that of CaMCa. It is evident the shift to higher energy of ΔE present in BTX.

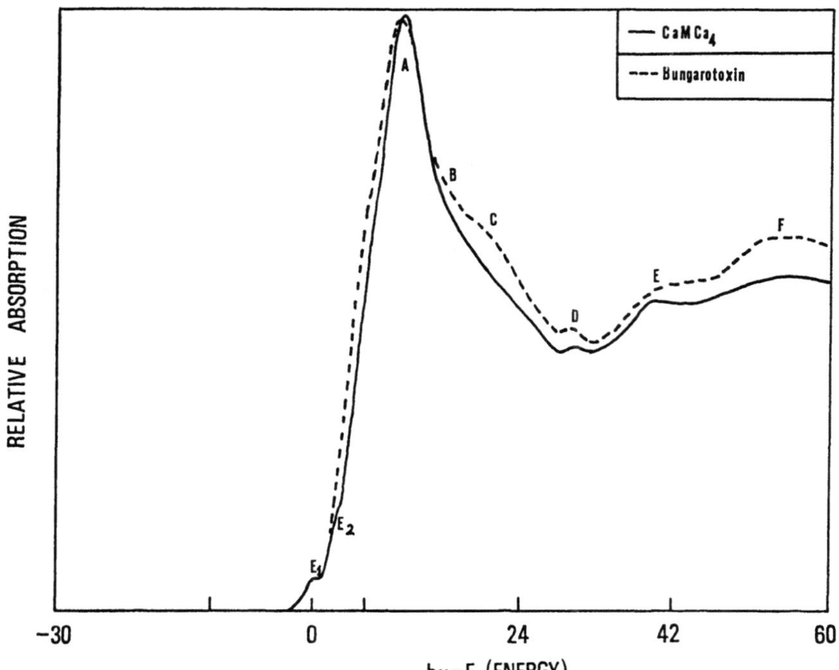

Fig. 5 The very close ΔE between BTX and $CaMCa_4$ reflects a similar mean Ca-O distance despite differences in the geomety of the sites.

DISCUSSION

The observation that the first shell (Ca-O) mean distance has been found to decrease in all samples in which the affinity is lower suggests a possible role in its control exterted by the microscopic environment. Confirmation of this can been found in BTX, which is believed to be relatively distant from EF hand proteins from an evolutionary point of view, that is, it has many substitutions in its primary structure, but these are compatible with certain chemical properties able to make the Ca binding a low affinity high specific one.

The spectra obtained by the specific Ca sites in which the affinity presents only small differences (CPa, S100 and CaM) show only small effects in the XANES peaks interpretable as produced by subtle geometrical distortions up to the third shell which affect the chemical bonding to only a minor degree.These data suggest that a particular microscopic binding site structure is necessary to achieve both the low and the high affinity of the binding site, even of thus cannot resolve the relative contributions to the affinity and specificity from the whole protein and from the local environment. Hence this means that, for example, the high affinity sites in CaM, reported to be III and IV, have structure different from the low affinity ones.

The results previously found in TnC, in which Ca is bound with an higher affinity in the sites III and IV, give a longer (R=2.43 Å) mean Ca-O distance. The comparison of the Ca binding properties of this protein with that of the others show that in TnC the high affinity sites have a lower specificity for Ca in spite of the high sequence homology with CaM. This suggests that the substitutions occurring in TnC have such an effect on the local environment of the site as to change the chemical properties of the chemical bond with Ca, and partially to affect the structure of the whole protein.

In conclusion this study provides evidence that the local environment of a Ca binding site has a fundamental part in controlling the Ca binding proprieties in Ca, that is the affinity and specificity. One of the key parameters in this control can be identified with the mean distance between the Ca ion and the oxygen atoms in the first coordination shell,which has been observed to decrease as the affinity decreases (TnC-CaM-CPa-S100-BTX, ordered from higher lower affinity), also if other factors, due to both the local geometry produced by the aminoacids constituting the site and to the whole protein, certainly have a role in establishing its chemical behaviour.

A limit of the method is the impossibility to determine what are the aminoacid residues that contribute fundamentally in providing such binding proprieties, due to the difficulty in performing theoretical calculation of the spectra able to give information on the contribution of a specific aminoacid to the binding.

A possible answer to this question can be found combining the X ray spectroscopy with gene manipulation techniques to evidence the contribution of the sequence to the properties of each Ca binding site. The introduction of specific point mutations can then, in principle, be correlated to the contribution of a single residue to the binding and to the structure.

REFERENCES

1) Kretsinger, R.H., (1980) *CRC Crit. Rev. Biochem.*, **8**, 119

2) Forsen S. , Anderson A. , Drakenberg T. , Teleman O. , Thulin E. and Vogel H.J. , (1983), ^{25}Mg , ^{43}Ca and ^{113}Cd NMR Studies of Regulatory Calcium Binding Proteins, in Calcium Binding Proteins (De Benard et al. Eds), Elsevier. North Holland, p. 121

3) Moews P.C. and Kretsinger R.H., (1975), *J. Mol. Biol.* **91,** 201

4) Szebenyi D.M.E., Obendorf S.K. and Moffat K. (1981), *Nature*, **294**, 327

5) Sudhakar B.Y., Sack J.S., Greenhough T.J., Bugg C.E., Means A.R., Cook W.J., 1985 , *Nature,* **315** , 37

6) Herzberg O. and James M.N.G. , (1985), *Nature,* **313,** *653*

7) Bianconi,A., XANES spectroscopy, (1985), *in "X Ray Absorption: Princip Application,Techniques of EXAFS, SEXAFS and XANES" edited by R.Pr and D.Koningsberger, J.Wiley and Sons, New York p. 15*

8) Kilhoffer , M.C. , Haiech , J. , Demaille , J.G. , (1983), *Mol. and Cell. Biochem* **51** , 33

9) Bianconi,A., Giovannelli,A., Alema',S., Castellani,L., Fasella,P., Mobilio,S., and Oesh,B., (1983) , *J. Mol. Biol* **165**, 125

10) Abe,T., Alema',S., and Miledi,R., (1980), *Eur. J. Biochem.,* **80**, 1

11) Laemmli U.K. , (1970) , *Nature* **227** , 680

12) Natoli C.R., (1984), EXAFS and Near Edge Structure III, Vol.2 of Springer Proceedings of Physics, ed K.O. Hodgson, B.Hedman, J.E. Penner - Hahn (Springer, Berlin, 1984) pag. 38.

X-RAY SCATTERING

Resonant X-Ray Scattering of Biological Systems

H.B.Stuhrmann[+], *K.S.Bartels*[+], *C.Boulin*[*], *F.Dauvergne*[*], *A.Gabriel*[*], *G.Goerigk*[+], *B.Munk*[+]

[+]University of Mainz and DESY HASYLAB, 2000 Hamburg 52
[*]European Molecular Biology Laboratory, Heidelberg and its Outstations

INTRODUCTION

Nearly all synchrotron radiation laboratories at high energy electron storage rings put enormous effort into the developement of resonant (anomalous) X-ray scattering techniques. So does the Hamburger Synchrotronstrahlungslabor (HASYLAB) at Hamburg. One of these intruments which is built in cooperation with the university of Mainz at the beam line A1 of the storage ring DORIS. How does the physics of resonant scattering enter into the design of the X-ray instrument.

Resonant scattering is a universal technique of non-destructive labelling in X-ray structure research. Strong resonant scattering, often called anomalous dispersion occurs in very narrow energy intervals, the X-ray absorption edges, which are specific for each element. A wide choice of elements then means the construction of an instrument which can cope with the strongly varying properties of the X-ray spectrum. - How can the necessary flexibility be obtained?

This question is of particular interest for the use of resonant scattering of light elements. Among them, phosphorus and sulfur are of particular interest in biological structure research and polymer science.

THE FORMALISM OF RESONANT SCATTERING

When the wavelength, λ, of the incident X-ray beam is close to the absorption edge of an atom, the atomic scattering factor, f, becomes complex to a greater extent

$$f(\lambda) = f_o + f'(\lambda) + if''(\lambda) \qquad (1)$$

f_o is the short wavelength limit of the scattering amplitude. Following the tradition of X-ray diffraction f is given in electrons, e.g. $f_o(0)$ of iron is 26. The scattering length b_e of one electron is 2.8 10^{-12} mm. f' and f" exhibit a wavelength dependence or dispersion. The absorption edge of X-rays represents the threshold frequency above which an electron can be ejected into the continuum. The resonance absorption also known as photoelectric absorption is an inelastic channel. The optical theorem of X-rays then is

$$\sigma_{(photoelectric)} = 2\,\lambda \frac{e^2}{m\,c^2} f''(0) = 2\lambda\, b''(0) \qquad (2)$$

Since σ exists only on the short wavelength region of the absorption edge, so also does f''(0). There is a weak dependence of f'' on the scattering angle which we shall neglect here. With this in mind we say that the imaginary part f'' of the scattering factor is completely controlled by the photoelectric absorption. For the K edge, σ is represented by fairly well by the empirical formula

$$\sigma = \left(\frac{\omega_k}{\omega}\right)^3 \sigma(\omega_k) \qquad \omega > \omega_k \qquad (3)$$

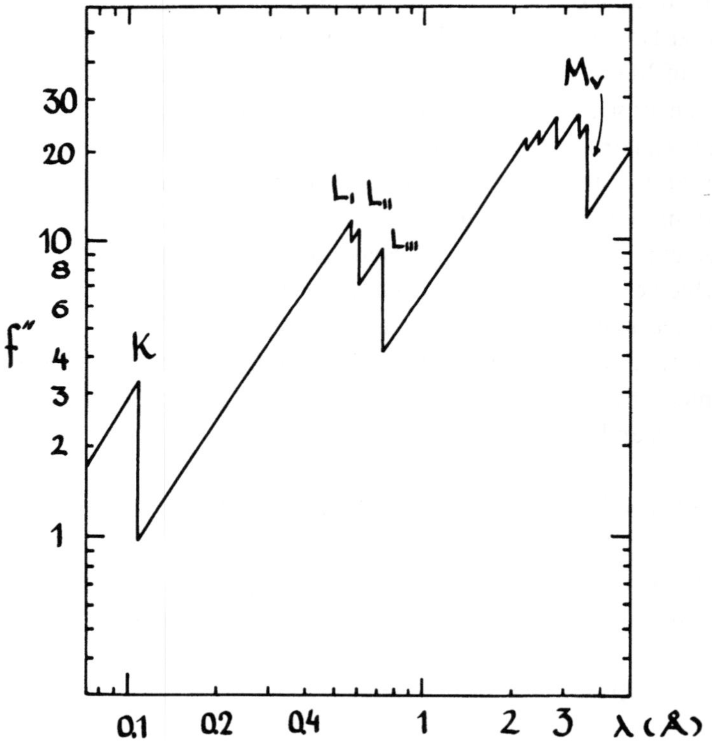

Fig.1 - Dispersion of the imaginary part f'' of the atomic form factor of uranium. f'' (in units of electrons) reaches about one third of the non resonant atomic form factor f=92. This schematic representation has been taken from "The International Tables of Crystallography, IV".

Since for the photon $k = \omega/c$, f'' should vary with the frequency roughly as $1/\omega^2$. Fig.1 shows that the dispersion of f" is slightly more complicated as ionisation of p and d electrons will give rise to additional changes of f. In fact, the schematic representation of the absorption edges in Fig.1 has to be replaced by more structured dispersion profiles at higher frequency resolution, as is shown in Fig.2.

The real part $f'(\omega)$ scattering depends on the dispersion of f" as described by the Kramers-Kronig relation:

$$f'(\omega) = \frac{2}{\pi} \int_{\omega'=0}^{\infty} \frac{\omega' f''(\omega')}{\omega^2 - \omega'^2} d\omega' \qquad (4)$$

Thus a knowledge of $f''(\omega)$ over a sufficiently wide frequency region permits the evaluation of $f'(\omega)$. With X-rays the frequency interval of strong anomalous dispersion may be very narrow (Fig.2). Then Eq.4 simplifies to

$$f'(\omega) = -\frac{1}{\pi} \int_{\varepsilon=-\delta}^{+\delta} \frac{f''(\omega+\varepsilon)}{\varepsilon} d\varepsilon \qquad (5)$$

where δ is of the order of 0.01.

Now we turn to resonance scattering from many atoms. Let us assume an assembly of M atoms each of them having fixed coordinates r with respect to an origin. Out of these a smaller number N of atoms is supposed to have a strong resonance scattering. The structure $\rho(r)$ is then represented by

$$\rho(\mathbf{r}) = \sum_{n=1}^{M} f_{\rho,n}\, \delta(\mathbf{r} - \mathbf{r}_n) + (f' + if'') \sum_{n=1}^{M} \delta(\mathbf{r} - \mathbf{r}_n) = u(\mathbf{r}) + (f' + f'')\, v(\mathbf{r}) \qquad (6)$$

It has been assumed that all strongly resonant scatterers have the same dispersion, i.e. they represent the same chemical element and they are assumed to have the same chemical environment. With $U(\mathbf{r})$ and $V(\mathbf{r})$ as the real and imaginary part of $\rho(\mathbf{r})$ respectively we obtain the scattering amplitude of the structure $\rho(\mathbf{r})$ as

$$F(\mathbf{h}) = \int [U(\mathbf{r}) + iV(\mathbf{r})] \exp(i\mathbf{h}\cdot\mathbf{r})\, d^3r \qquad (7)$$

where h is the scattering vector $h = |\mathbf{h}| = 4\pi \sin\theta/\lambda$.

The multiplication of F(h) with its complex conjugate yields the scattering intensity S(h)

$$\begin{aligned} S(\mathbf{h}) &= F(\mathbf{h})\, F^*(\mathbf{h}) \\ &\sim \iint [U(\mathbf{r}) + iV(\mathbf{r})]\, [U(\mathbf{r}) - iV(\mathbf{r})]\, e^{i\mathbf{h}\,.\,(\mathbf{r} - \mathbf{r}')}\, d^3r\, d^3r' \\ &= \iint [U(\mathbf{r})U(\mathbf{r}') + V(\mathbf{r})V(\mathbf{r}')] \cos[\mathbf{h}\cdot(\mathbf{r} - \mathbf{r}')]\, d^3r\, d^3r' \\ &\quad \iint [U(\mathbf{r})V(\mathbf{r}') - V(\mathbf{r})U(\mathbf{r}')] \sin[\mathbf{h}\cdot(\mathbf{r} - \mathbf{r}')]\, d^3r\, d^3r' \end{aligned} \qquad (8)$$

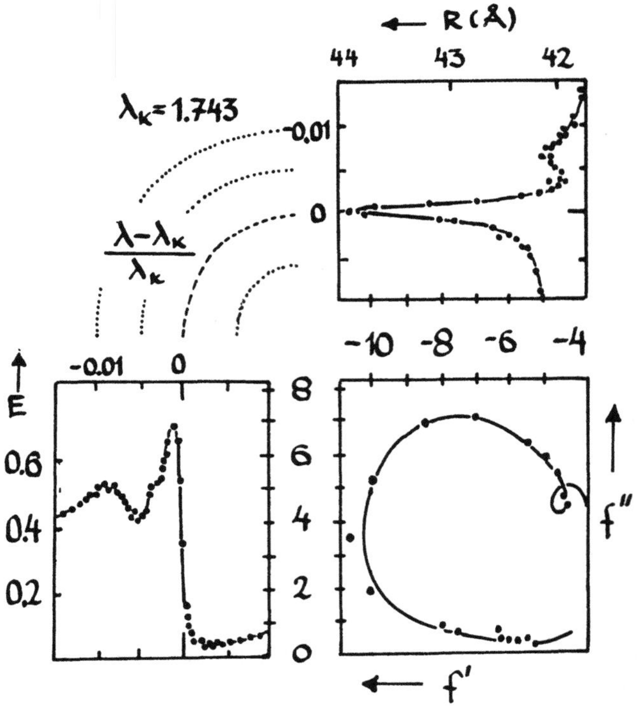

Fig.2 - The anomalous dispersion of iron in ferritin (Stuhrmann, 1980). Results obtained from X-ray small-angle scattering near the K absorption edge of iron at λ = 1.743 Å. The real resonant part is derived from the dispersion of the apparent radius of gyration of ferritin. - Eliminating the wavelength as a coordinate, f' is plotted as function of f'' (Argand diagram). The loop of this open circle at the short wavelength end is due to EXAFS. The variation in f'' is roughly as large as that in f'.

Separating the resonant real part we obtain

$$\begin{aligned} S(\mathbf{h}) \;\sim & \iint u(\mathbf{r})u(\mathbf{r}') \cos[\,\mathbf{h}\cdot(\mathbf{r}-\mathbf{r}')\,]\, d^3r\, d^3r' \\ & + 2 f' \iint u(\mathbf{r})v(\mathbf{r}') \cos[\,\mathbf{h}\cdot(\mathbf{r}-\mathbf{r}')\,]\, d^3r\, d^3r' \\ & + (f' + f'') \iint v(\mathbf{r})v(\mathbf{r}') \cos[\,\mathbf{h}\cdot(\mathbf{r}-\mathbf{r}')\,]\, d^3r\, d^3r' \\ & + f'' \iint [u(\mathbf{r})v(\mathbf{r}') - v(\mathbf{r})u(\mathbf{r}')] \sin[\,\mathbf{h}\cdot(\mathbf{r}-\mathbf{r}')\,]\, d^3r\, d^3r' \end{aligned} \tag{9}$$

$$= S_u(\mathbf{h}) + f' S_{uv}(\mathbf{h}) + (f' + f'') S_v(\mathbf{h}) + f''\varphi(\mathbf{h})$$

The overall effect of resonant scattering is to cause the break down of Friedel's law so that the Bijvoet pair of reflections S(**h**) and S(- **h**) are unequal. The difference

$$S(\mathbf{h}) - S(-\mathbf{h}) = 2 f'' \varphi(\mathbf{h}) \tag{10}$$

is used to determine the absolute configuration of a crystal structure. For randomly oriented microcrystals one has to consider the sum of S(h) and S(-h):

$$1/2\,[S(\mathbf{h}) + S(-\mathbf{h})] = S_u(\mathbf{h}) + f'\,S_{uv}(\mathbf{h}) + (f'^2 + f''^2)\,S_v(\mathbf{h}) \qquad (11)$$

Further averaging processes as they occur with scattering from randomly oriented macromolecules in solution do not change the dispersion given in Eq.11. Introducing the wavelength dependence of f' and f" explicitely, we obtain by integration of S(h) over the solid angle

$$\begin{aligned} I(h,\lambda) &= \int S(h,\lambda)\,d\Omega \\ &= \iint u(\mathbf{r})u(\mathbf{r}')\,\sin h|\mathbf{r}-\mathbf{r}'|\,/\,|\mathbf{r}-\mathbf{r}'|\,d^3r\,d^3r' \\ &\quad + 2f'(\ )\iint u(r)v(r')\,\sin h\,|\mathbf{r}-\mathbf{r}'|\,/\,|\mathbf{r}-\mathbf{r}'|\,d^3r\,d^3r' \\ &\quad + [f'^2(\ ) + f''^2(\ )]\iint v(r)v(r')\,\sin h\,|\mathbf{r}-\mathbf{r}'|\,/\,|\mathbf{r}-\mathbf{r}'|\,d^3r\,d^3r' \\ &= I_u(h) + f'(\lambda)\,I_{uv}(h) + [f'^2(\lambda) + f''^2(\lambda)]\,S_v(h) \end{aligned} \qquad (12)$$

Macromolecules in solution as described by Eq.6 give rise to three basic scattering functions. $I_u(h)$ is the off-resonance scattering, $I_v(h)$ originates from the structure of the resonant scatterers alone, and $I_{uv}(h)$ is a cross term containing the influence of both u(r) and v(r). In many cases the quadratic term in Eq.12 can be neglected and we obtain

$$I(h,\lambda) = I_u(h) + f'(\lambda)\,I_{uv}(h) \qquad (13)$$

The dominant role of f' in resonant scattering will be clarified in the case of forward scattering. According to Eq.12 the scattering cross section consists of three terms

$$\sigma_u = I_u(0);\quad \sigma_{uv} = f'\,I_{uv}(0);\quad \sigma_v = [f'^2 + f''^2]\,I_v(0) \qquad (14)$$

We are going to compare these cross sections of coherent scattering with those of absorption and fluorescence. The cross section of fluorescence σ_F is smaller than σ (photoelectric) given in Eq.2 by a factor c(c<1) taking into account the fluorescence yield. To simplify the discussion we assume the sample to be thin enough to show no appreciable absorption. Assuming a dilute solution of particles we can describe the resonant coherent cross section by

$$\sigma_v = 4\pi N^2\,(b'^2 + b''^2) == 4\pi\sqrt{2}\,N^2\,\frac{e^4}{m^2c^4}\,f''^2 \qquad (15)$$

This corresponds to $(f'^2 + f''^2)\,I_v(0)$ except that for the present pourposes we took into account the dispersion of f'^2 by an additional factor 1.4. N is the number of resonant atoms which form a definite structure either as a cluster or as definite binding sites of a macromolecule. The ratio of resonant zero angle scattering σ_v to fluorescence of σ_F in forward direction is

$$\frac{\sigma_v}{\sigma_F} = \frac{4\pi\sqrt{2}\ N^2 b''^2}{2\,N\,c\,\lambda\,b''} = \frac{2\pi\sqrt{2}\ N\,b''}{c\,\lambda} = \frac{N\,f''}{4000\ c\,\lambda(A)} \tag{16}$$

Clusters of some hundred atoms would give a σ_V comparable to the anomalous dispersion of fluorescence.

To derive a similar relation between the cross term σ_{uV} with respect to σ_F we introduce the number of electrons M of a macromolecule which will give rise to zero-angle scattering (see Eq.2). With N resonant scatterers per particle we obtain

$$\frac{\sigma_{uv}}{\sigma_F} = \frac{4\pi\,2\,M\,b_o\,N\,b'}{2\,N\,c\,\lambda\,b''} = \frac{0.00035\,M}{c\,\lambda(\text{Å})} \tag{17}$$

As the variation of f' and f" is of the same magnitude we have replaced f' by f". This should be kept in mind whenever a more detailed consideration becomes necessary. For instance with proteins in aqueous solution M is of the order of some thousand excess electrons with respect to the excluded volume of the solvent. The dispersion of fluorescence very often is of the order of $\sigma_{uV}(0)$.

The ratio between the f' term and the non-resonant scattering is

$$\frac{\sigma_{uv}}{\sigma_u} = \frac{2\,M\,b_o\,N\,b'}{(Mb_o)^2} = \frac{2\,N\,b'}{M\,b_e} = \frac{2\,N\,f'}{M\,f_e} = 2g \tag{18}$$

The fraction of resonant scattering due to f' dispersion is twice the ratio between resonant to non-resonant electrons. With biomolecules N/M is usually very small, e.g. 0.01 or less. Eq.13 then is a very good approximation.

As the fluorescence cross sections are known, the absolute intensity of the incoming beam can be determined readily from a measurement of the fluorescence intensity. On this basis all scattering data can be put on an absolute scale.

So far we were dealing with very thin samples that do not show any marked absorption. Contrary to the above relations, the dispersion of absorption can to some extent be adapted to the needs of the experiment. This is done either by change of the sample thickness or of the concentration of resonant scatterers is solution or a solid matrix. Usually very thin samples will be preferred in order to reduce the absorption dispersion.

$$\frac{2g}{d} = \frac{2\,Nb'/Mb_e}{z\,2\lambda\,N\,b''\,d} = \frac{1}{z\,Mb_e\,\lambda\,d} = \frac{1}{\rho\,\lambda\,\rho\,d\,v_{sp}} \tag{19}$$

Again we cancelled b' against b". z is the number density of macromolecules. Each particle has the excess scattering length of M b_e. This can be expressed by the product of the weight concentration ρ, the contrast ρ and the specific volume v_{sp}.

For a 5% hemoglobin solution at wavelengths near the K-absorption edge of iron we have $\lambda = 1.74\ 10^{-8}$cm, $\rho = 2\ 10^{10}$ cm^{-2}, $v_{sp} = 0,7$ cm^3/g and $\rho = 0.05$ g/cm^3. We obtain $2g/\mu d = 0.8$. The variation of forward scattering intensity due to f' dispersion nearly equals that of absorption. The variation of both is about 0.5% as one finds from Eq.17. Resonant scattering from metal proteins rarely exceeds 1% of the total forward scattering.

EXPERIMENTAL

The measurement of resonance scattering from disordered systems requires scattering experiments at wavelengths in the near vicinity of the absorption edge, i.e. in the threshold region of absorption or XANES region. The tunability of the X-ray source to any wavelength is a crucial demand. A continuous X-ray spectrum fulfils this requirement. The continuous spectrum emitted by X-ray tubes is too weak in order to allow high precision scattering experiments which are needed for the evaluation of resonance scattering. Synchrotron radiation is magnetic bremsstrahlung emitted by highly relativistic electrons circulating on macroscopic orbits with dimensions comparable to a footbal plane. Its spectral brilliance exceeds that of X-rays by about five orders of magnitude. Even rotating anodes are 10 000 times inferior to electron storage rings. This explains why diffuse resonant scattering techniques could not develope until X-ray synchrotron radiation became available. As we can deduce from the estimates on the resonante scattering the high intensity of sunchrotron radiation is absolutely needed. The small contribution of resonance scattering to the total scattering amounts to typically 0.001 to 0.05 and this fraction tends to be obscured by the much larger dispersion of the absorption and fluorescence (see optical theorem in Eq. 2.).

An instrument for resonance scattering has to fulfil two requirements:

- easy and highly reproducible selection of narrow wavelength intervalls
- high accuracy and resolution of the scattering intensity pattern

The hybridization of a rapidly tunable spectrometer with a diffractometer has been achieved to various extents. There are high performance spectrometers for the measurement of the fine structure of X-ray absorption edges on the one side and commercial diffractometers for crystallography or small-angle scattering on the other side. The existing instruments of resonance scattering are a modular arrangement of both. They may be classified according to the monochromator system which may consist of a

- double monochromator or
- focusing toroidal mirror and a double monochromator or
- focusing toroidal mirror, exchangeable plane mirror crystal monochromator.

We are not going to review the various designs and their perfomance but rather concentrate on an instrument which is capable of covering a very wide wavelength range from 0.8 Å to 7 Å.

The wavelenghts reflected by the monochromator crystal are given by Bragg's equation

$$n\lambda = 2d\sin\theta$$

d is the spacing of the reflection planes, e.g. 2d = 6.53 A for Ge (111).
Assuming a maximum deflection of the beam by 160° wavelengths up to 6.43 Å can be achieved. For still longer wavelengths indium antimonide crystals have proven to be useful. With d = 7.48 Å still longer wavelengths up to 7.36 Å be reached. These crystals would be suitable to produce resonance scattering at the K-edges of even rather light elements like chlorine, sulfur, phosphorus and silicon. For biological application the use of sulfur and phosphorus as labels is particularly interesting as these atoms are native components of most biomolecules. Phosporus is a regular constituent in nucleic acids and membranes. The aminoacids cystein and methionin contain sulfur. The cystin disulfid bridges stabilise protein structures. Consequently these elements are also found fossil fuels. Also in materials science these elements are of considerable interest: rubber contains sulfur. Chlorine is found in certain classes of polymers. Semiconductors are based on silicon.

The transmission function for a perfect crystal is strictly calculable from the dynamic theory, and if good crystals are used the experimental results follow closely the theoretical predictions (Fig. 3).

The choice of a perfect crystal is particularly important if one uses a double reflection geometry, e.g. a channel cut, because in this case, if imperfect crystals are used, not only will the wavelength resolution be poor but also large losses in flux occur. On the other hand if the crystal is perfect the use of a double reflection geometry leads to a marked improvement in the ratio of the peak to tail intensity, since within the confines of the rocking curve the reflectivity is almost 100% at wavelengths shorter than 2 Å (Fig. 3). At longer wavelengths there is a considerable decrease of peak reflectivity> Using the 111 reflection of a germanium crystal the peak reflectivity will drop to 15% at a wavelength of 6 Å. Reflection by the second crystal of a double monochromator then reduces the maximum transmission of 6 Å wavelengths to 0.15 * 0.15 = 0.023, - It is for this reason why a single reflection monochromator is preferred in the soft X-ray spectrum.

Using the continous spectrum of sunchrotron radiation higher harmonics will be reflected by the monochromator as well. These have to suppressed by appropriate premonochromatisation of the synchrotron radiation spectrum. This is usually done by grazing incidence mirrors which will reflect the long wavelength part of a spectrum from a critical wavelength onwards (Fig. 4).

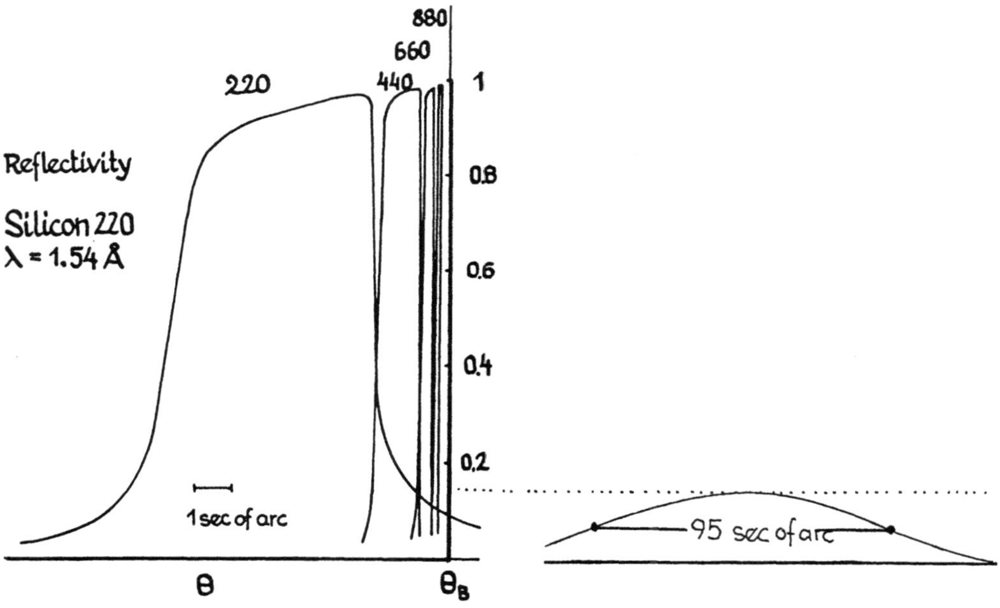

Fig. 3 - The reflection characteristics of a silicon crystal 220 reflection at a wavelength of 1.54 Å (left) and of a germanium crystal 111 reflection at a wavelength of 6 Å. Observe the departure from the simple law of Bragg due to refractive index corrections.

Using a diamond type crystal as a monochromator (e.g. Si, Ge) no second order harmonics are encountered. With the premonochromatisation given in Fig. 4 third order harmonics then will be transimitted to a marked extent From 3.5 Awavelengths onwards only. For the use of wavelenghs beyond 3.5 Å the premonochromater has to be modified. - We do this by using mirrors in double reflecting geometry (Fig. 5).

- The first gold coated, focusing mirror will deflect the beam by either
- - 10 mrad to the monochromator (reflection stars from λ = 0.8 A) or by
- - 14 mard to the second mirror
- The two halves of the second mirror have different surfaces:
- - gold coated (reflectivity given in Fig. 4) and
- - quartz (reflectivity starting from 3 A onwards).

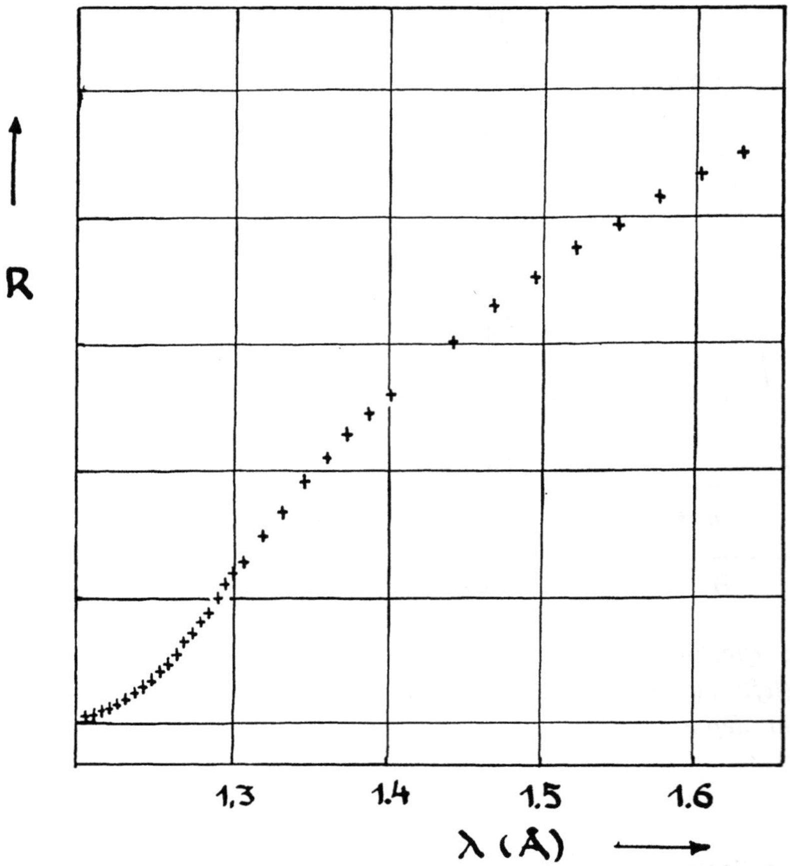

Fig. 4. Reflectivity of two plane gold mirrors in double reflection geometry at a glancing angle of 7 mrad. - The current of the ionisation chamber produced by the transmitted beam is plotted against the wavelength. The data were taken with the resonance scattering instrument at beam line A1 of HASYLAB.

It is reflection of synchrotron radiation under a glancing angle of 7 mrad by a quartz mirror, which offers the right spectrum for the monochromator to operate in the long wavelength spectrum. As wavelengths shorter than 3 Å are cut off by the quartz mirror only the fundamental wavelength will be reflected by the 111 plane of the germanium (or silicon) crystal.

As synchrotron radiation is highly polarised in the plane of the orbit, the vertical reflection of the beam by monochromator crystals is preferred. Using one monochromator crystal only, the diffractometer has to follow any change of direction of the monochromatic beam by 2 θ on rotation of the crystal by θ. As this camera is also ment to meet the requirements of small-angle scattering as well the maximum length of the camera (from axis of the monochromator to the detector) can reach 4.3 m. For reasons of mechanical stability however, resonance scattering experiments can performed at a total camera length of not more than 3.1 m. The maximum distance between sample and detector then is 1.85 m. When the primary beam is deflected by more than 48° the camera length has to be reduced by another 0.6 m in order to avoid collision with the bottom of the 0.8 m deep grove (Fig. 6). Both the monochromator crystals (three can be kept resident on a common holder) and the camera are rotated by 1.4 m long arms driven by stepping motors. The angular resolution in θ is 0.7 seconds of arc. The perfomance of this instrument as an EXAFS spectrometer has been tested at various absorption edges. Using the 111 reflection of a germanium crystal the resolution is about 3 eV at the K-absorption edge of titanium (Fig. 7).

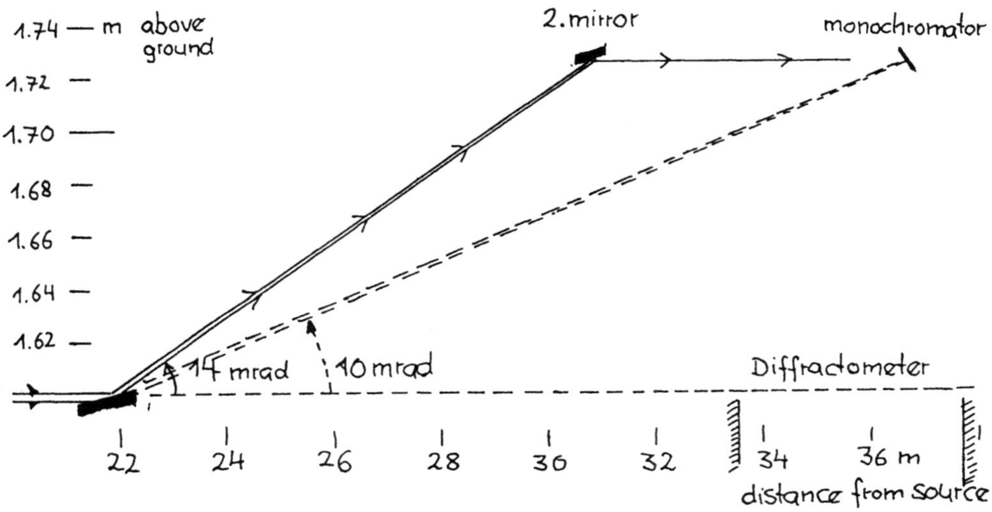

Fig. 5 The premonochromator consists of two mirrors in double reflecting geometry. The two mirrors at 21.84 m and 30.80 m from the source point produce a vertical off-set of the beam by 126 mm. The ultra high vacuum section of the of the premonochromator is separated by a 50 µm kapton foil from the high vacuum section of the diffractometer shown in Fig. 6. For further details see text.

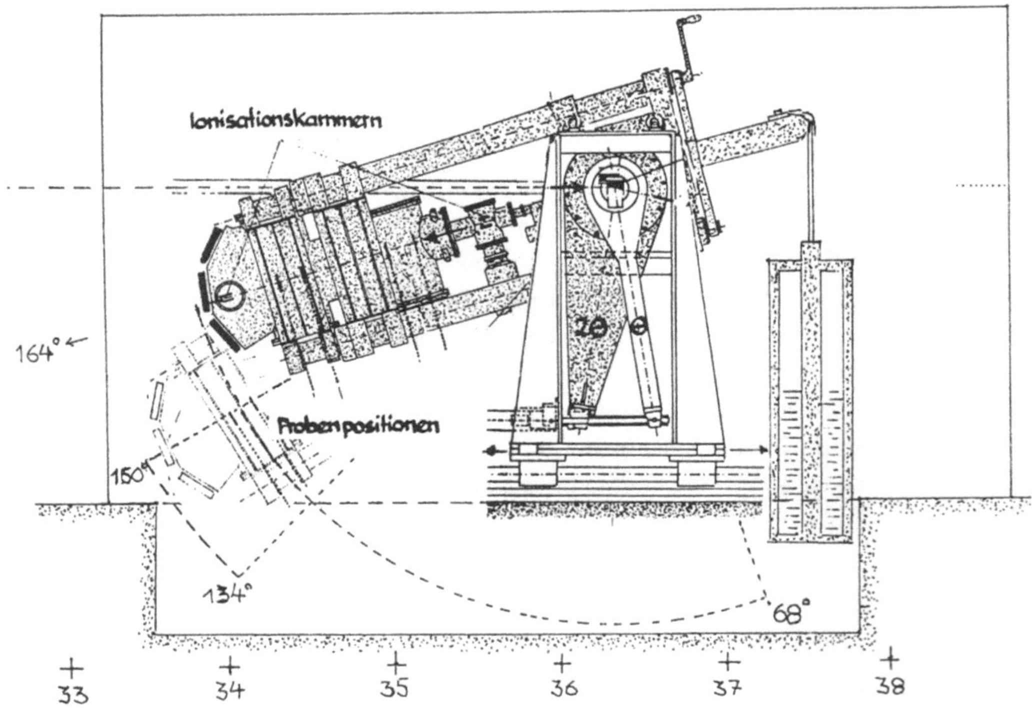

Fig. 6. Side view of the new instrument for the measurement of resonance scattering at the beam line A1 of HASYLAB. The camera (shaded) is balanced by 400 Kg lead on the left side. - In order to use long wavelengths obtained in back reflection by one of the three resident monochromator crystals, the counter balance is removed, the whole diffractometer driven to the right side. Then deflection angles up to 165° can be reached. The distance between sample and detector can vary between 0.35 m and 1.85 m. The flux is measured by ionisation chambers. One is half way between the monochromator and the first sample position, the second one is in the beam stop. - The vacuum inside the camera is 0.001 mbar at present. The large X-ray windows at the end of the camera are made of two 10 µm mylar foils supported by a steel grid. Small leaks are likely to occur here.

The resolution of the scattering pattern is dicated by the specifications of the detectorsystem.We are using three multiwire proportional counters wich were constructed by A. Gabriel and F. Dauvergene (EMBL, Grenoble). The spatial resolution is 1 mm in the direction of the anode wires which have a spacing of 2mm. The sensitive area is 200 mm * 200 mm out of which 180 mm * 180 mm are used. The three detectors are mounted in an asymmetric way so as to cover the whole angular range up to scattering angles of 60°. The structural resolution then is equal to the wavelenght used (Fig. 9). - The performance of the detector is best demonstrated by a scattering pattern of purple membrane which has been taken recently at our instrument by G. Buldt and coworkes (Fig. 8).

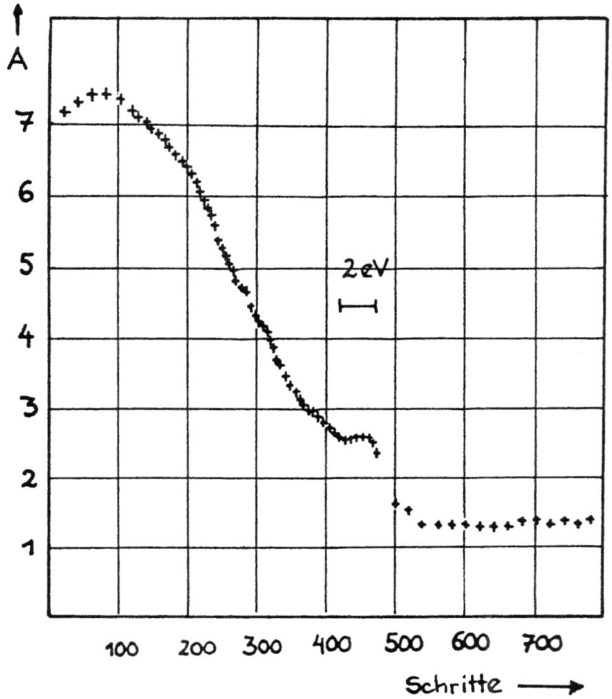

Fig. 7 - K - Absorption edge of a titanium foil near $\lambda = 2.5$ Å The pre-edge is just being resolved. The abscissa is given in steps of the motor.

X-ray resonant scattering has been measured from the dysprosium derivative of purple membrane in collaboration with G.Büldt (Biophysics Department of the Free University of Berlin) and from an iron-copper alloy (2% copper) at the K-absorption edges of iron and copper. The analysis of the data is in progress.

*Fig. 8 Scattering of purple membrane from halobacterium halobium. this pattern has been taken with an area detector from A. Gabriel (EMBL Grenobel) and displayed on a television screen by a data transfer system developed by C. Boulin (EMBL, Heidelberg). The picture is turned by 90°. In the lower part the shadow of the beam stop with outer dimensions of 12 mm * 50 mm and an opening of 3 mm * 20 mm to let the beam enter into the ionisation chamber. The horizontal lines trace anode wires,. The resolution of 1 mm of the delay lines is good enough to see each wire as well separated lines. The resolution of the picture by 256 * 256 words of 16 bit length makes the representation of the 90 visible anode wired slighly more complicated. - Each event on the detector leads to an incrementation +1 in an intermediate storage memory (from Sension Ltd). About 200 000 cps can be handled without significant dead time losses. After each measurement the picture is read by a minicomputer (VAX 750 from DEC), which is busy with data reduction while the next picture is taken. Further data analysis is done on the Mikro-VAX II while the following scan through a series of programmed wavelengths is going on. The development of the data aquisition system is in progress. It relies to a considerable extent on the cooperation with the EMBL Outstation at DESY.*

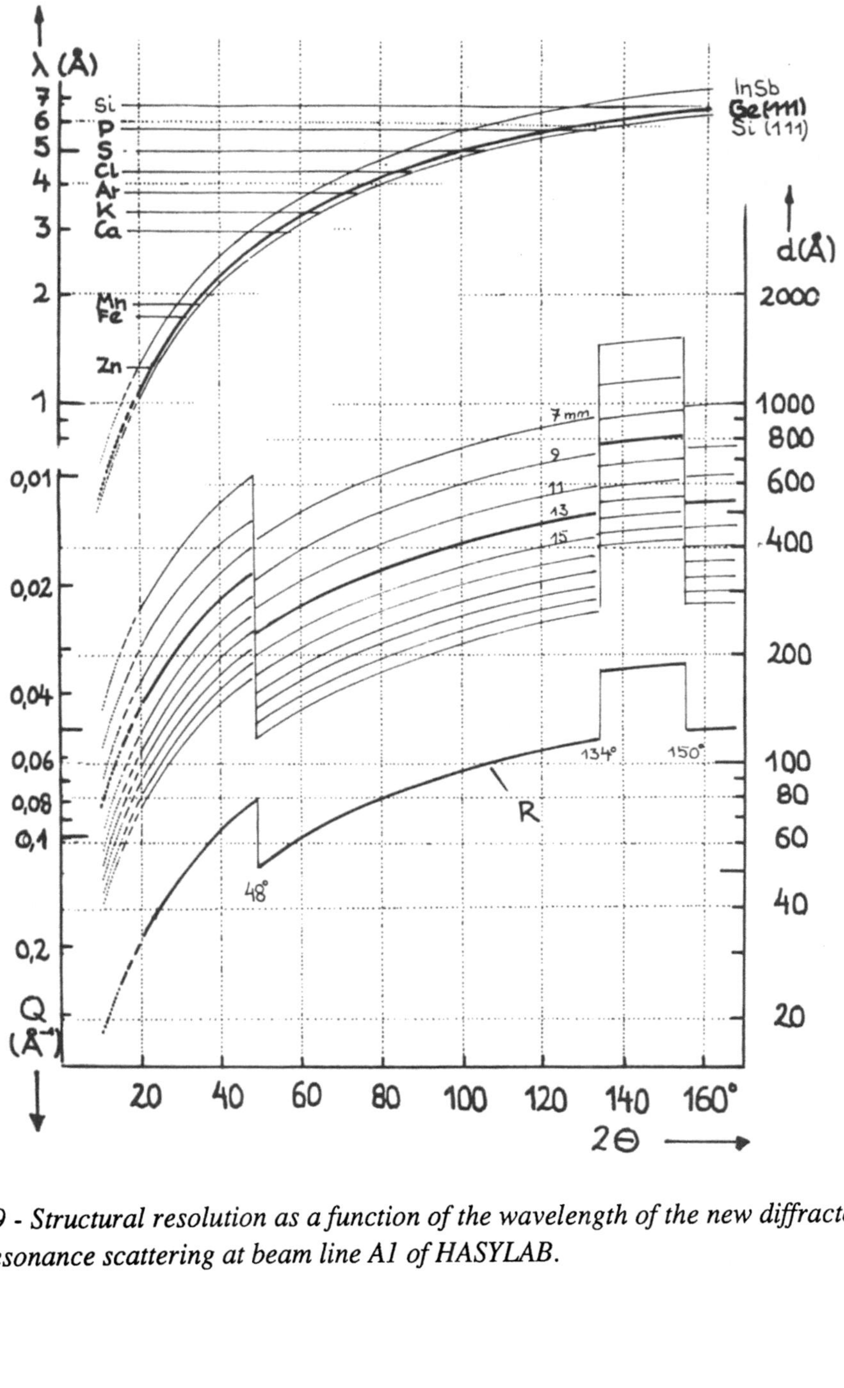

Fig. 9 - Structural resolution as a function of the wavelength of the new diffractometer for resonance scattering at beam line A1 of HASYLAB.

REFERENCES

For a review, which presents some results of resonance scattering of biological macromolecules are:

- H.B.Stuhrmann, Advances in Polymer Science 67, 124-163 and citations therein
- H.B.Stuhrmann, Acta Cryst A36, 996 (1980).

Chromatin, Solution Scattering and Perception

M.H.J. KOCH and Z. SAYERS

European Molecular Biology Laboratory
Hamburg Outstation
c/o DESY Notkestrasse 85D-2000 Hamburg 52

In the present paper we draw a parallel between small angle scattering (SAS) experiments and visual perception and illustrate the origin and importance of the "personal equation" (see Gregory ref.1) in the design of experiments (i.e. manipulations designed to test a hypothesis based on *a priori* knowledge) as opposed to measurements (i.e. manipulations designed to obtain a numerical value without reference to its meaning). Similar considerations probably apply to most experimental techniques.

Small angle scattering shares with visual perception [(2)] the property of being an ill-posed problem in the sense that there is no unique solution which depends continuously on the initial data leading from a SAS curve to the structure. Ill-posed problems are usually solved by a regularization process, which transforms them into well-posed ones by using *a priori* knowledge (models) to limit the class of admissible solutions. In the case of SAS interpretation, one uses simple real space models (e.g. spheres or cylinders) and *a priori* knowledge to select the experimental data (e.g. by selecting the region for determination of the radius of gyration).

Although the use of *a priori* knowledge is clearly indispensable to make any interpretation of a SAS pattern, one has to be cautious, as illustrated by the evolution of the interpretation of the SAS results for chromatin during the last decade.

To clarify matters, the scattering patterns for various types of chromatin published since 1976, are shown in figure 1 as plots of $\log(s \cdot I(s))$ vs . s^2, where $s=2\sin\theta/\lambda$ 2θ is the scattering angle and λ is the wavelength. Clearly there are differences between the experimental data which are most likely attributable to improved preparative methods; the data obtained since 1979 consistently display a broad maximum centred around 0.05 nm^{-1}.Synchrotron radiation sources have been applied in the most recent experimental work [(3-6)].

The present interpretation by Perez-Grau et al. [(3)], by Bordas et al.[(4,5)] and by Koch et al. [(6)] is that even at the lowest ionic strength the nucleofilament exists as a loose helix-like structure with a diameter of 25 - 35 nm. The maximum at 0.05 nm^{-1}

is attributed to the interference between nucleosomes. At increasing ionic strength the structure condenses in a continuous manner into the so-called 30 nm fibre and the interference maximum disappears as the distance between nucleosomes decreases.

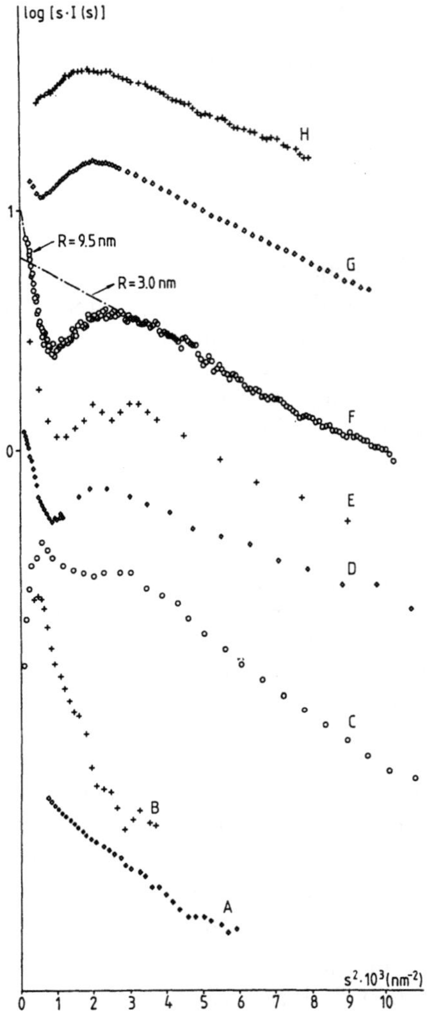

Small angle scattering data for chromatin at low ionic strength

Fig.1 Chromatin small angle scattering data published since 1976.
A: (1976) Sperling and Tardieu(7), **B**: (1978) Baudy and Bram,(9)
C: (1978) Hollandt et al.(10), **D**: (1979),Suau et al.(11)
E: (1981) Brust and Harbers(12), **F**:(1984) Perez-Grau et al.(3)
G: (1986) Notbohm (13), **H**: (1986) Greulich et al.(14).

Previous interpretations entailed a sharp transition from a 10 nm filament at low ionic strength to the 30 nm filament and did not explain the presence of the maximum at 0.05 nm^{-1}. The notion of a 10 nm filament was introduced on the basis of the early electron microscopy and SAS results. Sperling and Tardieu [7] measured a mass per unit length (M/L) of 257000 daltons/11 nm and a radius of gyration of the cross-section (Rx) of 3.8 nm, consistent with a filament of 11 nm diameter for rat liver chromatin in solution. With hindsight, it appears that their preparation, which was exposed to X-rays for several hours, must have contained very short chromatin. The scattering pattern is then essentially that of mononucleosomes, which has a slope of the cross-section plot corresponding to Rx = 3.8 nm and does not display an interference maximum. Finch and Klug [8] described their electron microscopy (EM) observations on rat liver chromatin in 0.2 mM EDTA as follows:

"Chromatin appears as filaments about 100 Å in diameter, which often tend to coil up loosely. There is no visible repeat along the length of the filament. The filamentous character agrees with Sperling and Tardieu's result on low angle X-ray scattering.... In a few instances the coiling of the nucleofilament is so pronounced as to lead to a more condensed state, but the majority of filaments are extended. In 0.2 mM $MgCl_2$, almost all material has this condensed character".

The micrographs correspond to a higher resolution than the SAS data. The latter are, however, incompatible with a loosely coiled filament, which would give rise to an interference between nucleosomes and have an apparent outer diameter at very low resolution of about 25 nm.

Later the very plausible result of Sperling and Tardieu was never reproduced but in several instances, gross differences in experimental data were overlocked in order to interpret the results in terms of a transition between a 10 nm and a 30 nm filament. Although this does not detract from the validity of much of the work referred to, which is mostly concerned with the structure of the "30 nm" condensed chromatin filament occurring under physiological conditions, it is relevant to the condensation mechanism of chromatin and to a better understanding of its hydrodynamic properties. Interpretation of the scattering patterns as resulting from a 10 nm filament leads logically to the conclusion that there would be a sharp transition to the 30 nm filament with increasing ionic strength rather than a continuous condensation process. This interpretation is, however, based upon different ranges of the data chosen to determine Rx: from $s>0.05$ nm^{-1} at low ionic strength to $s<0.05$ nm^{-1} in the condensed state.Figure 1 shows that this will clearly give different results.

Baudy and Bram [9] in their neutron scattering study on calf thymus chromatin in 1 mM salt solutions obtained a value of Rx = 8.0 nm which they interpreted as originating from a loose coil. From the value of the slope above s = 0.025 nm^{-1} they determined a cross-section with an Rx = 4.1 nm and an M/L of 253000 daltons/11 nm. They attributed the difference between their results and those of Sperling and Tardieu to the fact that in the latter study "...30 - 35% mass contribution of NHP (non histone proteins) was neglected...".

Hollandt et al. [10] measured an Rx = 2.9 nm and an M/L of 233200 daltons/11 nm. They considered the difference between their results and those of Sperling and Tardieu to be due to a stronger shearing of their material giving a "higher proportion of 'unwound' beads". Concerning the results of Baudy and Bram [(9)] they write: "A possible expalanation might be that fragments of quaternary structure were involved in the measurements of Baudy and Bram. In fact, we could observe such an effect in our scattering curves just in the angle range used by Baudy and Bram for their analysis.".

In their neutron scattering study, Suau et al. [(11)] observed that for chicken erythrocyte chromatin in 10 mM Tris HCl 1 mM EDTA at pH 7.8: "The down turn of the Guinier plots at the lowest k ($k = 2\pi$ s) we attribute to the persistence of chromatin segments of 2 - 3 nucleosomes at low ionic strength." and measured an M/L of 286000 daltons/11 nm and an Rx = 2.7 nm in H_2O in the range $0.045<s<0.10$ nm^{-1} and Rx = 3.4 nm when the scattering is dominated by DNA. The noticed the difference with Sperling and Tardieu [(7)] and write: "One major difference between chicken erythrocyte chromatin and rat liver chromatin is that the latter contains a much higher amount of non-histone proteins. If the location of the non-histone proteins was on the outside of the 10 nm filament it would increase the apparent cross-section radius of gyration".

As the scattering pattern at low ionic strength of Suau et al. [(11)] is shown only in the range $0.035<s<0.12$ nm^{-1} the corresponding curve in figure 1 was reconstituted by combining this data with those given in the same paper for chromatin at 0.4 mM $MgCl_2$ in the range $8.7 \cdot 10^{-3}<s<0.035$ nm^{-1}. This is justified by the fact that we find little difference between the scattering patterns in absence of salt and in the presence of 0.4 mM $MgCl_2$ at typical chromatin concentrations used for solution scattering experiments (about 3 mg DNA/ml)[(3)].

At 0.4 mM $MgCl_2$ Suau et al.[(11)] found an Rx of 9.2 nm using the range $8.7 \cdot 10^{-3}<s<0.022$ nm^{-1} and an M/L of 2.2 nucleosomes/11 nm and concluded from the increase in M/L and the small changes in Rx at increasing $MgCl_2$ concentrations that: "This suggests that in solution there is a family of supercoils of nucleosomes of similar radius of gyration but with very different degrees of compaction.".

The persistence of the "10 nm to 30 nm transition" interpretation is interesting, since, as shown above, the existence of a loose helix had been suggested by several authors. Further electron microscopy suggested a similar interpretation. A particularly clear account is given by Thoma, Koller and Klug [(15)] in their EM study of salt dependent superstructures of chromatin: "When fixation is performed in 5 mM TEACl and 0.2 mM EDTA the fibres are about 200 - 250 Å in diameter .... Such findings suggest that in solution the nucleosomes may be arranged in helical turns, but on the grid there is a tendency for the nucleosomes to fall on their flat faces, giving rise to the zig-zag appearance. Allowing for the bias, one concludes that in solution there is a helix with about two to three nucleosomes per turn. ...When 10 mM NaCl is added to the 5 mM TEACl buffer before fixation, the zig-zag disappear in favor of a more acompact fiber with a diameter of about 250 Å.

This is even more evident at 20 mM NaCl. Again in certain regions the nucleosomes approach closely to give the impression of a loose super-helix that could have three to four nucleosomes per turn. ...Between 20 and 40 mM NaCl a more striking condensation of the chromatin fibers is observed. At 40 mM the fibers are irregular and inhomogeneous in shape, about 250 Å in diameter and individual nucleosomes are hard to see". It should be noted that the micrographs for chromatin at low ionic strength in this study are quite different from the early ones in Finch and Klug [8].

Brust and Harbers [12] using rabbit thymus chromatin noted that in 0.2 mM EDTA "...chromatin is usually considered as having exclusively tertiary structure, organized as a chain of nucleosomes likes 'beads on a string'. However, our small angle X-ray scattering data do not correlate perfectly to such an interpretation".

Very recently, Greulich et al. [14] attempted to model the structure at low ionic strength by the compaction of a wormlike chain and showed that the downturn of the scattering pattern, which was also observed previously by this group [16] cannot be interpreted in this way. They measured values of M/L of 220000 daltons/11 nm and Rx = 2.5 nm. In a way it is rather amusing that the values of M/L are largely independent even of the gross shape of the primary scattering patterns! The values of Rx, on the contrary, cluster around 3 nm and 10 nm. The reason is that as a result of the particular shape of the scattering curve, the M/L is not very sensitive, compared to Rx, to the region taken for extrapolation as illustrated in curve F in figure 1.

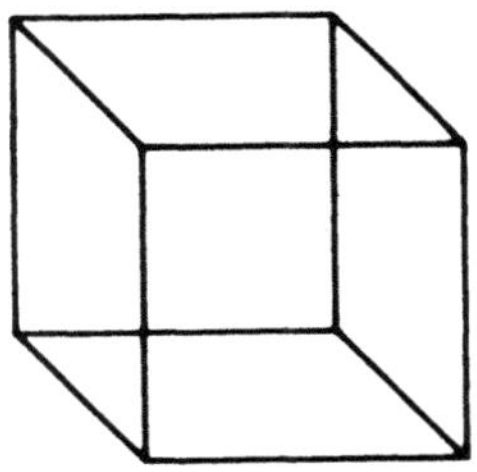

Fig. 2 The cube developed by L.A. Necker in 1832. One can readily perceive two interpretations: A cube seen from above or from below. In fact, the most 'realistic' interpretation (i.e. lines connecting eight points in the plane of the paper) is the one which is most difficult to see.

There are many cases of visual images such as Necker's cube, shown in figure 2, where the observer actually becomes aware of the ill-posed nature of perception and can fit several hypothesis to an image. The important thing is that although it is possible to switch from one perception to the other, the various possibilities are never perceived simultaneously - one excludes the other. The same applies obviously to the interpretation of SAS patterns.

How easily one can be fooled by solution scattering data is very well illustrated by the following anecdote which also shows that awareness of a danger offers little protection. Recently we made a few preliminary experiments with L. Wijns and G. Meersseman from the University of Brussels on samples of nuclei and solutions of chromatin from rye embryos. The solution scattering patterns of long chromatin fragments displayed no maximum at 0.05 nm^{-1} and had a radius of gyration of the cross-section of 3.8 nm, exactly the value reported by Sperling and Tardieu! Moreover, the patterns had a maximum at about $(11\ nm)^{-1}$ as expected. Addition of up to 150 mM did not result in any significant change in the scattering pattern. The obvious reaction in such case is to suspect the preparation. However, the perfect histone and DNA gels obtained from the samples only added to the confusion. Having exhausted our experimental skills and having run out of beam time, we decided to look at the scattering from Percoll, the medium used to isolate the rather fragile rye nuclei. The material (Pharmacia) is extensively used by biochemists and consists of PVP-coated silica beads about 20 nm in diameter. At dilutions of 1:100 (v/v) it gives rise to very nice scattering patterns with slopes in cross-section plots corresponding to an Rx of 3.8 nm. So, all this time we had been looking at the residual Percoll in the solutions. In fact, as shown in figure 3, the material provides a schoolbook example to illustrate interparticle interference and is an excellent calibration standard.

Fig. 3. Scattering pattern of pure Percoll (A), of a 1:100 (v/v) dilution (B) and interference pattern (C) obtained from (A) and (B). (Unpublished data).

The survey of experimental data from SAS studies of chromatin and their interpretation discussed in the literature is a good example of an attempt at a solution of an ill-posed problem and leads one to agree with the assertion of Whevell(17) in 1847, namely that "There is a mask of theory over the whole face of nature" or that expectation largely determines the selection of data, be it in everyday perception or in the contrived case of an experiment.

Statistical and information theoretical methods are of little help in getting round this problem, since, although they allow statements as to how well the experimental data can be fitted using a particular model, they do not give any indication as to the validity of the - usually unstated - hypothesis underlying the selection of data.

As stated by A. Toynbee [18] in a different context: "Facts are, in truth exactly what is meant by the latin word facta from which the English word is derived. They are 'things that have been made' - that is to say 'fictitious' rather than 'factual' things - and this truth cannot be evaded by calling them 'data' ('gifts') instead.". Thus, "It is the power of expectation rather than the power of conceptual knowledge that molds what we see in life no less than in art" (Gombrich, ref.19) - or for that matter in scattering patterns!

REFERENCES

1) Gregory R., (1984) *Mind in Science.* Penguin (Peregrine) Books, Harmondsworth, UK.
2) Poggio T., Torre V., Koch C., (1985) Computational vision and regularization theory. *Nature* **317** 314-319.
3) Perez-Grau L., Bordas J., Koch M.H.J., (1984) Synchrotron radiation X-ray scattering study on solutions and gels. *Nucleic Acid Res* **12**, 2987-2995.
4) Bordas J., Perez-Grau L., Koch M.H.J., Nave C., Vega M.C., (1986) The superstructure of chromatin and its condensation mechanism: I Synchrotron radiation X-ray scattering results. *Eur J. Biophys.* **13**, 157-174.
5) Bordas J., Perez-Grau L., Koch M.H.J., Nave C., Vega M.C., (1986) The superstructure of chromatin and its condensation mechanism: II Theoretical analysis of the X-ray scattering patterns and model calculations. *Eur. J. Biophys.* **13**, 175-186.
6) Koch M.H.J., Vega M.C., Sayers Z., Michon A.M., (1986) The superstructure of chromatin and its condensation mechanism. III: Effect of monovalent and divalent cations, X-ray solution scattering and hydrodynamic studies. *Eur. J. Biophys.* (in press).
7) Sperling L., Tardieu A., (1976) The mass per unit lenght of chromatin by low-angle X-ray scattering. *FEBS Lett* . **64**, 89-91.
8) Finch J.T., Klug A., (1976) Solenoidal model for superstructure in chromatin. *Proc. Natl. Acad. Sci. Usa* **73** ,1879-1901.
9) Baudy P., Bram S., (1978) Chromatin fiber dimensions and nucleosome orientation: a neutron scattering investigation. *Nucleic Acid Res* **5,** 3698 - 3713.
10) Hollandt H., Notbohm H., Riedel F., Harbers E., (1979) Studies of the structure of isolated chromatin in three different solvents. *Nucleic Acid Res* **6**, 2017-2027.
11) Suau P., Bradbury E.M., Baldwin J.P., (1979) Higher-order structures of chromatin in solution. *Eur . J. Biochem.* **97**: 593-6602.
12) Brust R., Harbers E., (1981) Structural investigations on Isolated Chromatin of Higher-Order Organization. *Eur J. Biochem.* **117,** 609-615.
13) Notbohm H., (1986) Comparative Studies on the structure of soluble and insoluble chromatin from chicken erythrocytes. *Int. J. Biol. Macromol.* **8,** 114-120.
14) Greulich K.O., Wachtel E., Ausio J., Seger D., Eisenberg H., (1986)

Transition of chromatin from the "10 nm" Lower Order Structure, to the "30 nm" Higher Order Structure, as followed by Small Angle X-ray Scattering. *J. Mol. Biol.* **193**: 709-721.`
15) Thoma F., Koller T., Klug A. (1979) Involvement of histone H1 in the organization of the nucleosome and of the salt dependent superstructures of chromatin. *J. Cell. Biol.* **83**, 403-427.
16) Reich M., (1982) PhD Thesis Weizmann Institute of Science - Rehovoth.
17) Whevell (1847) *The Philosophy of the Inductive Sciences* , 2nd Ed., London.
18) Toynbee A., (1972) *A study of History* Weathervane Books: New York p 486
19) Gombrich E.H., (1960) *"Art and Illusion"* Phaidon Press, London, p. 225.

Pre-Nucleation and Nucleation Events in Microtubule Assembly: A Time-Resolved X-Ray Scattering Study

U.Spann*, E.M.Mandelkow+, J.Bordas# and E.Mandelkow+

*Max-Planck-Institute for Medical Research, Department of Biophysics
Jahnstrasse 29, D-6900 Heidelberg, FRG

+Max-Planck-Unit for Structural Molecular Biology
Ohnhorststrasse 18, D-2000 Hamburg 52, FRG

#MRC-SERC Biology Support Laboratory
Daresbury Laboratory, Warrington WA4 4AD, U.K.

Microtubules are hollow cylinders of about 25-30nm diameter which are built from protein subunits, tubulin and microtubule-associated proteins (MAPs). They play a role in a variety of cellular processes, for example mitosis, axonal transport, or beating of flagella. The subunit protein can be isolated (e.g. from mammalian brain) and induced to form microtubules in vitro.

In earlier studies we have described the process of temperature-induced microtubule assembly as seen by time-resolved X-ray scattering (Mandelkow et al., 1980; Bordas et al., 1983; Mandelkow et al., 1983). The observed states could be broadly subdivided into the initial (cold) state, pre-nucleation events, nucleation, elongation, the final (warm) state, and postassembly events. The initial cold state was characterized by a mixture of rings, oligomers, and dimers. Pre-nucleation events were seen as an undershoot in scattering which reflected the breakdown of rings and oligomers into smaller units. Nucleation was interpreted in terms of an association of oligomers, while the final warm state was dominated by the scattering from microtubules. Oligomers were observed not only with MAP-containing microtubule protein but also with purified tubulin which lacked the rings typical of cold microtubule protein. Both the ring and non-ring oligomers appeared to consist of short stretches of protofilaments, and the structural and kinetic evidence suggested that oligomeric states of tubulin were important for microtubule assembly.

Under the experimental conditions employed previously there was an overlap of the pre-nucleation and nucleation events following a temperature jump from 4° C to 37° C, such that the magnitudes and rates of the individual phase could not be analyzed separately. In this report we describe experiments designed to separate the pre-nucleation events from the subsequent microtubule assembly. This can be achieved, for example, by lowering the pH and the ionic strength. Within the time resolution of the temperature jump (a few seconds) the oligomeric forms appear to be in a rapid temperature-dependent equilibrium, independent of whether microtubule assembly is rapid or slow. The results suggest that the transitions between pre-nucleation oligomers occur independently of the interactions between the nucleating oligomers (for details see Spann et al., 1987).

MATERIALS AND METHODS

(a) time-resolved X-ray scattering

Microtubule protein was prepared by one of the temperature cycle methods as described previously (Mandelkow et al., 1985). The reassembly buffer was usually 0.1 M PIPES, pH between 6.9 and 6.5, containing 1 mM each of $MgSO_4$, DTT, EGTA, and GTP. For certain experiments the ionic strength was varied by reducing the buffers and/or adding salts. Protein concentrations ranged up to 3o mg/ml. The X-ray experiments were performed at the EMBL Outstation at DESY, Hamburg, as described (Mandelkow et al., 1980; Bordas et al., 1983), using instruments X31 or X11 (Koch & Bordas, 1983; Hendrix et al., 1979), position sensitive detectors (linear detector, see Gabriel 1977; or a quadrant detector integrating the circularly symmetric solution pattern over an angle of 90°; J.Hendricks & J. Bordas, unpublished), and a fast data acquisition system (Bordas et al., 1980). Assembly and disassembly were induced by temperature jumps between about 0^oC and 37^oC (half time 4-10 sec) using the T-jump device described elsewhere (Renner et al., 1983).

(b) data analysis

The observed intensities were corrected for background (measured from a buffer-filled cell), detector response, and variations in incident intensity due to the decay of the current in the storage ring. The scattering data were interpreted as described (Mandelkow et al., 1983; Bordas et al., 1983).

Fig. 1 shows calculated scattering curves of several representative model structures, including microtubules, rings, protofilament fragments, and tubulin dimers. The curves are normalized to the same total protein concentration so that the forward scattering (S=0) represents the degree of polymerization. Microtubules show the highest central scatter and several subsidiary maxima separated by clear minima. Rings have a lower central scatter and side maxima that are approximately out of phase with respect to those of microtubules. Since microtubules have smaller diameters than rings their subsidiary peaks occur at higher S-values. By contrast, both tubulin dimers and short protofilament fragments show smoothly decaying scattering curves. In practice one usually observes a mixture of several types of aggregates, especially at low temperature where rings, oligomers, and dimers are in equilibrium. The minimum preceding the first maximum of microtubules is particularly sensitive to the presence of oligomeric structures.

Because of the beam stop and parasitic scatter the lowest observable S-value are around 0.01 nm^{-1}. In the following the intensity integrated between about 0.01 and 0.015 nm^{-1} will be referred to as "central scatter" for short (note that this is only an approximation to the forward scattering); integration between about 0.025 and 0.035 nm^{-1} yields the "ring scatter" since this overlaps with the first side maximum of the form factor of rings, and the region around 0.045 to 0.05 nm^{-1} will be called "microtubule scatter" (see arrows C, R and M in Fig.1). The time-dependent changes in these regions allow one to assess the main features of assembly and disassembly (e.g., Fig. 2a, c). For example, a rise in the central scatter indicates overall assembly, a rise in the microtubule scatter usually means formation of microtubules, and a rise in the microtubule scatter usually means formation of microtubules, and a rise in the ring scatter may mean ring formation and/or other structures (e.g. oligomers). The actual structures present may be analyzed from the scattering curves or difference plots at given time points (e.g. Fig. 2b).

RESULTS

(a) pre-nucleation and assembly phases in standard buffer conditions

Fig. 2 illustrates one cycle of assembly and disassembly of microtubule protein in standard buffer conditions (0.1 M PIPES reassembly buffer pH 6.9). The projection plot (Fig. 2a) shows the time dependence of the scattering traces. Representative curves from the initial (cold) and the polymerized state (warm) are shown in Fig. 2b. The initial state has a low central scatter (left) and the side maxima typical of rings of diameter 36 nm; the polymerized state has a higher central scatter and the side maxima characteristic of microtubules (weight average mean diameter 25 nm). A detailed comparison with the calculated scattering of model structures shows that most observed traces can only be explained if one allows for contributions from other structures (e.g. oligomers, dimers, MAPs; see Bordas et al., 1983). At high temperatures the patterns are dominated by the scattering from microtubules, with a minor contribution from species that are best explained in terms of oligomers and/or MAPs that contribute mainly at very small scattering angles and fill in the minima of the microtubule trace. The scattering at low temperature can be modelled by mixtures of single and double concentric rings plus oligomers. The scattering from double rings has a higher first subsidiary peak but decays more quickly than the scattering from single rings. As before, oligomeric species contribute mainly at small angles, with radii of gyration on the order of 10 nm (determined from the slope of Guinier plots). The contribution

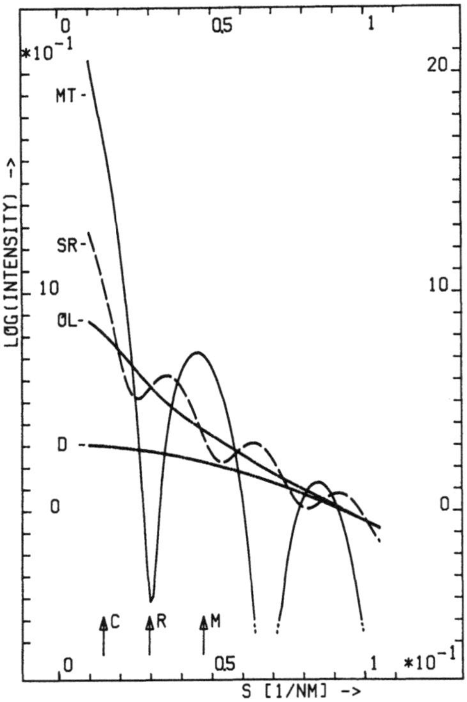

Fig. 1

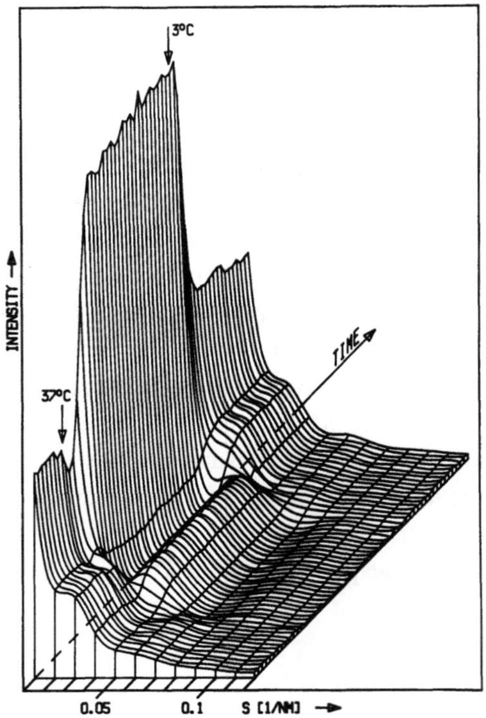

Fig. 2a

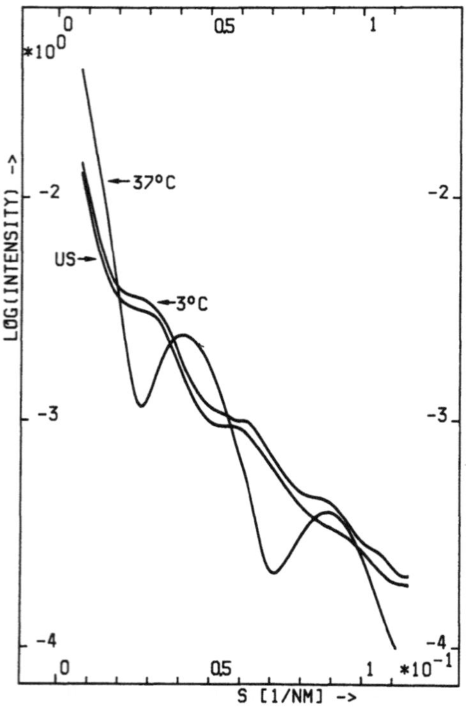

Fig. 2b

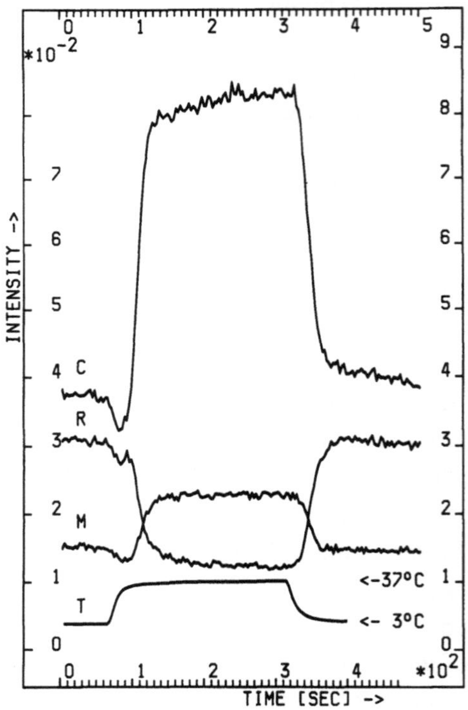

Fig. 2c

from tubulin dimers is generally small and thus negligible within noise until one reaches higher scattering angles (second side maximum and beyond, see Fig. 1) However, the presence of species with unknown form factors allows only an approximate quantitation of the individual componentsm especially at low temperature and during the assembly phase.

Fig. 2c shows the time courses of the intensities in the regions designated in Fig. 1 as central scatter (c), ring scatter (R), and microtubule scatter (M); the temperature (T) is shown in the bottom trace. They show the following features: After the upward

---

Fig. 1: Theoretical scattering curves of model structures. The curves were calculated on the basis of Debye's formula, using spheres of 4 nm diameter to represent the protein subunits. The scattering of mixtures of aggregates can be calculated from $I(S,t) = \Sigma_k \; x_k(t) \, p_k \, i_k(s)$ where the index k refers to the species of aggregate, $x_k(t)$ is the fraction of subunits assembled into the species k, p_k is the degree of polymerization, $i_k(s)$ is the form factor shown in the Figure, and $S=2\sin\theta/\lambda$ is the Bragg scattering vector (for details see Mandelkow et al ., 1983).

From top to bottom on left-hand side: Microtubule with 13 protofilaments of mean diameter 26 nm (MT); ring of 36 nm mean diameter (SR), oligomer consisting of 4 dimers (=protofilament fragment, OL), and dimer (D). Note that the central scatter (left) increases with the degree of polymerization; the side maxima of rings and microtubules are roughly in antiphase; oligomers and dimers have smoothly decaying scattering curves, with the former contributing noticeably at small angles. The arrows marked C, R, and M indicate the ragions sensitive to overall assembly ("central scatter", C), rings (R) or microtubules (M). They are used to monitor the time-dependence of the reactions.

Fig. 2: Microtubule assembly in 0.1M PIPES reassembly buffer pH 6.9.

(a) Projection plot of control experiment. EMBL instrument X33, camera length 3 m, linear position-sensitive detector with 256 channels, 256 time frames of 3 sec. per run (only every third shown), temperature jumps from 3°C (initial cold) to 37°C and return (arrows). Note the increase in central scatter during assembly and the change in side maxima. The scattering curves in this Figure and elsewhere have been smoothed by cubic spline functions.

(b) Selected scattering traces of experiment shown in (a). Initial state at 3°C (arrow) showing side maxima typical of rings (36 nm mean diameter). Just below this trace we show the undershoot before onset of assembly, showing a decrease in the fraction of rings. The pattern at 37°C (arrow) shows side maxima typical of microtubules (25 nm weight average mean diameter).

(c) Time course of scattered intensities. From top to bottom: Central scatter (integrated from S=0.011 to 0.015 nm^{-1}, C) ring scatter (0.023 to 0.031 nm^{-1}, R); microtubule scatter (0.041 to 0.051 nm^{-1}, M), and temperature (initial and final values 3 and 37°C, T).

temperature jump the central scatter drops briefly, then rises rapidly to twice its initial value, followed by a second slow rise, and after the reverse T-jump the intensity drops in a biphasic manner and equilibrates slightly above its initial value. The microtubule scatter shows a similar behavior but with different amplitudes since the rise in polymarization is due to the formation of microtubules. The time course of the ring scatter is roughly the mirror image of the microtubule scatter since rings disappear when microtubules are formed, and since the microtubule scatter has a minimun where the ring scatter has a peak (Fig. 2b).

Here we are particularly concerned with the small deviations from the initial level before and after assembly. The undershoot after the T-jump up can ben accounted for by the partial dissolution of rings; this feature is seen in all therr time courses shown. The increase in the level after on cycle is seen mainly in the central scatter and is explained by non-ring oligomers which re-assemble into rings slowly or not at all.

Several lessons can be learned from experiments such as that of Fig. 2: (1) The apparent duration and magnitude of the undershoot is determined by two reactions affecting the scattering in opposite directions. (2) Structurally the first phase of the undershoot may be interpreted in terms of partial ring breakdown, while the beginning of the rising phase corresponds to the association of oligomers. (3) The observed phases are closely reminiscent of the assembly induced by the slow T-scan described previously (Bordas et al., 1983), in spite of the 60-fold difference in heating rate (2^{0}C/sec and 2^{0} C/min, resp.). This suggests that similar structural transitions take place in the two conditions. (4) The effective temperature measured at the onset of assembly is variable reflecting the lag due to prenucleation and nucleation events. By contrast, disassembly sets in rather reproducibly at a temperature just below 20^{0}C, indicating a major change in microtubule stability around this temperature.

(b) separation of pre-nucleation and nucleation events

When surveying different assembly conditions we found that a clear separation of pre-nucleation and nucleation events could be achieved by lowering ionic strenght and pH. The experiments to be described now have been perfomed with the same protein preparation as in Fig. 2, carried through another cycle in order to change buffer conditions. Fig. 3a shows the time courses of the intensities at various scattering angles of a sample in 0.05 M PIPES buffer pH 6.5 In these conditions the undershoot reaches equilibrium before the onset of assembly. The difference curve between the initial state and bottom of the undershoot indicates that the intensity decay is due to the disappearance of rings.

Microtubule assembly starts about 16 sec after initiating the T-jump. The longer delay compared with the previous experiment (Fig. 2c) is one of the factors explaining the large undershoot. The second factor is the higher initial concentration of rings. This can be shown by scaling the initial cold patterns with respect to the patterns after microtubule assembly. This means the lower ionic strenght and pH favor the formation of rings at low temperature but do not protect them against temperature-induced disassembly. The experiment also supports the view that ring formation and microtubule formation represent antagonistic reactions.

The apparent half-time of ring breakdown is about 10 sec. Close inspection of the intensity decay shows that is is nearly linearly related to the temperature, both in fast temperature jumps and slow temperature scans (this is best show by correlation plots, see Bordas et al ., 1983) Thus the apparent rate does not represent a reaction rate but rather a temperature-dependent equilibrium between rings, oligomers, and dimers whose interconversion is faster then the T-jump employed here. This explains why the breakdown of rings produced by the temperature rise itself is only partial in the temperature range studied here.

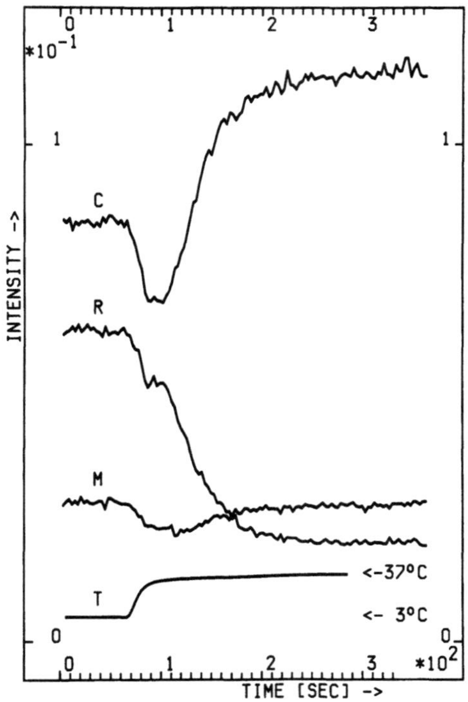

Fig. 3a

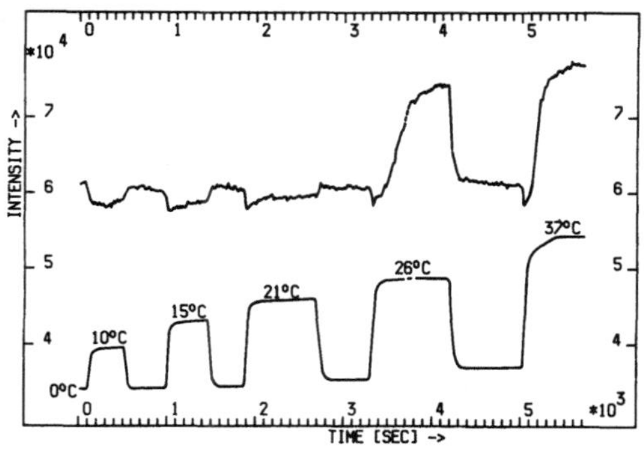

Fig. 3b

Fig. 3a: Assembly of microtubule protein prepared from the protein used for Fig. 2 by another cycle of assembly and resuspending the pellet in 0.05 M PIPES reassembly buffer at pH 6.5. From top to bottom: Time courses of central scatter "ring" peak, "microtubule" peak, and temperature.

Fig. 3b: Experiment showing the time course of the central scatter (top) during limited temperature jumps (bottom). The low temperature was kept at 0°C, the high temperature was set at 10, 15, 21, 26 and 37°C during successive cycles. Every T-jump induces the reversible disassembly of rings, but only the last two cycles cause microtubule assembly.

Another way to separate pre-nucleation events from assembly is to apply limited temperature jumps, as illustrated in Fig. 3b. In this experiment the protein was cycled several times. The low temperature was kept at 0°C, while the high temperature was initially set at 10°C and increased in several steps during successive cycles (i.e. 10, 15, 21, 26, and 37°C). In this example the temperatures of the first three cycles are not sufficient to initiate assembly so that one observes only the pre-nucleation events. Microtubules are formed during the fourth and fifth cycle, preceded as before by the undershoot due to prenucleation breakdown of oligomers.

In summary, the experiments of Fig. 3 illustrate that the different reactions of the proteins following a temperature jump may be separated in several ways whose common feature is to prevent or retard the onset of nucleation.

(c) effect of pH and cycling on oligomeric and polymeric states

The distribution of oligomers of microtubule protein depends on solution conditions. Using analytical ultracentrifugation Marcum & Borisy (1978) showed that low pH, low ionic strength, and high protein concentration favored a 30S species (rings), whereas higher pH (>7) and ionic strength cause its disintegration into 18S oligomers and 6S dimers. Since we were interested in the role of oligomers during assembly we perfomed X-ray experiments in similar solution conditions. The protein was prepared by 2 cycles in standard conditions (0.1 M PIPES pH 6.9) and adjusted to the desiderd pH during a third cycle.

Fig. 4 (top) shows the scattering traces after assembly at pH 6.5, 6.9, and 7.7 in 0.1 M PIPES reassembly buffer. The clearest pattern (in terms of contrast between maxima and minima) is obtained at pH 6.9; it is dominated by the scattering of microtubules of 28 nm mean diameter. The other patterns are less contrasty and indicate an increasing contribution from non-microtubular aggregates. This is particularly apparent at pH 7.7.

In the case of the cold solutions (Fig. 4, bottom) the curves at PH 6.9 and 6.5 show clear maxima arising from rings. The middle curve (pH 6.5) decays more rapidly, as expected from a larger fraction of double rings, compared to pH 6.9 (in agreement with electron microscopy). Note that the single and double rings have similar mean diameters (about 38 nm). By contrast, the maxima at pH 7.7 are strongly reduced, indicating that rings break apart at higher pH-values and give rise to other oligomeric structures, as seen from the slope of the traces (as before, the fact that some rings are observed is due to the high protein concentration).

In spite of the differences in assembly forms the kinetics are similar at the three pH values studied; little undershoot is visible, and the half time of assembly is about 15 sec. In other words, the high pH at 0.1 M PIPES reduces the extent of microtubule assembly but not the rate of nucleation. This is in contrast to the effect of lowering the buffer strenght to 0.05 M which slows down nucleation but has a lesser effect on the extent of assembly (see Fig. 3).

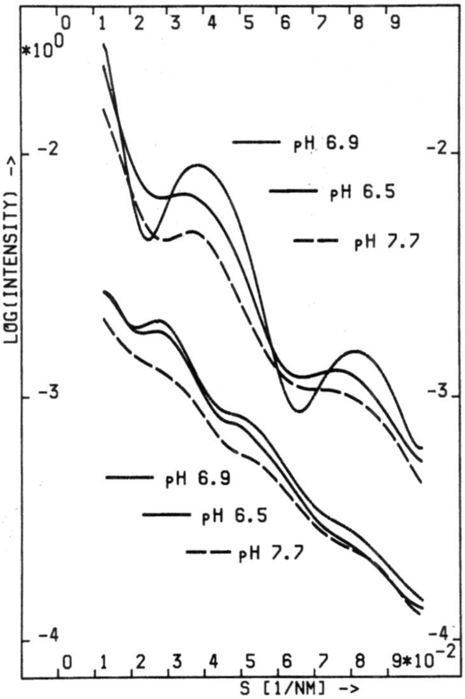

Fig. 4

Fig. 4: Effects of pH on the X-ray scattering. Bottom: the three curves show the scattering of solutions at 4°C at pH 6.9, 6.5, and 7.7 (in descending order). The subsidiary maxima typical of rings of about 38 nm diameter are seen most clearly at pH 6.9. At pH 7.7 the ring scattering is greatly reduced, and the maxima are shifted to the right, corresponding to a mean diameter around 36 nm. Top: Scattering of same three preparations after microtubule assembly, same descending order in the region of the subsidiary maximum typical of microtubules. Note that the maxima and the contrast between the minima is most clearly expressed at pH 6.9 and least well at pH 7.7. The top curves have been shifted up by 0.75 log units (factor 5.6) for better visibility.

The Figures shown thus far illustrate that the equilibrium between oligomeric states is affected by a variety of solvent conditions such as temperature, ionic strength, or pH, a complete map of which would result in a phase diagram. It is clear that other factors may be equally important, as illustrated by the following two examples:

When microtubule protein is stored frozen in reassembly buffer lacking GTP the concentration of rings after unfreezing is very low, even when GTP is added; instead one observes a higher contribution from non-ring oligomers. However, a cycle of assembly and disassembly restores the usual equilibrium between rings and oligomers (comparable to that shown in Fig. 2b). Secondly, purified tubulin does not contain rings in the standard buffer, and only few are formed when the colde protein is mixed with MAPs. However, the usual fraction of rings reappears after one cycle. These examples indicate that the history of the sample must be taken into account; the species generated by microtubule disassembly are not necessarily the same as those initiating assembly.

DISCUSSION

Despite much work on microtubule assembly in vitro the nature of the entities that contribute to nucleation and elongation has remained rather elusive. Studies with brain microtubule protein showed that in conditions conducive to assembly one observes a variety of polymorphic forms (rings, sheets, helical ribbons, etc.). These led to models of nucleation based on the assumption that the polymorphic forms were precursors or intermediates (reviewed by Kirschner, 1978). On the other hand, kinetic data supported models of elongation based on the endwise addition of subunits (e.g. Gaskin et al ., 1974). Other workers investigated the assembly of purified tubulin and showed that it did not require the polymorphic intermediates. This poitend to the influence of MAPs which not only stimulate assembly but also cause polymorphism, indicating that the relationship between the two effects may only be indirect. Finally, the demonstration that microtubule assembly was preceded by the disassembly of rings led to the view that oligomeric species smaller than rings contributed to nucleation and possibly to elongation (Engelborghs et al., 1977; Mandelkow, et al 1980 Weisenberg, 1980). Present discussion on microtubule assembly are largely centered on the mechanism of elongation or shortening, in particular the role of nucletoides and their possibile implicatoins in treadmilling (Margolis & Wilson, 1978) or dinamic instability (Mitchison & Kirschner, 1984). By contrast, the mechanism of nucleation and the events leading to it have received less attention, probably for want of a reliable assay. In this report we use the method of time-resolved X-ray scattering to study the fate of oligomers before and during assembly.

In earlier experiments employing a moderate temperature jump we found that the assembly of microtubule protein was preceded by pre-nucleation events consisting largely of ring disassembly (Mandelkow et al., 1980). Near-equilibrium T-scan experiments showed that the subsequent nucleation could be described in terms of oligomer association (Bordas et al., 1983). It was not clear however what the relationship between the breakdown products of rings and the nucleating oligomers was. These experiments were perfomed in reassembly buffer with 0.1 M PIPES pH 6.9 in which the pre-nucleation events largely overlapped. We therefore attempted to find conditions where the two phases can ben observed separately. This can be achieved in several way, for example by lowering the pH and the ionic strenght, by reducing GTP, or by lowering the temperature jump. The main findings can be summarized as follows:

(1) Temperature dependence: Under all conditions studied (including slow T-scan and fast T-jump) the pre-nucleation events were almost linearly correlated with temperature, i.e. the half time of equilibration between rings and their disassembly

products is in the range seconds or less. Thus, the ring pattern can be used as a thermometer to monitor the temperature of the solution. Thus the formation of rings from their subunits is an exothermic process.

(2) Extent of the reaction : Between 3°C and 37°C the ring fraction decreases only by about 20%. Thus the rings appear surprisingly stable, even at elevated temperature. Since the initial ring fraction is about 50-70% of the total protein in our concentration range the enthalpy of ring formation from their subunits is on the order of a few kcal/mole. The complete disappearance of rings during microtubule assembly therefore requires processes other than temperature-induced ring breakdown. The simplest explanation would be the depletion of the subunit pool (small oligomers and dimers) by the nucleating species. This implies that there is an equilibrium of the form

$$\text{rings} \leftrightarrow (\text{oligomers} + \text{dimers}) \leftrightarrow \text{nuclei},$$

such that both rings and nuclei draw on the same pool (see Fig. 5).

(3) Reversibility: Prior to assembly the equilibrium between rings and their break-down products appears to be reversibile. By contrast, once nucleation sets in the assembly cycle shows hysteresis (see Bordas et al., 1983).

(4) Nucleotides: Their role and their dependence on Mg^{++} will be described in more detail elsewhere, but we note some key features here: In the absence of GTP one obsevers the reversible dissociation and association of rings. In the presence of GTP the same effect can be observed, provided that the temperature jumps remain below the point where nucleation sets in. This suggests that neither binding nor hydrolysis of GTP plays a role at the pre-nucleation stage.

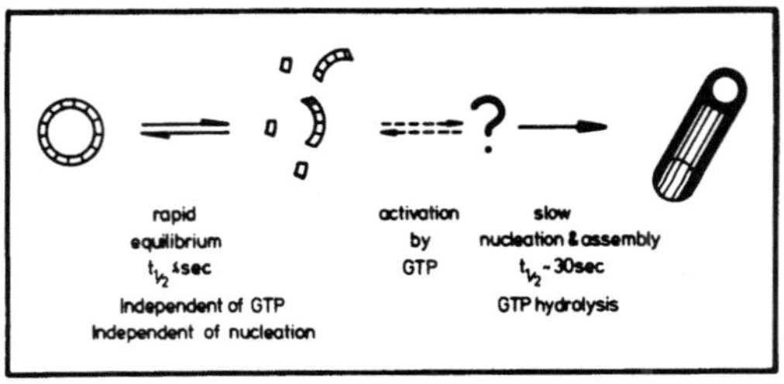

Fig.5

Fig. 5: Diagram summarizing the differences between pre-nucleation and nucleation events. There is a rapid and reversible temperature-dependent equilibrium between rings smaller oligomers, and dimers that is not directly related to microtubule assembly (left part of diagram). By contrast, microtubule nucleation and assembly is slow and takes palce above a threshold temperature (right part of diagram).
Since the ring-forming and microtubule-forming modes of assembly draw on the same pool of building blocks (oligomers and dimers) one requires one or more intermediate steps during which the subunits are activated for assembly, e.g. by temperature and GTP. The structure (s) of the activated oligomers and dimers are unknown but could correspond to straightened protofilament fragments, in contrast to the inactive coiled ones.

(5) Role of MAPs: MAPs from mammalian brain support the formation of ring oligomers, but the efficiency is low unless the mixture the tubulin and MAPs is cycled. This indicates that MAPs bind to microtubules in a specific fashion such that subsequent disassembly generates preformed building blocks which then reassociate into rings (e.g. the coiled protofilament fragments observed by cryo-electron microscopy, see Mandelkow & Mandelkow, 1985). Although complete rings are not observed in cells the tubulin -MAP building blocks may be functionally importante, e.g. as storage and/or assembly units.

(6) Role of microtubule assembly inhibitors: There are a variety of conditions which prevent microtubule assembly without a major effect on ring stability. Two of these have already been mentioned, namely low temperature and absence of GTP. Other examples are the presence of mM amounts of Ca^{++} or GDP which leave rings largely intact even 37°C (data not shown). It appears that conditions that are conducive to ring formation are detrimental for microtubule formation, and vice versa. This fact may be expressed more simply in structural terms: protofilament coiling (rings) is antagonistic to protofilament straightening (microtubules). One could turn the argument around by postulating that microtubule disassembly is caused by factors whose primary effect is to induce a coiled protofilament conformation, even when there is no ring-like closure of the protofilaments. Evidence in support of this comes from disassembly experiments with colchicine (Kirschner et al., 1974) or rapid freezing (Mandelkow & Mandelkow, 1985). Note that this interpretation rests on the view that tubulin oligomers consist of short protofilament fragments.

Oligomeric species other than rings are almost invariably observed by negative stain electron microscopy of cold microtubule protein. Analytical ultracentrifugation tends to emphasize bimodal distributions between tubulin dimers (6S) and rings (30S and up), but intermediate species are observed as well. Some of these may be the result of isodesmic self-association (especially in the case of purified tubulin, see Frigon & Timasheff, 1975) while others appear to represent discrete assembly states that can be fixed and observed by electron microscopy (e.g. the 18S species of microtubule protein, Marcum & Borisy, 1978). Small oligomers are particularly enriched during microtubule nucleation, as seen both by analytical ultracentrifugation (Weisenberg, 1974) and electron microscopy (Bordas et al., 1983). As described here, non -ring oligomers can also be enriched at the expense of rings by certain solution conditions suche as high pH, high salt, storage of the protein at low GTP or without MAPs. Finally, the participation of tubulin-MAP oligomers in the nucleation of microtubule protein is suggested by the low apparent cooperativity (n=2, Engelborghs et al., 1977), compared with that of purified tubulin (n=12, Carlier & Pantaloni 1978).

The sum of these observations is somewhat contradictory: On the one hand, non-ring oligomers appear to play a role in microtubule nucleation, on the other hand conditions that inhibit microtubule assembly favor the formation of non-ring oligomers. This is best explained by assuming that there are different types of oligomers, active and inactive ones. Active oligomers would contribute to microtubule nucleation and possibly elongation. Inactive oligomers could consist of partially denatured protein, as suggested by Correia & Williams (1985). Alternatively, oligomers may be temporarily inactivated, e.g. by binding of assembly inhibitors such as GDP or Ca^{++} or at low temperature, but may be reactivated by removing these ligands, replacing them by activators (e.g. GTP), or raising the temperature. Structurally, an inactive oligomer ma be viewed as a coiled protofilament fragment which would be straightened as a result of activation.

The continuity between the pre-nucleation and nucleation events observed previously (Mandelkow et. al., 1980) suggested that there was a direct conversion from ring breakdown products to microtubule nuclei. Since the two phases can be separated it appears now that there must be additional intermediate steps, i.e. breakdown of rings into oligomers or dimers is not sufficient to cause microtubule assembly (and vice versa). This is best explained if we make the following assumptions:

- The ring-forming and microtubule-forming modes of assembly are not directly linked to one another (this would be in analogy with the disk-forming and helix-forming modes of assembly of the well-known example of tobacco mosaic virus protein see Lauffer, 1975).
- Both assembly forms draw on the same pool of intermediates, i.e. oligomers and dimers.
- The intermediates can be activated or inactivated for the respective modes of assembly by binding of ligands (e.g. GDP, GTP, etc.).

Thus a minimal reaction scheme would be

$$\text{rings} \leftrightarrow (\text{oligomers + dimers})_C \leftrightarrow (\text{oligomers + dimers})_S \leftrightarrow \text{nuclei},$$

in which the subscripts C ("coiled") refers to the oligomeric state that is active for ring formation but inactive for microtubule assembly, and S ("straight") refers to the state that is active for microtubule assembly (Fig. 5).

ACKNOWLEDGEMENTS

We gratefully acknowledge the important contributions of our colleagues to the project: P. Maier (MPI) for building the T-jump cell, E Dorrington (EMBL) for the electronic control modules, J. Hendricks (EMBL) for providing the position-sensitive detector, U.Rühl (MPI) for protein preparations, and M.Koch (EMBL) for discussions and computer programs for data processing. This work was supported by the Max-Planck-Gesellschaft (MGP), the European Molecular Biology Laboratory (EMBL), the Deutsche Forschungsgemeinschaft (DFG), and the Commision of the European Comunities (CODEST).

REFERENCES

Bordas, J., Koch, M.H.J., Clout, P.N., Dorrington, E., Boulin, C., & Gabriel, A. (1980). J. Phys. E: Sci. Instrum. 13, 938-944.
Bordal, J., Mandelkow, E.-M. & Mandelkow, E. (1983). J. Mol. Biol. 164, 89-135.
Carlier, M.F. & Pantaloni, D. (1978). Biochemistry 17, 1908-1915.
Correia, J.J. & Williams, R.C. (1985). Arch. Biochem. Biophys. 239, 120-129.
Engelborghs, Y., De Mayer, L.C.M & Overbergh, N. (1977). FEBS Letters 80, 81-85.
Frigon, R.P. & Timasheff, S.N. (1975). Biochem. 14, 4559-4566.
Gabriel, A. (1977). Rev. Sci. Instrum. 48, 1303-1305.
Gaskin, F., Cantor, C.R. & Shelanski, M.L. (1974). J. Mol. Biol. 89, 737-758.
Hendrix, J., Koch, M.H. & Bordas, J. (1979). J. appl. Cryst. 12, 467-472.
Kirschner, M.W., Williams, R.C., Weingarten, M. & Gerhart, J. (1974). Proc. Natl. Acad. Sci. U.S.A. 71, 1159-1163.
Kirschner, M. W. (1978). Int. Rev. Cytol. 54, 1 - 71.
Koch, M.H.J. & Bordas, J. (1983). Nucl. Instrum. Meth. 208, 461-469.
Lauffer, M.A. (1975). Entropy-driven processes in biology. Springer-Verlag Berlin.
Mandelkow, E.-M.,Harmsen, A.,Mandelkow, E. & Bordas, J. (1980). Nature 287, 595-599.
Mandelkow, E., Mandelkow, E.-M. & Bordas, J. (1983). J. Mol. Biol. 167, 179-196.
Mandelkow, E.-M. & Mandelkow, E. (1985). J. Mol, Biol. 181, 123-135.
Mandelkow, E.-M., Herrmann, M. & Ruhl, U. (1985). J. Mol. Bio. 185, 311-327.
Marcum, J.M. & Borisy G.G. (1978). J. Biol. Chem. 253, 2825-2833.
Margolis, R.L. & Wilson, L. (1978). Cell 13, 1-8.
Mitchison, T. & Kirschner, M. (1984). Nature 312, 237-242.
Renner, W., Mandelkow, E.-M., Mandelkow, E. & Bordas J. (1983). Nucl. Instr. Meth. 208, 535-540.
Spann, U., Renner, W., Mandelkow, E.-., Bordas, J. & Mandelkow, E. (1987). Biochemistry, in press.
Weisenberg, R.C. (1974). J. Supramol. Struct. 2, 451-465.
Weiswnberg, R.C. (1980). J. Mol. Biol. 139, 660-677.

Sinusoidal Length Change Study of Muscle Contraction and Self-Induced Translation Model of Myosin Motion

Toshio MITSUI*, Katsuzo WAKABAYASHI*, Hidehiro TANAKA**, Takakazu KOBAYASHI**, Yutaka UENO*, Yoshiyuki AMEMIYA***, Hiroyuki IWAMOTO**, Toshiaki HAMANAKA* and Haruo SUGI**

* Department of Biophysical Engineering, Faculty of Engineering Science, Osaka University, Toyonaka, Osaka 560, Japan; ** Department of Physiology, School of Medicine, Teikyo University, Itabashi-ku, Tokyo 173, Japan; *** Photon Factory, National Laboratory for High Energy Physics, Oho-machi, Tsukuba-gun, Ibaraki 305 Japan

1. Introduction

By using synchrotron radiation, it was observed that X-ray diffraction pattern changed markedly when sinusoidal length changes were applied to isometrically contracting frog skeletal muscle [1]. We proposed a model for the filament sliding mechanism in order to explain the obtained data as well as various other data. Basic idea of the model was given briefly in a report [2]. Here we discuss the model in more detail and present experimental data which are in agreement with what the model predicts.

2. Model

Neccesity of a new model

It is generally believed that the force for filament sliding in contracting muscle is generated by a cyclic interaction of myosin heads with actin molecules coupled with ATP hydrolysis and that a power stroke (a change in orientation) of the myosin head takes place during the cyclic interaction [3,4]. So far, however, no direct evidence has been found for the power stroke. For instance, by polarized fluorescence study, Yanagida found that about 40% of myosin heads were attached to the actin filament in isometrically contracting rabbit psoas muscle [5], and optical probe bound to the attached myosin head oriented relative to the actin filament axis at nearly the same angle as that in rigor [6]. This fact suggests that there is no change in the orientation of the probed region of the myosin molecule during the attached phase. Cooke et al. [7] also observed that spin-labeled domain does not change orientation in contractile interaction. Further Yanagida [5] observed that the orientation of the optical probe bound to myosin molecule in the attached phase was not altered during sudden changes in length of isometrically contracting muscle. On the other hand, with synchrotron radiation, Huxley et al. found that X-ray diffraction pattern from frog skeletal muscle showed very significant changes during tension development [8] and for rapid change of muscle length [9]. As mentioned above, Wakabayashi et al. observed that X-ray diffraction pattern changed markedly with sinusoidal change of muscle length [1]. Their data are cited in Fig.1, which shows that, for the sinusoidal length change (a), pronounced changes took place in intensities of the 14.3 nm meridional reflection (b) and the equatorial (1,0) and (1,1) reflections (c). Figure 1 shows also intensity changes during tension development, which are similar to those observed by other authors [8, 10, 11]. These X-ray results indicated that there occurs large molecular motion associated with length changes of isometrically contracting muscle, whereas the fluorescence and EPR studies did not provide evidence of orientational change of myosin head attached to the actin filament as mentioned above. This apparent

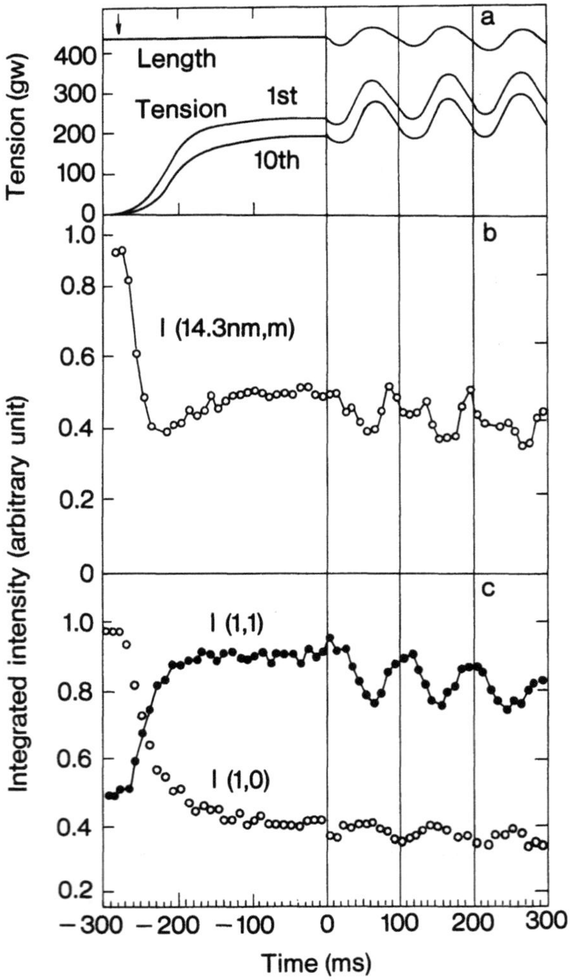

Fig.1. Effect of sinusoidal length changes of frog skeletal muscles on the intensities of the myosin meridional reflection of the spacing of 14.3 nm and the equatorial (1,0) and (1.1) reflections, at 17°C [1]. (a) Changes of muscle length and tension. Arrow indicates the onset of stimulation (t=-275ms). When the steady tentanic tension was reached, sinusoidal length change (1%, 10 Hz) was applied at t=0. Two curves of tension correspond to the first and the tenth experiments. (b) intensity of the 14.3 nm meridional reflection, I (14.3 nm, m). (c) Integrated intensities of the equatorial (1,0) and (1,1) reflections, I(1,0) and I(1,1).

contradiction was our first motive to proposing a new model.

Recently Yanagida et al. [12] measured velocity of sliding of free thin filament on the thick filament and rate of ATP hydrolysis, and estimated sliding distance of the thin filament during one ATP hydrolysis cycle as more than 60 nm. They suggested the possibility that the range of movement over which myosin molecule can attach to the thin filament during one ATP hydrolysis cycle under an unloaded condition is more than 60 nm. We shall explain how such long distance movement is possible with the following model.

Self-induced translation model

The model is based upon the assumption that myosin molecule moves along the thin filament by self-induced motive force. More specifically the assumptions are as follows.
(1) When a myosin head activated by ATP hydrolysis binds to actin monomer in the thin filament, the attached and neighbouring actin monomers change their conformations.
(2) The conformational changes cause increase of affinity to myosin heads in the upward actin monomers and decrease of the affinity in the downward actin monomers, where the upward direction is defined

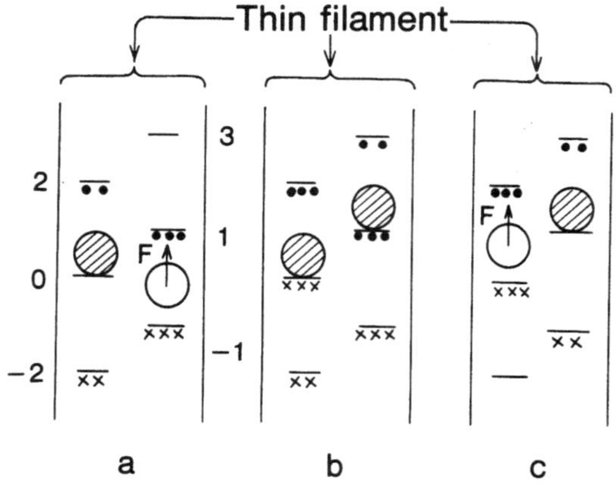

Fig.2. Explanatory diagram of the self-induced translation model. Bar stands for binding site of actin monomer to myosin head: • and × symbolize imcrease and decrease of the affinity to myosin head, respectively. Two rows of bars represent the two strands, neglecting the helical structure. Shaded and white circles represent bound and unbound myosin heads, respectively.

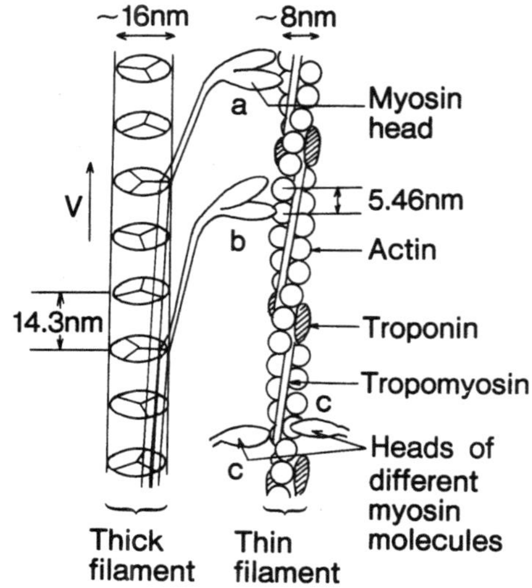

Fig.3. Illustration of the self-induced translation model. Myosin molecules climb up the thin filament causing the thick filament to slide.

as the direction to which myosin molecules move during muscle contraction.

(3)The conformational changes modify distribution of electric field in the vicinity of the thin filament and exert an upward electric force upon near-by myosin heads. The thin filament is polar in structure and it seems reasonable to expect such an effect as piezoelectricity.

Figure 2 illustrated the model. The head bound to actin is depicted by a shaded circle and unbound head by a white circle. The unbound head will move upward by the force F assumed in (3) as shown in Fig.2a and will attach to the upper actin monomer 1 or 2 (Fig.2b). Then affinity of the actin monomer 0 to the myosin head will decrease by the assumption (2) and the head will detach from it (Fig.2c). The detached head will move upward by the force F (Fig.2c). Repeating these steps, myosin molecule will climb up the thin filament resulting in sliding of the thin filament.

In this model, formation of a pair of two heads is essential and the pairing mate does not necessarily belong to the same myosin molecule. It seems, however, reasonable to expect that the pair will be formed most frequently by two heads of the same myosin molecule in actual muscle. Then, considering matching of polarity of myosin and actin molecules, the actin monomers to be bound will be on the same strand as indicated by 0, 2 in Fig.2a. Figure 3 illustrates two myosin molecules climbing up the thin filament with their two heads by a and b. Two heads which belong to different myosin molecules may climb up along different strands as shown by c in Fig.3.

The above scheme implies that there is the following feedback loop.

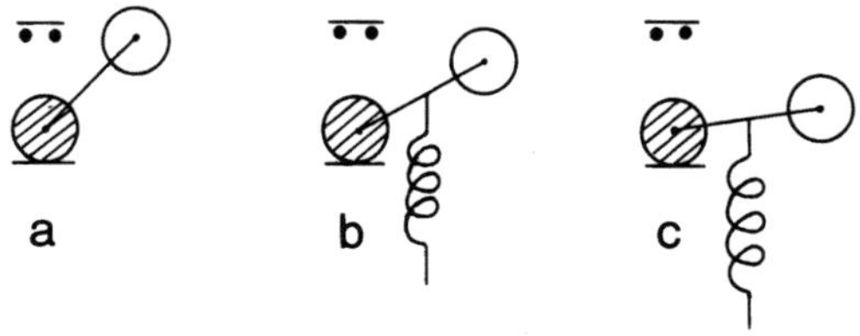

Fig.4. The mean position of unbound head is lowered with increasing load. (a) no load, (b) loaded shortening, (c) isometric contraction.

Binding of myosin head to actin
↓
Local conformational changes of the thin filament
↓
→ Upward translation of myosin molecule
(feedback loop) ↓
— Upward translation of the local conformational changes

This scheme is similar to the self-induced oscillation in electric circuit, and we call the model a self-induced translation model.

3. Explanation of Experimental Results

According to our model, the power stroke of the bound head is not needed for filament sliding, and thus the bound heads may appear immobile during tension development or during changes of muscle length as was reported by Yanagida [5,6] and Cooke et al. [7].

In our model, two myosin heads may attach to neighbouring two actin monomers at the same moment as shown in Fig.2b, but life time of this state will be very short according to the assumption (2). Therefore, it is expected that the population of myosin heads near the thin filament will be about twice the population of bound heads. Matsubara et al. [10] estimated that about 80% of myosin heads exist in the vicinity of the thin filament in isometrically contracting frog skeletal muscle, based upon the marked changes of the equatorial (1,0) and (1,1) reflections during tension development (as seen in Fig.1c). On the other hand, Yanagida [5] reported that about 40% of the myosin heads were bound to actin filament in isometrically contracting rabbit psoas muscle. The ratio of these values is 80/40=2, as expected above.

The mean position of unbound head will be determined by balance between the electric force assumed in (3) and load on muscle. Figure 4 illustrates the situation, where the force by load is assumed to be exerted on myosin heads through a spring. The mean position of the unbound head will be lowered with increasing load as shown from a to c in Fig.4. Now let us consider how the X-ray intensity I(14.3 nm, m) changes for the sinusoidal length change if the position of unbound head is at the same level as bound head at isometric contraction. When muscle length changes sinusoidally as shown in Fig.5a, the unbound head will oscillate as in Fig.5b. Then electron density projection of the two heads onto the fibre axis will change as shown in Fig.5c, and simple calculations gave variation of I(14.3 nm, m) as shown in Fig.5d, which demonstrates that period of oscillation of I (14.3 nm, m) is twice the period of tension variation in Fig.5a. The actual situation is more complicated as will be discussed below, but this result suggests that variation of I(14.3 nm, m) should contain

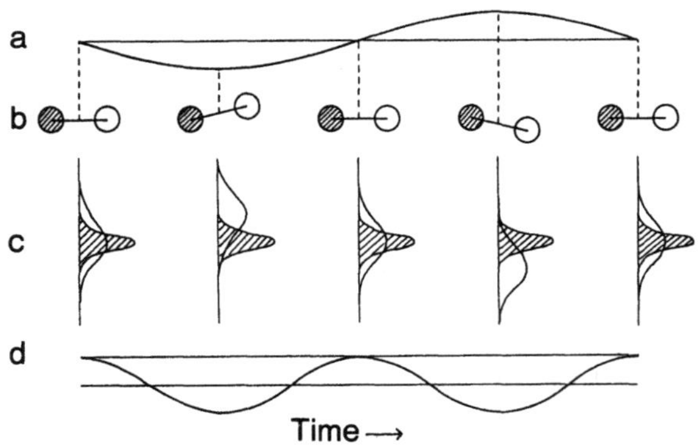

Fig.5. Interpretation of the large second harmonic in the time course of I (14.3 nm, m) in response to the sinusoidal length change. (a) sinusoidal length change, (b) tilting oscillation of two heads of myosin molecule, (c) projection of electron density of the two heads onto the filament axis, (d) calculated I(14.3 nm, m)

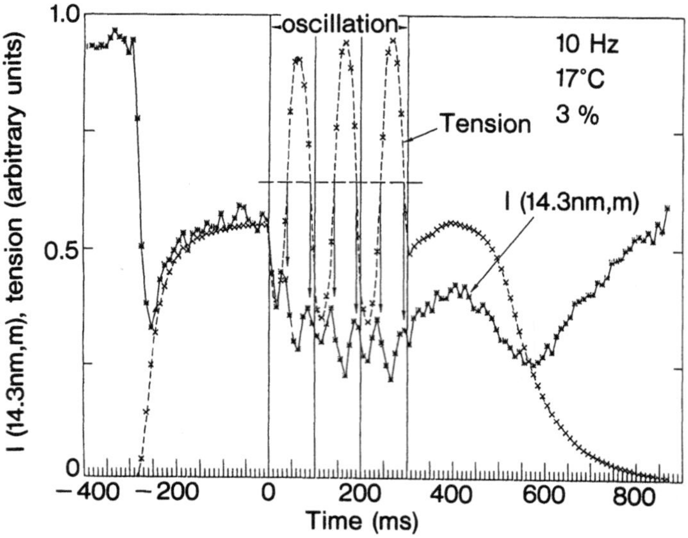

Fig.6. Observed time courses of tension and I(14.3 nm, m) for sinusoidal length change of 3%. The peaks of I(14.3 nm, m) appear around midpoints of the tension oscillation.

second or higher harmonics in response to the sinusoidal length change, as actually observed (cf. Fig.1b). This reasoning led us further to expect that two peaks would appear more evidently per one period of tension oscillation for larger length change. Figure 6 shows observed time courses of tension and I(14.3 nm, m) for amplitude of the length change of 3%, and demonstrates that two peaks actually appeared in I (14.3 nm, m) per one period of tention oscillation.

One characteristic feature of the tension curve in Fig.6 is that the mean level of oscillation (indicated by dashed line) is higher than the isometric tension level. We were afraid that this effect was artifact due to some defect in instrumentation, and checked apparatus by replacing the muscle specimen with rubber string. The rise of oscillation level was not observed with rubber, indicating that the rising level effect is intrinsic with muscle. This effect can be explained by our model as illustrated in Fig.7a. In the experiment shown in Fig.6, the length oscillation started from the release phase.

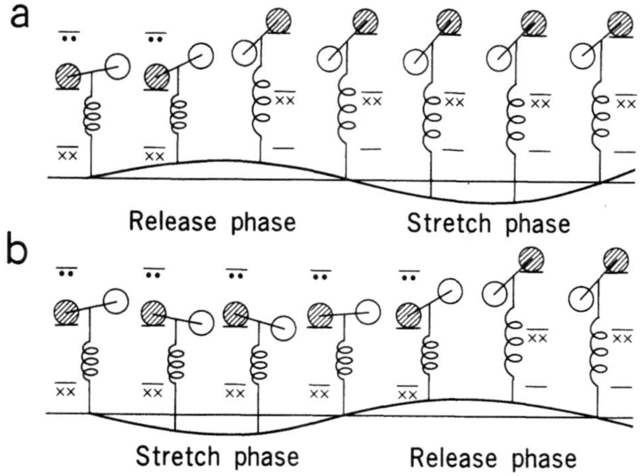

Fig.7. Explanation of the rise of the mean level of tension oscillation by the self-induced translation model. (a) Length oscillation was started from the release phase. Hanging-up of myosin head takes place in the first half of the cycle. (b) Started from the stretch phase. Hanging-up does not take place in the first half of the cycle.

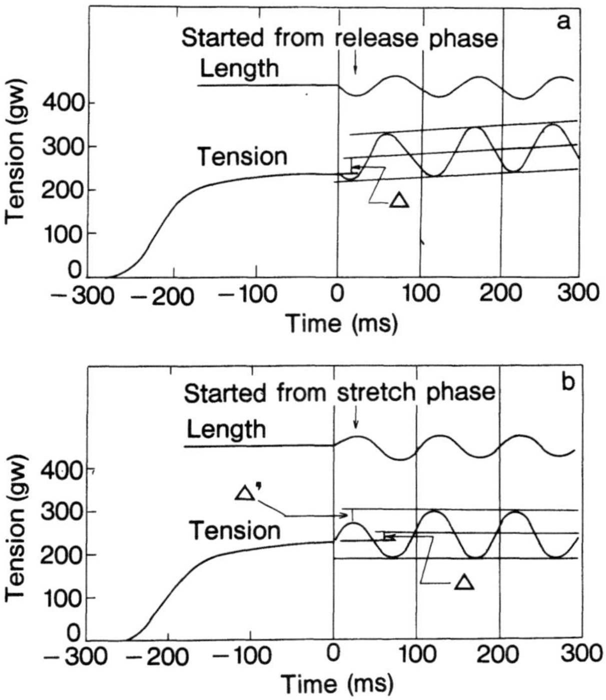

Fig.8. Observed time courses of tension for sinusoidal length changes. Stimulation was started at t=-280 ms, and sinusoidal length changes (1%, 10 Hz) were applied from t=0 after the tetanic tension became steady. Temperature was 17°C. (a) Sinusoidal length change was started from the release phase. Δ indicates rise of the mean level of tension oscillation relative to the isometric tension level. (b) Sinusoidal length change was started from the stretch phase. Rise of the mean level of tension oscillation (Δ) became evident after the first cycle. Δ' indicates that rise of the peak value of tension took place from the second cycle.

Then some myosin molecules will move upward by changing binding site to upper actin monomers during the release phase as shown in Fig.7a since the upper actin monomers have large affinity. The myosin molecules will remain hanged up at the position during the stretch phase since the lower actin monomers have much less affinity, resulting in the rise of the tension oscillation level.

If this explanation is correct, different time course of tension variation is expected when a sinusoidal length change starts from the stretch phase, because such hanging up of myosin molecule will not take place during the stretch phase. That is, the mean level of tension oscillation is expected to remain at the isometric tension level

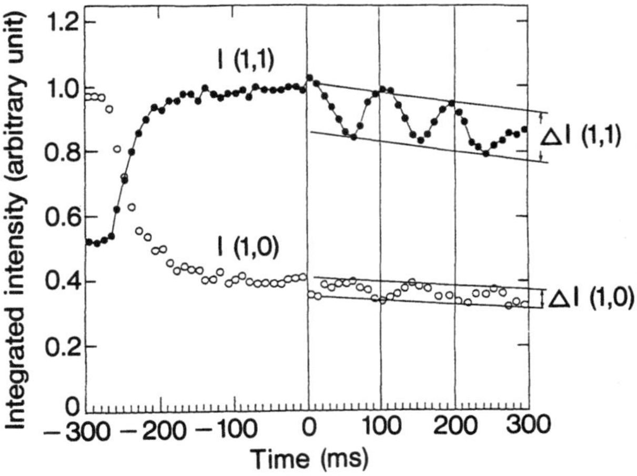

Fig.9. Illustration of the quantities in Eqs.(1) and (2). The data are the same as in Fig.1c. The sinusoidal length change was started at t=0, and the values of I(1,0) and I(1,1) around t=-50ms were used for the calculation in Eq. (2).

in the first cycle of the oscillation. We examined time course of tension against sinusoidal length changes either by starting from the release phase or from the stretch phase. Figs.8a and b show the results. The oscillation level remained at the same level as the isometric contraction level for the initial cycle of oscillation when oscillation was started from the stretch phase as seen in Fig.8b, and such differences as indicated by Δ and Δ' were observed reproducibly.

our model suggests that variation of electron density in muscle will take place mainly near the thin filament during the sinusoidal length change. Then simple calculation gives the following relation

$$\frac{\Delta I(1,0)}{\Delta I(1,1)} = -\frac{1}{2}\left\{\frac{I(1,0)}{I(1,1)}\right\}^{1/2}, \tag{1}$$

where ΔI(1,0) and ΔI(1,1) are intensity changes shown in Fig.9 and minus sign means that their variations should be antiphase, as seen in Fig.9. From the experimental data given in Fig.9, we have

$$\left|\frac{\Delta I(1,0)}{\Delta I(1,1)}\right| = 0.37, \qquad \frac{1}{2}\left\{\frac{I(1,0)}{I(1,1)}\right\}^{1/2} = 0.32, \tag{2}$$

in fairly good agreement with expectation.

We have estimated the distance L over which a myosin molecule moves along the thin filament by using energy liberated by hydrolysis of two ATP molecules, by the following equation:

$$L = \frac{2\varepsilon(ATP)\cdot W/(W+H)}{f(isom)\cdot P/P(isom)}. \tag{3}$$

Here ε(ATP) is the energy produced by hydrolysis of one ATP molecule, W and H are work and heat liberated during contraction, respectively, f(isom) is a force produced by one myosin head in isometric contraction, P is load upon muscle and P(isom) is P for isometric contraction. We determined W/(W+H) by using Hill's data [13]. f(isom) was estimated on the assumption that 80% of the myosin molecules continuously contribute to produce force for sliding, and this value of 80% remains constant for P/P(isom) > 0.14, as was discussed by Podolsky et al. [14]. Obtained values of L are given in Fig.10, which indicates that L > 120 nm for free shortening (P=0). The traveling distance of myosin molecule per hydrolysis of one ATP molecule is thus L/2 > 60 nm, in accordance with the obsevation by Yanagida et al. [12] that the average sliding distance of the free actin filament is more than 60 nm

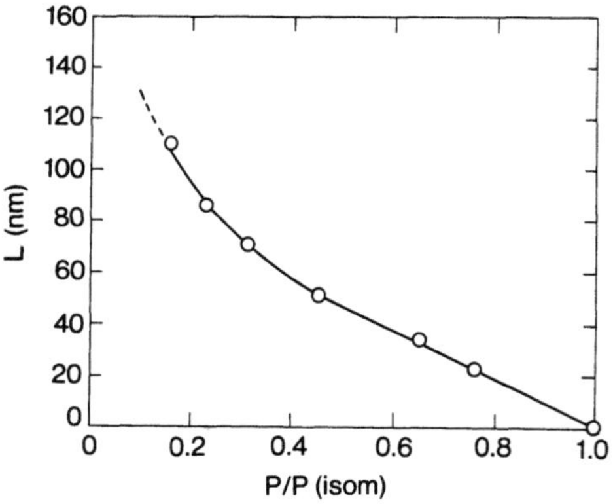

Fig.10. Traveling distance L of myosin molecule along the thin filament by using energy liberated by hydrolysis of two ATP molecules, calculated with Eq.(3).

during one ATP cycle.

According to our model, the two-headed structure of myosin molecule can be looked upon as a device by nature to produce force effectively. Due to the helical structure of the thin filament, however, there is a limit (say, ~ 30 nm) in the self-induced translation by two heads of one myosin molecule along one strand, since myosin molecule cannot go round the thin filament (cf. a and b in Fig.3). There seem to be two possibilities to get over this limit. One is the cooperation of two myosin heads of different myosin molecules as shown by c in Fig. 3. Another possibility is that the detachment of a bound head from actin is made by itself in a self-induced manner (without help of the pairing mate) when the bound head delays behind mean translation of other head on sliding filament and thus effective load on it becomes small. This single head process may still contribute to force generation for the sliding if the self-induced force F remains to exist for a while after detachment of the head.

We thank Dr. T. Yanagida for his helpful discussion.

References

1. K. Wakabayashi, H. Tanaka, T. Kobayashi, Y. Amemiya, T. Hamanaka, S. Nishizawa, H. Sugi and T. Mitsui: Biophys. J. 49 (1986) 581-584.
2. T. Mitsui, H. Sugi, K. Wakabayashi, H. Tanaka, T. Kobayashi, Y. Ueno, Y. Amemiya, H. Iwamoto and T. Hamanaka: Photon Factory Activity Report 3 (1984/85) p.249.
3. H.E. Huxley: Science 164 (1969) 1356-1366.
4. A.F. Huxley and R.M. Simmons: Nature 233 (1971) 533-538.
5. T. Yanagida: J. Muscle Res. Cell Motility 6 (1985) 43-52.
6. T. Yanagida: J. Mol. Biol. 146 (1981) 539-560.
7. R. Cooke, M.S. Crowder and D.D. Thomas: Nature 300 (1982) 776-778.
8. H.E. Huxley, A.R. Faruqi, M. Kress, J. Bordas and M.H.J. Koch: J. Mol. Biol. 158 (1982) 637-684.
9. H.E. Huxley, R.M. Simmons, A.R. Faruqi, M. Kress, J. Bordas and M.H.J. Koch: J. Mol. Biol. 169 (1983) 469-506.
10. I. Matsubara, N. Yagi and H. Hashizume: Nature 255 (1975) 728-729.
11. H. Sugi, Y. Amemiya and H. Hashizume: Proc. Japan Adad. 54 (1978) 559-564.
12. T. Yanagida, T. Arata and F. Oosawa: Nature 316 (1985) 366-369.
13. A.V. Hill: Proc. Roy. Soc. London B159 (1964) 297-318.
14. R.J. Podolsky, R.St. Onge, L. Yu and R.W. Lymn: Proc. Nat. Acad. Sci. USA 73 (1976) 813-817.

Small Angle X-Ray Diffraction of Collagen Fibrils Using a Three-Dimensional Imaging Gas Detector

A. LA MONACA
INFN, Laboratori Nazionali di Frascati, 00044 Frascati, Roma (ITALY)

A. BIGI, A. RIPAMONTI, N. ROVERI
Istituto di Chimica "G. Ciamician", Università di Bologna, 40126 Bologna (ITALY)

M. IANNUZZI, R. MESSI
Dipartimento di Fisica, II Università di Roma, 00173 Roma (ITALY)

ABSTRACT - Small angle X-ray diffraction investigation of collagen fibrils in cornea and in rat-tail tendon have been carried out. The experiments have been performed using synchrotron radiation and a three-dimensional imaging gas detector.

The preliminary results on anisotropic effects at ultra-small angles are discussed.

With the present availability of synchrotron radiation as high brilliance X-ray source, the whole information extractable from a small-angle scattering experiment could be obtainable. However, this is actually true only when the performances of the X-ray detectors used in the experiment are suitable for an unambiguous identification of the reflections arising from the sample, and for a quantitative analysis of the shape, and of the time evolution of the scattered radiation intensity. For anisotropic scattering measurements, two-dimensional operation is also a main point. A detector satisfying all these requirements has to be of electronic type, with high spatial resolution on both coordinates, with high time resolution, with high count-rate capability.

For isotropic scattering, all the required information can be obtained by measuring the radiation scattered in one direction. In this case, the linear position sensitive detectors (PSD) are suitable for SAXS experiments.

For anisotropic scattering, the performances of the most used two-dimensional detectors (TV systems; multiwire proportional chambers: MWPC; charge-coupled devices: CCD; multi-channel plates: MCP)[(1), (2), (3), (4)], are far from those of an ideal detector.

We propose to use the self-explanatory diagram shown in fig. 1 to plot the significant figures of such detectors.

The present commercial matrices of CCD or MPC have serious limitations. CCD matrices have low dynamic range, are noisy and suffer from radiation damage. MCP have low efficiency, low count-rate capability and their response is not constant for different incidence angles of the photons. Moreover, both devices have a small detection area, $\sim$1 cm^2. TV systems can be useful only for high scattered intensity, because their response is noisy and not uniform on the whole detecting area. Multiwire proportional chambers are

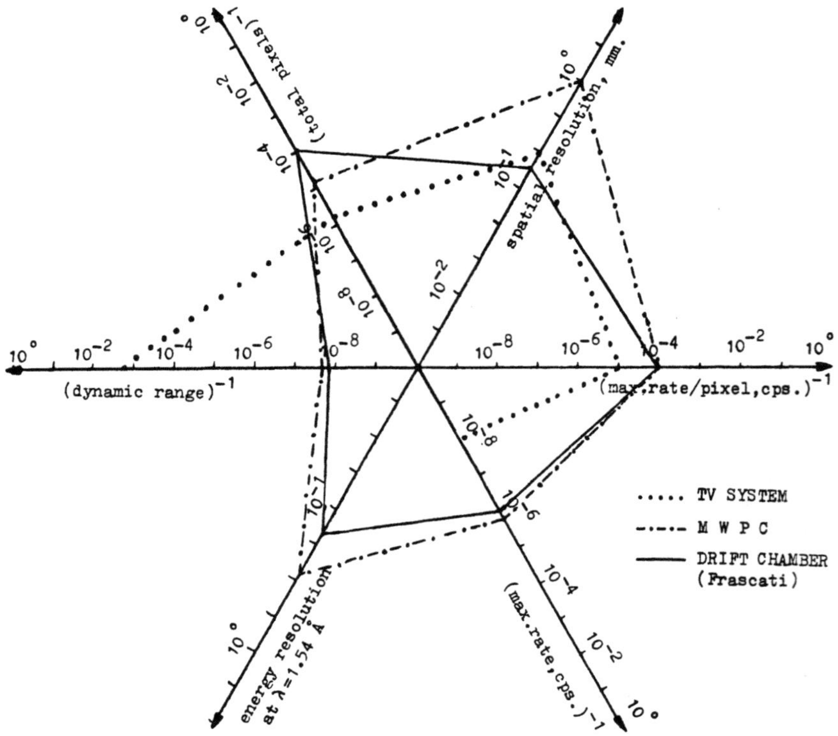

Fig. 1. - Diagram of significant parameters of the present detectors for SAXS experiments.

versatile and potentially fine devices, but at present their response is not uniform on the whole detecting area because of their wire structure. Because their spatial resolution is not high, in order to increase the angular resolution, they have to be located at a large distance from the sample. The three-dimensional imaging "drift chamber" developed at Frascati is a proportional detector in the class of the multiwire chambers, with better performances especially in the spatial resolution and in the uniform response on the whole area.

EXPERIMENTAL

The experimental apparatus operating at Frascati [4] has been used to study collagen fibrillar organization in cornea and in rat-tail tendon.

Cornea (stroma) is made up of lamellae, arranged almost parallel to the surface of the tissue, where collagen fibrils lie with their axes oriented nearly parallel to each other.

Collagen fibrils in one lamella lie approximately orthogonal to those in the successive lamella (fig. 2). The diffraction effects due to the interfibrillar distance in cornea were previously observed using small-angle cameras provided with slit collimation of focussed diffraction beam [5], [6]. However, these apparatus do not allow for accurate determination of the degree of orientation and spread of the reflections.

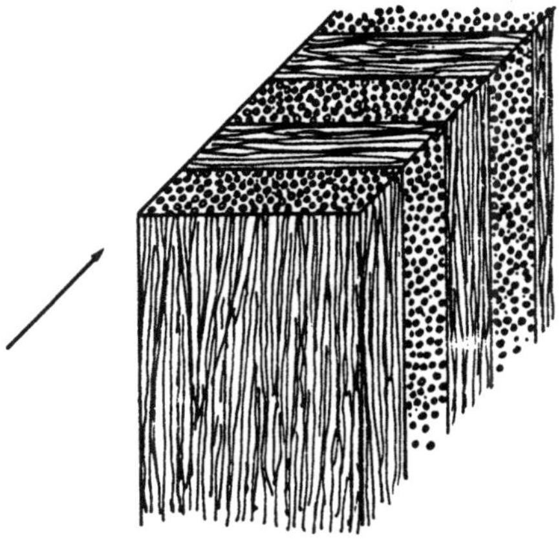

Fig. 2. - Drawing giving a conceptual view of the cornea stroma where fibrils in two successive lamellae make an angle of nearly 90°. The arrow indicates the direction of the visual axis of the eye.

The experimental apparatus (optical system and detector) used for the experiment has been described elsewhere [4]. In table 1 we summarize its more recent figures. We wish to notice the present improvement of the detector's spatial resolution along both directions, the increase of the pixel number, and the high optical collimation. All the measurements have been taken at Adone with an illuminance $E = 5 \times 10^6$ phot/sec at $\lambda = 1.54$ Å on the samples. For comparison, this value is $\sim 10^5$ smaller than the illuminance obtained at Desy for SAXS experiments using photographic films.

Samples of bovine cornea were washed in distilled water and dried under stress (50% of uniaxial elongation). Strip shaped samples were examined with the incident X-ray beam parallel to the visual axis of the eye (fig. 2).

RESULTS

The pattern reported in fig. 3 and 4 show a diffraction ring corresponding to the 6th order of the collagen axial periodicity of 65 ± 1 nm, and several diffraction peaks corresponding to the superior orders of a periodicity of 88 ± 9 nm, which appear oriented parallel to the stretching direction.

These reflections, which are observed within angles of 500 - 2100 arcsec from the primary beam, are superimposed upon the diffuse scattering. The length of the reflections along the layer lines is approximately constant and their angular spread decreases from 70° (2nd order) to 25° (5th order). This pattern can be interpreted by a two-dimensional model of paracrystal lattice [7]. The observed periodicity of about 88 nm can be attributed to the average distance between collagen fibrils distributed inside the lamellae almost

Table 1: Experimental parameters of the diffractometer.

Cross section of pin-hole collimated primary beam	0.2 mm^2
Primary beam angular divergence (2α)	16.5 arcsec
Maximum angle covered by the beam-stopper	412 arcsec
Angle covered by residual parassitic scattering (2β)	495 arcsec
Total angle subtended by the detector	2°
Total angle covered by the goniometer	10°
Power incident on specimen at λ=1.54 Å, with $\Delta\lambda/\lambda=10^{-4}$ and I=50mA (Adone in single mode operation)	10^6 phot/sec
Quantum efficiency of detector operating in Argon flow at λ=1.54 Å	17%
Detector size	17x20 mm^2
Angular resolution in the plane of the detector: x-direction	70 arcsec
y-direction	70 arcsec
Pixel size	$3.6\times10^{-2} mm^2$
Number of detector pixels	9714
Maximum pixel number shown on the MCA	8192
Maximum count rate per pixel	10^4 cps
Maximum ultimate counting capability of detector	10^6 cps
Maximum ultimate recording rate of data	9×10^5 cps

orthogonally with respect to the stretching direction. The value of 88 nm is slightly higher than the values observed by other authors [(5), (6), (8)] in unstretched samples at various degrees of hydration. It can be interpreted on the basis of the stretching effect which induces an increase of the interfibrillar distance in lamellae where fibrils are oriented orthogonal to the stretching direction.

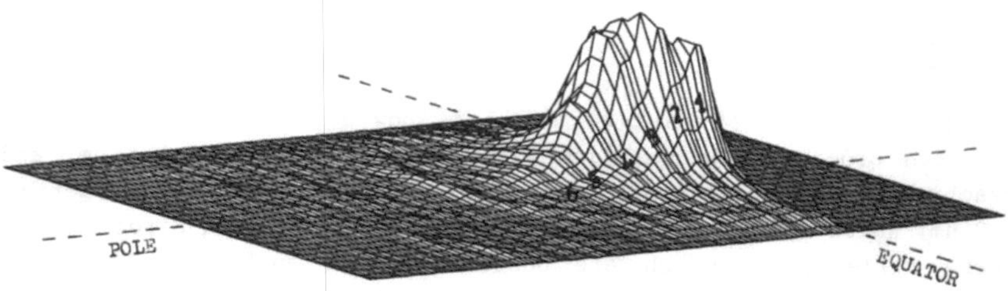

Fig. 3. - Cornea three-dimensional diffraction semi-pattern in the region 500-2100 arcsec from the primary beam. The semi-ring outside this region corresponds to the 6th order of the collagen axial periodicity.

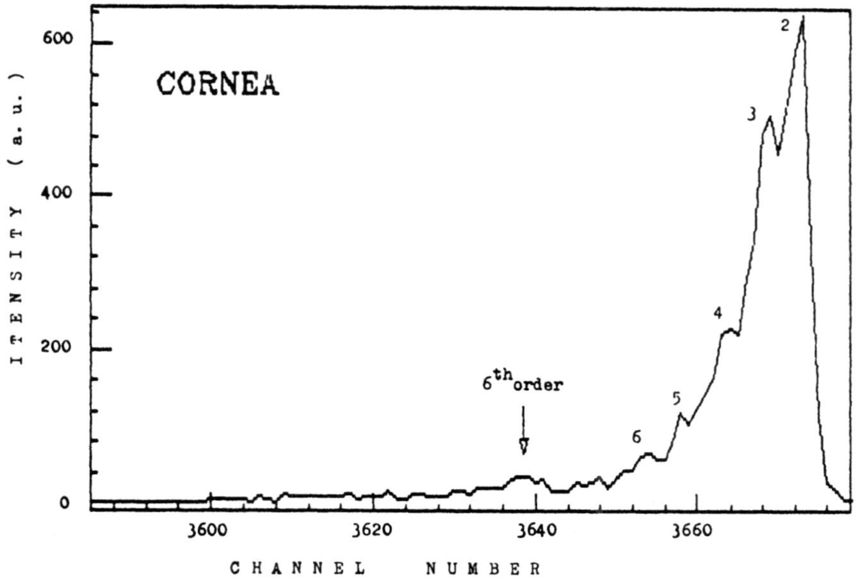

Fig. 4. - Cornea unidimensional diffraction semi-pattern along the meridional direction. Periodicity: 88.5 ± 9 nm.

Thus, the high angular resolution of the detector and the high pin-hole collimation of the beam appear the most suitable requirements to be satysfied in order to investigate the degree of orientation and the spread of ultra-small angle reflections. Such experimental conditions have allowed us to observe also the well-known crimping effect (9) on the small-angle meridional reflections of rat-tail tendon collagen (fig. 5).

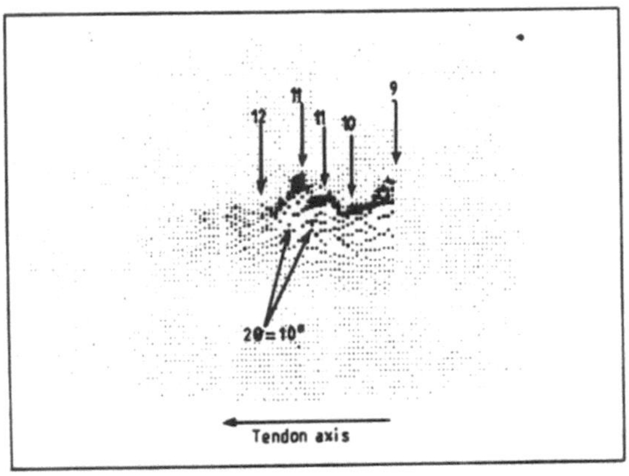

Fig. 5. - Meridional splitting (crimping affect) of the 11th order reflection of stretched rat-tail tendon collagen.

ACKNOWLEDGMENTS

We wish to thank Mr. R. BOLLI for his continuous technical assistance.

REFERENCES

(1) B. Buras and S. Tazzari, Report of the ESRP, CERN (1984)
(2) L.N. Koppel, Rev. Sci. Instr. 48 (1977) 669; R.H. Dyck, Fairchild CCD Catalog (1984) 188
(3) G.W. Fraser, Nucl. Instr. and Meth. 195 (1982)
(4) L. Allocca, M. Iannuzzi and A. La Monaca, Nucl. Instr. and Meth. 219 (1984) 277
(5) C.R. Worthington and H. Inouye, Int. J. Biol. Macromol. 7 (1985) 2
(6) Z. Sayers, M.H.J. Koch, S.B. Whitburn, K.M. Meek, G.F. Elliott and A. Harmsen, J. Mol. Biol. 160 (1982) 593
(7) R. Hosemann, S.N. Bagchi, "Direct Analysis of Diffraction by Matter", North Holl. Publ. Comp., Amsterdam, (1962)
(8) J.M. Goodfellow, G.F. Elliott and A.E. Woolgar, J. Mol. Biol. 119 (1978) 237
(9) S.P. Nicholls, L.J. Gathercole, A. Keller and J.S. Shah, Int. J. Biol. Macromol. 5 (1983) 283

X-RAY MICROSCOPY

X-Ray Microscopy at the Electron Storage Ring BESSY

G.Schmahl and D.Rudolph

Forschungsgruppe Röntgenmikroskopie, Universität Göttingen
Geismarlandstrasse 11, D-3400 Göttingen, Fed.Rep. of Germany

1. Introduction

The limitations of the imaging techniques which dominate submicron biological imaging today, namely optical microscopy, diffraction of x-rays and electron microscopy, are due to the imaging particles themselves [1]. In optical microscopy the wavelength of the visible-light photon limits the resolution to about 0.2 μm. In x-ray diffraction the relatively low coherent-scattering cross-section of the hard x-ray photon makes it necessary to use macroscopic specimens to obtain sufficient interaction for imaging. In electron microscopy the large cross-section of the scattering electron makes it necessary to disassamble the cell. In addition, the weakness of the dependence of cross-section on atomic number makes staining and dehydration necessary.

Soft x-ray photons as imaging particles have interesting and basic attractions for microscopy. It is possible to examinate, in transmission, individual intact cells up to several microns thick. It is not necessary to remove water or to stain because the interaction of soft x-rays depend strongly on atomic number. Because of this x-ray microscopy was started with considerable enthusiasm already in the 1950's. This initial enthusiasm, however, slowed down after it became evident that the achievement of high resolution and short ex-exposure time was beset with considerable experimental difficulties. During the last years advances in soft x-ray sources and optics have led to a renewed interest in this field. For detailed discussions of the possibilities of x-ray microscopy also in comparison to electron and light microscopy see [1,2,3,4,5 and 6].

X-ray microscopy requires high resolution x-ray optics and intense x-ray sources of high spectral brilliance. Concerning x-ray optics condenser optics to illuminate the object with quasimonochromatic x-radiation as well as high resolution optics are necessary. Up to now for condenser optics zone plates [7,8] or grazing incidence optics and zone plates [5] were used. In future in addition multilayered mirrors [9,10] possibly in combination with zone plates and/or gratings may be applied. As high resolution optics up to now only micro zone plates are used [7,11].

Suitable x-ray sources have to be of high spectral brilliance. Tunability to different wavelengths may be advantageous though not necessary in any case. Bending magnet synchrotron radiation from 0.5 - 1.0 GeV electrons in dedicated storage rings is currently the most widely used radiation [1,6].

2. The Beamlines of the Microscopic Area at BESSY

At the electron storage ring BESSY in Berlin two beamlines are dedicated to x-ray microscopy. Figure 1 shows the arrangement of the beamlines of the x-ray microscopic area.

Both beamlines emerge from the teefitting behind the valve V1 with which the beamlines can be separated from the outlet of the storage ring. In the mirror chamber MC the beam of the second line can be reflected into a third line provided for an experimental chamber of testing x-ray optical elements etc. This line as well as the second line itself which is dedicated to the scanning x-ray microscope [12] are not subjects of this paper. The first beamline is dedicated to the imaging x-ray microscope which has been in operation since April 1983.

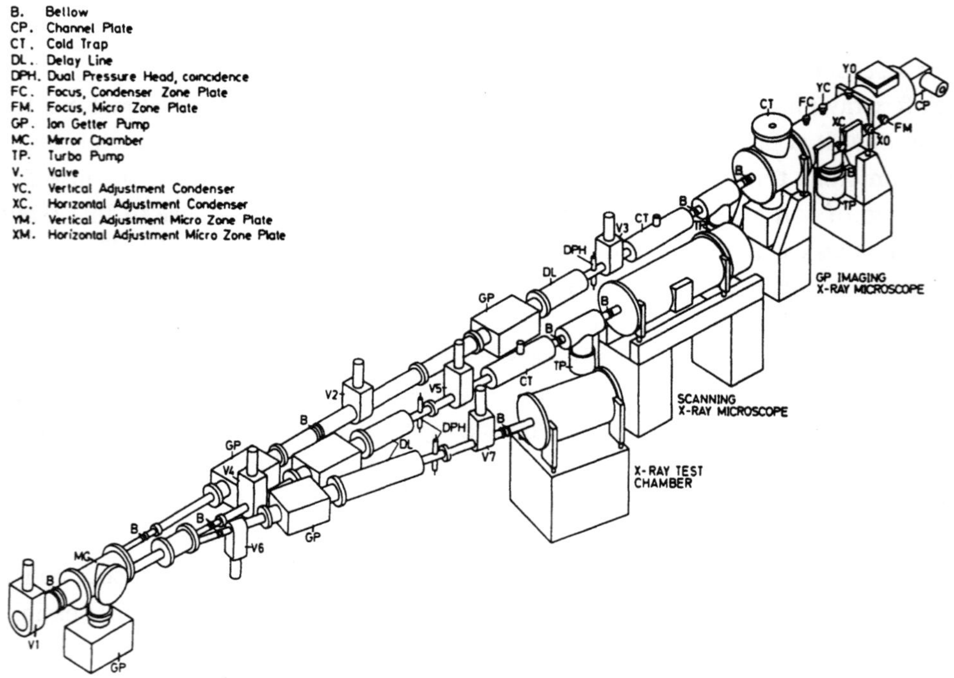

Fig. 1 - The arrangement of the beamlines of the microscopic area at the BESSY storage ring

3. The X-Ray Microscope

The first beamline ends up in the imaging x-ray microscope. Figure 2 shows the schematic arrangment. The polychromatic x-radiation meets the condenser zone plate at a distance of 15 m of the source which is the tangent point to the electron storage ring. The condenser zone plate has a diameter of 9 mm and a focal length of 3o4 mm for 4.5 nm radiation. A micro zone plate of 55.6 μm diameter and a focal length of 0.69 mm at 4.5 nm generates an enlarged image of the object in the image field. A central stop at the condenser zone plate prevents contrast reducing zero order radiation of the condenser zone plate and of the micro zone plate reaching the image field. The enlarged image can be viewed using a microchannel plate for converting the x-radiation into the visible or it can be photographed directly. The x-ray microscope is operated under vacuum. For examination of wet biological specimens in their natural state an environmental object chamber has been developed in which the specimen is separated from vacuum by thin polyimide foil transparent to x-radiation. Detailed descriptions of the x-ray optics are given in [7,11] and of the microscope and the object chamber in [6].

Figure 3 shows the imaging x-ray microscope at the storage ring BESSY in Berlin.

4. X-Ray Images

Figure 4 shows an image of the cell nucleus of a human fibroblast made with the x-ray microscope with an x-ray magnification of 330 x. The cells were critical point dried without any staining. Images of wet cells, e.g. a human epithelial cheek cell and spores of the Australian moss Dawsonia superba were made [13].

5. Radiation Damage - Spore Viability

Investigations of viability of spores of Dawsonia superba under x-ray irradiation were started [14]. The results indicate that for 10-50% viability of the spores an exposure of $5x10^6$ photons/μm^2 were applied during 130 s exposure time. Assuming the spores absorb 5o% of the photons per μm and the density is 1 g/cm^3, this leads to a dose of $5x10^4$ Gy or $5x10^6$ rad. This number indicates that roughly 500 photons in a resolution element of 10nm x 10nm can be applied according to this viability test. In a scanning microscope with a 50% efficient detector this will be enough to detect a 20% contrast with a signal to noise ratio of 2 at 10nm resolution.

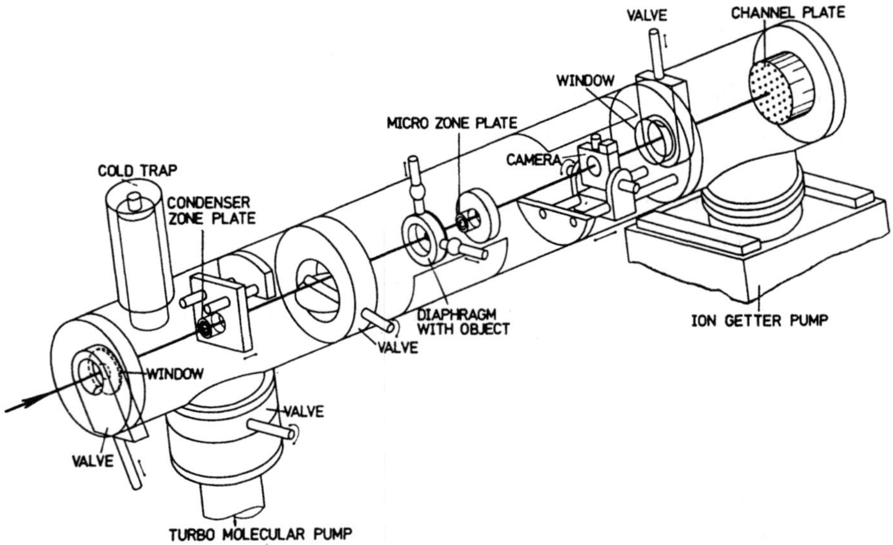

Fig. 2 - The schematic arrangement of the imaging x-ray microscope

Fig. 3 - The imaging x-ray microscope

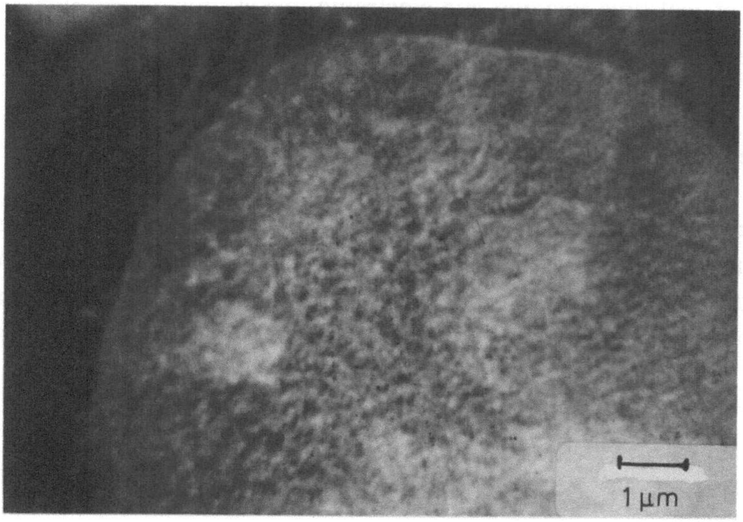

Fig. 4 - Part of a human fibroblast, critical point dried, imaged with x-rays at a wavelength of 4.5 nm and at a primary x-ray magnification of x 330. The exposure time was 4 s.

6. Aspects and Future Developments

Up to now for x-ray imaging amplitude zone plates are used with a theoretical maximum first order diffraction efficiency of 10%; in practice values of about 5% are achieved. Higher values of diffraction efficiency can be obtained by using phase zone plates (up to about 20% for laminar phase structures). Phase zone plates with laminar phase structures have been made and will soon be tested and used.

Zone plates with higher diffraction efficiency enable shorter exposure times. In addition a larger amount of the x-radiation which loads the object is used for image generation, leading to a reduced radiation dose applied to the specimens.

Up to now a coarse focussing of the object was done looking at the x-radiation converted to the visible by the microchannel plate and subsequently taking a few line focussing photos with slightly different focus setting. An improvement was done by introducing an optical microscope into the x-ray microscope for prefocussing. Prefocussing with visible light reduces the x-radiation load to the object to the amount necessary for taking just only one x-ray image. This improvement is of crucial importance to any investigation of living cells.

In the existing x-ray microscope at BESSY the image is directly viewed by converting the x-radiation to the visible using a microchannel plate or it is directly photographed. Using semiconductor imaging devices such as CCD's will provide the possibility of real time imaging and electronic image processing. Such devices are under development; preliminary results are given in [15].

Acknowledgements

The x-ray microscopic experiments at BESSY in Berlin are supported by the Bundesministerium für Forschung und Technologie. For a joint effort to compare the results of different microscopic methods Dr.P.C.Cheng, IBM Th.J.Watson Research Center, Yorktown Heights, NY, USA, provided us with critical point dried human fibroblasts (see figure 4).

References

[1] Howells, M., J.Kirz, D.Sayre and G.Schmahl, Soft X-ray microscopes, Physics Today, 1985, vol.38, No.8, p.22-32.

[2] Niemann, B., G.Schmahl, D.Rudolph, X-ray microscopy: recent developments and practical applications, in: SPIE Vol.368 - Techniques and Capabilities (Sira), (1982) p.2-8.

[3] Kirz, J. and D.Sayre, Soft X-ray microscopy of biological specimens, in: Synchrotron radiation research, H.Winnick and S.Doniach, eds., Plenum Press, 1980, p.277-322.

[4] Duke, P.J., X-ray microscopy: Recent developments and future prospects, Proceedings RMS, Vol.16/3, 1981, p.186-192.

[5] Schmahl, G., D.Rudolph, B.Niemann and O.Christ, X-Ray Microscopy of biological Specimens with a Zone Plate Microscope, in: Ultrasoft X-Ray Microscopy: Its Application to Biological and Physical Sciences, D.P.Parsons, ed., Annals of the New York Academy of Sciences, Vol.342, 1980, p.368-386.

[6] Rudolph, D., B.Niemann, G.Schmahl and O.Christ, The Göttingen X-ray Microscope and X-Ray Microscopy Experiments at the BESSY Storage Ring, in: X-Ray Microscopy, G.Schmahl and D.Rudolph, eds., Springer Series in optical Sciences, Vol.43, Springer-Verlag Heidelberg, 1984, p.192-202.

[7] Schmahl, G., D.Rudolph, P.Guttmann and O.Christ, Zone Plates for X-ray Microscopy, ibid., p.63-74.

[8] Thieme, J., Construction of Condenser Zone Plates for a Scanning X-Ray Microscope, ibid., p.91-96.

[9] Barbee, T.W., Sputtered layered synthetic microstructure (LSM) dispersion elements, in: Low Energy X-Ray diagnostic - 1981, D.T.Attwood and B.L.Henke, eds., AIP Conference Proceedings No.75, 1981, p.131-145.

[10] Dhez, P. and J.M.Esteva, Les multicouches metalliques, Direction des Recherches, Etudes et Techniques, Rapport AEPA 80/043, 1980.

[11] Guttmann, P., Construction of a Micro Zone Plate and Evaluation of Imaging Properties, ibid., p.75-90.

[12] Niemann, B., The Göttingen Scanning X-Ray Microscope, ibid., p.217-225.

[13] Schmahl, G., B.Niemann, D.Rudolph, P.Guttmann and V.Sarafis. 1985. X-Ray Microscopy: experimental results with the Göttingen X-ray microscope at the electron storage ring BESSY in Berlin, Journal of Microscopy, Vol.138, p.279-284.

[14] Niemann, B., V.Sarafis, D.Rudolph, G.Schmahl, W.Meyer-Ilse and P.Guttmann. 1986. X-ray Microscopy with Synchrotron Radiation at the Electron Storage Ring BESSY in Berlin, Proceedings of the "International Conference X-Ray and VUV Synchrotron Radiation Instrumentation", Stanford 1985, Nuclear Instruments and Methods, Physics Research A 246, p.675-680.

[15] Germer, R. and W.Meyer-Ilse. 1986. X-Ray TV-Camera at 4.5 nm, Rev.Sci. Instrum., Vol.57, No.3, p.426-427.

TIME-RESOLVED FLUORESCENCE

Quaternary Structure Changes in Aspartokinase-Homoserine Dehydrogenase Studied by Time-Resolved Fluorescence

Magali JULLIEN, Sylvie BAUDET and Francis RODIER
Laboratoire de Biologie Physicochimique, Bâtiment 433,
Université Paris-Sud, 91405, Orsay, France.

INTRODUCTION

Time-resolved fluorescence spectroscopy provides a non perturbing probe that permits simultaneous monitoring of multiple protein domains. These domains may be subdivided according to their accessibility to various types of quenchers i.e. according to whether they are exposed to the external solvent or buried within the core of the protein.

The purpose of this work has been to study by time-resolved fluorescence the allosteric transition of a bifunctional enzyme Aspartokinase-Homoserine dehydrogenase I (AK-HDH).

AK-HDH from E.coli catalyzes two non-consecutive reactions in the biosynthetic pathway leading from L-aspartate to L-threonine ; both activities are regulated allosterically by L-threonine. The regulatory properties of the enzyme have been interpreted in terms of a two-state model involving cooperative changes of the subunits ; the preexisting equilibrium would be shifted towards an R active form in presence of potassium ions or L-aspartate and a T inactive form in their absence or in the presence of L-threonine (**Janin & Cohen**, 1969). It is made up of four identical subunits of molecular weight 89 000 daltons (**Falcoz-Kelly** et al., 1972), the amino acid sequence of which has been determined (**Katinka** et al., 1980). Limited proteolysis (**Veron** et al.) and refolding experiments (**Dautry-Varsat & Garel**, 1981) have shown that the two catalytic activities are carried out by two independent regions of the polypeptide chain.

THEORY

The intrinsic fluorescence of proteins originates from aromatic amino acids and is usually dominated by the emission from tryptophan residues. The fluorescence decay can be described by a sum of exponentials :

$$I(t) = \sum_i a_i \exp(-t/\tau_i),$$

where a_i is a preexponential factor representing the fractional contribution of the component with a lifetime τ_i.

Fluorescence can be quenched by diverse external molecules (**Lakowicz**, 1983).

For acrylamide, a neutral molecule which quenches efficiently tryptophan fluorescence, the quenching proceeds via a diffusion - controlled process. A bimolecular quenching rate constant k can obtained from the Stern-Volmer equation :

$$\frac{\tau_o}{\tau} = 1 + k\,\tau_o\,[Q],$$

where τ_o and τ are the fluorescence lifetimes in the absence and presence of quencher respectively and [Q] is the quencher concentration.

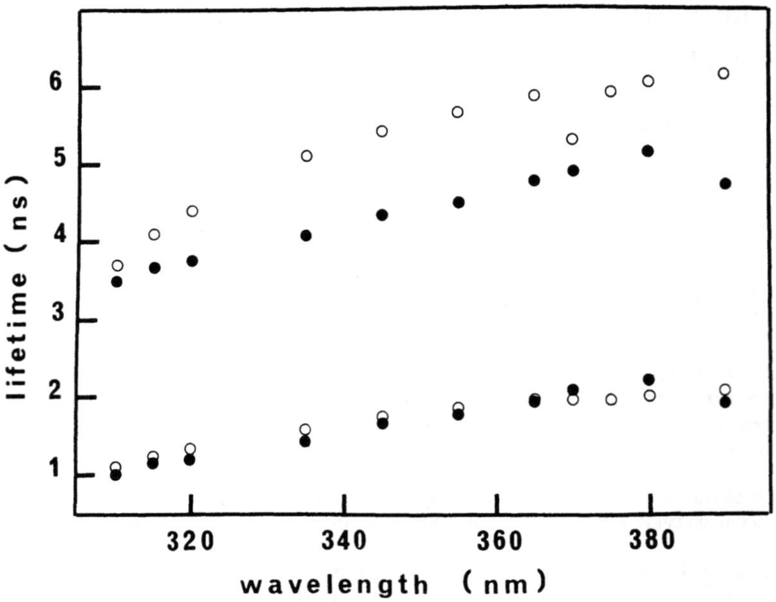

Figure 1. Plots of the two fitted lifetimes of AK-HDH as a function of emission wavelength in absence (•) and in presence (o) of threonine.

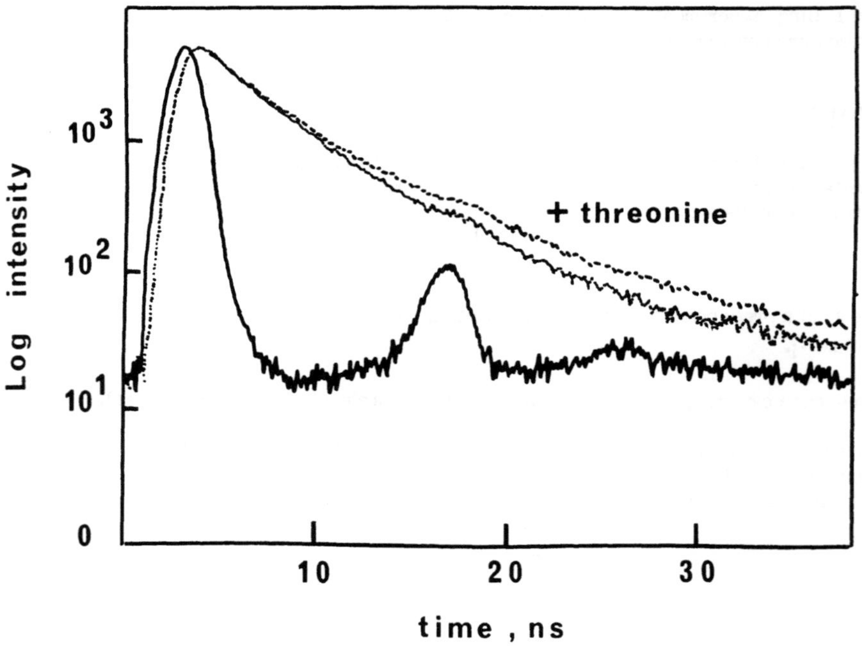

Figure 2. Fluorescence decay curves for the active (....) and inactive (----) forms of AK - HDH recorded at 350 nm ; the continuous curve is the apparatus response function.

In the case of a coenzyme, like NADPH (Nicotinamide adenine dinucleotide phosphate, reduced form), the quenching proceeds via a non-radiative energy transfer, since there is an overlap between the emission spectrum of tryptophan residues and the absorption spectrum of the coenzyme. The transfer efficiency is :

$$E = \frac{\tau_o - \tau}{\tau},$$

where τ_o and τ are the fluorescence lifetimes of the donor in absence and in presence of acceptor respectively.

MATERIALS and METHODS

AK-HDH was prepared according to the procedure of **Truffa-Bachi** et al. (1968).

Fluorescence decays were measured at **LURE** (Laboratoire pour l'Utilisation du Rayonnement Electromagnétique), Orsay, using synchrotron radiation from **ACO** (Anneau de Collisions d'Orsay) as a light source ; the pulse frequency was 13.6 MHz and the pulse width at half-height was about 1.4 ns. The signal obtained from an SLM 8 000 spectrofluorometer was analyzed by the single photon counting technique with an apparatus previously described (**Brochon**, 1980).

The numerical values of relaxation times and amplitude factors were determined by a non-linear least-squares procedure (**Provencher** & **Vogel**, 1983), taking into account the measured instrumental response function.

RESULTS

1 - Fluorescence lifetimes

When excited at 295 nm, the fluorescence decay of AK-HDH was correctly fitted with two exponentials : a fast component with a lifetime of about 1.5 ns and a slow component with a lifetime of about 4.5 ns. The longer lifetime increases significantly with emission wavelength (figure 1). Addition of threonine does not change the fast component ; it increases the lifetime of the slow component by about 20 % (figure 2) and slightly increases its dependence upon emission wavelength (figure 1). The effect of threonine follows a sigmoidal curve as in an allosteric transition (figure 3).

The contribution of each component in the total intensity changes along the emission spectrum, indicating that the two different lifetimes do not exhibit the same emission spectra, especially different emission maxima. The relative intensity spectra of the two components indicate that, in absence of threonine, the fast component has an emission maximum at 340 nm whereas the maximum of the slow component is near 335 nm ; adddition of threonine results in a red shift of the slow component maximum from 335 nm to 345 nm (figure 4).

2 -Acrylamide quenching

Acrylamide quenches the two tryptophan components of AK-HDH with two different quenching rates (figure 5) . The fast component is highly quenched with a quenching rate constant of $(8.0 \pm 1.0).10^{8}\ M^{-1}.s^{-1}$. The slow component is less quenched with a quenching rate constant of $(2.0 \pm 0.3).10^{8}\ M^{-1}.s^{-1}$; in the active form of the enzyme, the slow component is less accessible to an external quencher than the fast component. In presence of threonine, only the quenching of the slow component is modified ; a 35% increase for the quenching rate constant is observed if the change in lifetime is taken into account (figure 5). Therefore, in the inactive form, the accessibility of the slow component is increased, but this accessibility is still less important than the accessibility of the fast component.

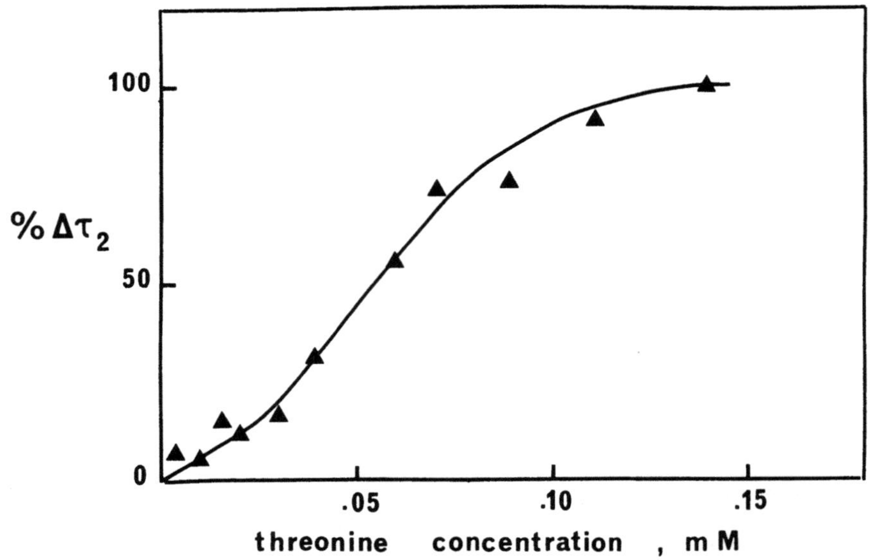

Figure 3. Relative variation of the lifetime of the slow component as a function of inhibitor concentration.

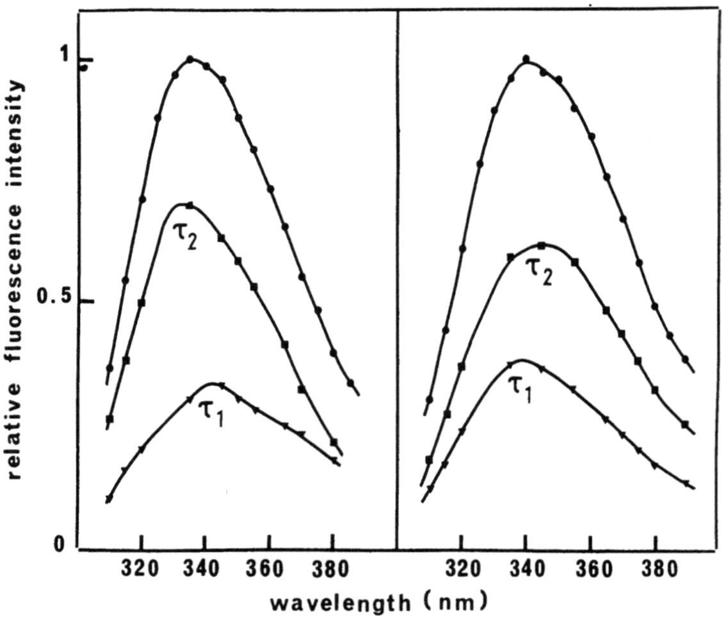

Figure 4. Steady-state fluorescence spectra of AK-HDH resolved into fractional contributions of the fast (τ_1) and slow (τ_2) components ; left = active form ; right = inactive form.

3 - NADPH quenching

NADPH is a coenzyme of dehydrogenase activity ; binding of NADPH results in a decrease of the two lifetimes (table 1) ; this means that there is some non radiative energy transfer between the tryptophan residues of the enzyme and the coenzyme. The transfer efficiencies are 27 % for the fast component and 14 % for the slow one. Addition of threonine decreases significantly the transfer efficiency of the slow component. The transfer efficiency may have been reduced either by a larger distance between donor and acceptor or by a change in their mutual orientations.

DISCUSSION and CONCLUSION

The AK-HDH fluorescence decay can be resolved into two distinct classes of tryptophan emitters which are characterized by different lifetimes and spectra, and which behave differently upon addition of inhibitor, coenzyme or quencher.

The fast component, with a lifetime of 1.5 ns, is insensitive to the allosteric transition of the enzyme : its emission spectrum, its accessibility to a quencher, its transfer efficiency to NADPH are unmodified in the presence of L-threonine . On the contrary, all the properties of the 4.5 ns lifetime are affected by the allosteric transition.

Addition of threonine increases the longer lifetime by about 20 % ; a similar increase has been obtained for the R-T transition of human hemoglobin (**Itoh** et al.,1981) ; this effect is accompanied by a strong dependence of the lifetime upon the wavelength (figure 1) and a red-shift of the spectrum (figure 4). These observations suggest a more polar environment for this component in the T-form than in the R-form. Since lifetimes of exposed residues never exceed 4 ns (**Grinvald** & **Steinberg**, 1976), this change in polarity may not be attributed to an exposure to the aqueous solvent. Moreover such a wavelength dependence indicates that the dipolar relaxation of the surrounding environment is comparable to the fluorescence lifetime, i.e. that the time constant of the reorganization around the excited state dipole is not smaller than the fluorescence lifetime ; in this case, the composite emission displays a broader spectral distribution. **Lakowicz** & **Cherek** (1980) have pointed out that this phenomenon is observed only for fluorophores located in very viscous media or buried in proteins. Therefore, although in a more polar environment, tryptophan residues responsible for the slow component would still be buried inside the protein matrix.

Spectral shifts may also be due to specific interactions between the fluorophore and its environment (**Beechem** & **Brand**, 1985) ; an unambiguous interpretation would require a detailed understanding of the types of specific interactions which are likely to occur with the polar groups present on the macromolecule and the time scale over which these interactions occur. In absence of the knowledge of the crystallographic structure of the enzyme, no hypothesis can be made concerning these specific effects.

Acrylamide is an efficient quencher of indole derivatives (**Eftink** & **Ghiron**, 1984), and the bimolecular rate constant between the probe and the indole side chain is a good measure of the degree of exposure of a tryptophan residue. Quenching results suggest that the slow component corresponds to fluorophores buried in the protein, partially shielded from collisions with acrylamide. Addition of threonine slightly increases the quenching rate constant which still remains smaller by a factor of 3 with respect to the quenching rate of the fast component. Therefore, the increased value of the lifetime, its wavelength dependence and its relatively small quenching rate corroborate that the slow component is not completely exposed to the external solvent and still partially buried in the T-form.

Assignment of the two components to specific residues of the protein sequence looks hazardous, since each subunit contains four tryptophan residues, one located in the AK domain and three located in the HDH domain. Nevertheless, if one supposes quantum yields proportional to lifetimes, the relative intensity of the fast component corresponds to 2.4 tryptophan residues. Since the transfer efficiency of these residues to NADPH is around 30 % and unaffected by the allosteric

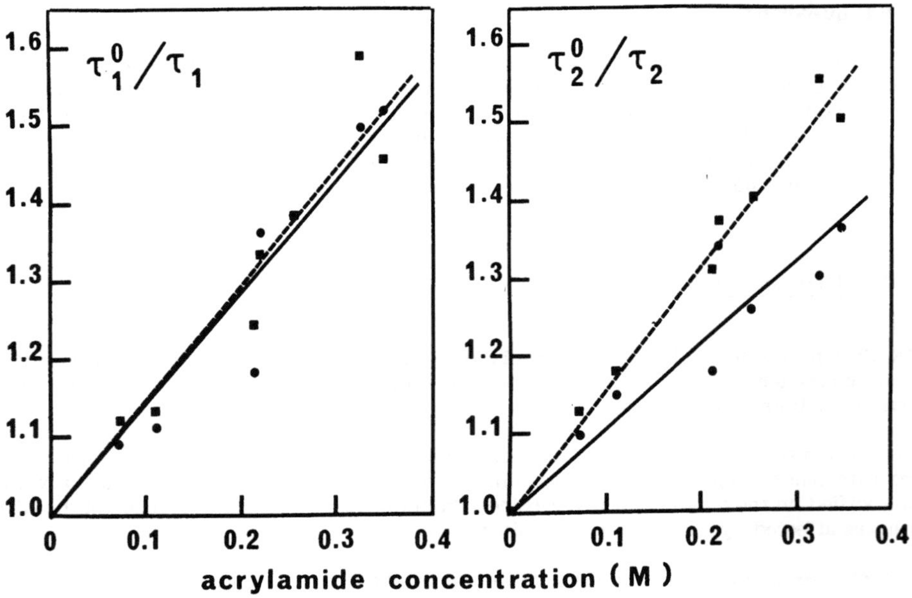

Figure 5. Stern-Volmer plots for the acrylamide quenching of the fast (left) and the slow (right) components in absence (●) and in presence (■) of threonine.

Table I. Fluorescence lifetimes and transfer efficiencies in presence of NADPH.

	a_1	τ_1 (ns)	E_1 (%)	a_1	τ_2 (ns)	E_2 (%)
AK-HDH	.53	1.51		.47	4.26	
+ NADPH	.69	1.10	27	.31	3.68	14
+ Threonine	.63	1.62		.37	5.40	
+ Threonine and NADPH	.69	1.11	32	.31	5.02	7

a_1, a_2 = normalized amplitudes
τ_1, τ_2 = fluorescence lifetimes
E_1 = transfer efficiency for τ_1
E_2 = transfer efficiency for τ_2

transition, they may be located at less than 25 Å from the HDH active site (**Steinberg**, 1971) in both conformations. Therefore, the fast component might include at least two of the three tryptophans of the HDH domain. The unique tryptophan of the AK domain looks like a good candidate for the slow component : it is located in a polar amino sequence of the protein (**Katinka** et, al., 1980) between two cysteinyl residues connected by a disulfide bridge (**Falcoz-Kelly** et al., 1972) . Finally, each of the two components of the fluorescence decay could be a reflect of one functional domain of the protein and the present results would indicate the esssential role played by the AK domain in the allosteric transition.

ACKNOWLEDGEMENTS

The authors would like to thank **Dr. J-C. Brochon** who kindly gave them the opportunity for using his apparatus .

REFERENCES

Beechem, J.M. & Brand,L. (1985) Ann. Rev. Biochem. 54, 43-71.

Brochon, J-C. (1980) in Proceedings of the sixth Taniguchi International Symposium (Ishiwata,S. ed.) pp. 163-189.

Dautry-Varsat,A. & Garel, J-R. (1981) Biochemistry 20, 1396-1401.

Eftink,M.R. & Ghiron,C.A. (1984) Biochemistry 23, 3891-3899.

Falcoz-Kelly,F., Janin,J., Sarri,J.C., Veron,M., Truffa-Buchi,P. & Cohen,G.N. (1972) Eur. J. Biochem. 28, 507-519.

Grinvald,A. & Steinberg,I.Z. (1976) Biochim. Biophys. Acta 427, 663-678.

Itoh,M., Mizukoshi,H., Fuke,K., Matsukawa,S. & Mawatari (1981) Biochem. Biophys. Res. Commun. 100, 1259-1265.

Janin,J. & Cohen,G.N. (1969) Eur. J. Biochem. 11, 520-529.

Katinka, M. , Cossart,P., Sibilli,L., Saint-Girons,I., Calvignac,M.A., Le Bras,G., Cohen,G.N. & Yaniv,M. (1980) Proc. Natl. Acad. Sci. USA, 77, 5730-5733.

Lakowicz, J.R. (1983) in Principles of fluorescence spectroscopy, pp.258-295, Plenum Press, New-York and London.

Lakowicz, J.R. & Cherek,H. (1980) J. Biol. Chem. 255, 831-834.

Provencher,S.W. & Vogel,R.H. (1983) in Progress in Scientific Computing (Deuflhard,P. & Hairer,E. Eds) vol.2, pp.304-319.

Steinberg,I.Z. (1971) Ann. Rev. Biochem. 40, 83-114.

Truffa-Bachi,P., van Rapenbusch,R., Janin,J. , Gros,C. & Cohen,G.N. (1968) Eur.J. Biochem. 5, 73-80.

Veron,M., Falcoz-Kelly,F. & Cohen,G.N. (1972) Eur. J. Biochem. 28, 520-527.

Lifetime of Fluorophores Excited State in Biological Membranes and Their Model Systems

Filippo Conti and Tiziana Parasassi

Dipartimento di Chimica, Università degli studi di Roma"La Sapienza", Roma and Istituto di Medicina Sperimentale, CNR, 00185 Roma, Italy

During the last years our research interest has been devoted to the investigation of the structural and dynamical modifications of biological membranes during physiological processes such as the cell duplication cycle and the differentiation. Our research has initially considered membrane model systems constituted of phospholipid vesicles of various composition.

We utilized different spectroscopic techniques. Interesting information can be achieved by H, ^{2}H, ^{13}C, ^{31}P Nuclear Magnetic Resonance (NMR) (1). However, NMR spectroscopy is restricted in the study of natural membranes, due to the complexity of the spectra obtained in these systems. We also utilized Electron Spin Resonance (ESR) spectroscopy with the aim of determining structural and dynamic parameters of model phospholipid vesicles. However from the preliminary stage of our research we observed that the high required paramagnetic probe concentration (around 10^{-4} - 10^{-5} M) could introduce relevant alterations to the bilayer structure.

Considerable advantages with respect to both NMR and ESR are offered by fluorescence spectroscopy: i) the high sensitivity of the phenomenon allows the required probe concentration to be reduced to 10^{-7} M; ii) the intrinsic timescale of the fluorescence emission is of the order of 10^{-9} sec. This timescale is also considered to be representative of the dynamics of biological membranes and of many other biological processes such as the enzyme activity and the dynamics of nucleic acids (2). NMR and ESR spectroscopy are principally characterized by longer timescale, of around 10^{-5} and 10^{-7} sec., respectively; iii) the possible resolution of the parameters arising from biological samples; iiii) the possibility of studying the interactions between fluorescent probes inserted in the lipid domains and the natural fluorescent residues of membrane proteins.

The most commonly used steady-state parameter to describe the degree of order in membranes is the fluorescence polarization, P, or the derived anisotropy, r, value:

$$P = (I_{\parallel} - I_{\perp}) / (I_{\parallel} + I_{\perp})$$
$$r = (I_{\parallel} - I_{\perp}) / (I_{\parallel} + 2I_{\perp}) = 2P/(3-P)$$

where $I_{\parallel}$ and $I_{\perp}$ refer to the fluorescence intensity parallel and perpendicular, respectively, to the polarization of the exciting light. By observing the polarization or the anisotropy variation vs. temperature of the 1,6-diphenyl-1,3,5-hexatriene (DPH) fluorescence in phospholipid vesicles, gel to liquid-crystalline phase transitions have been determined and phase diagrams have been derived (3).

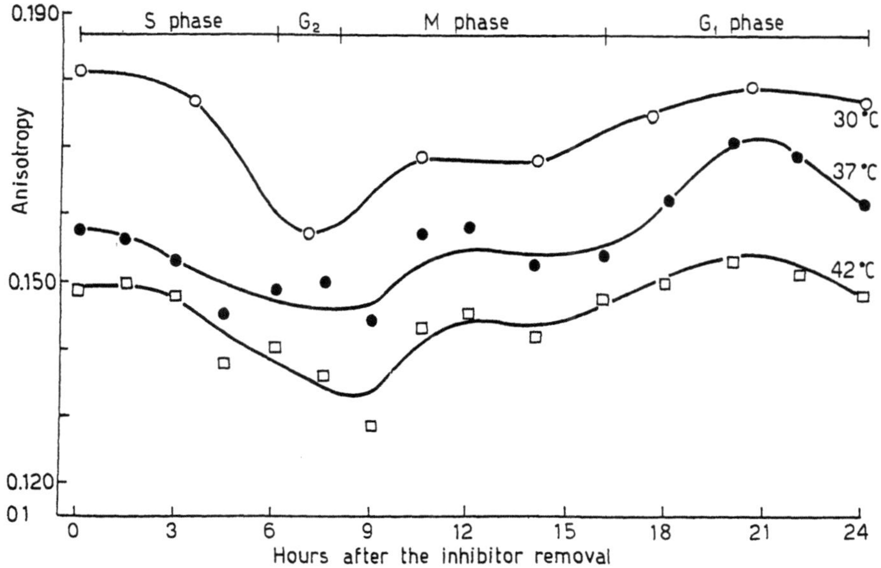

Fig.1 DHP steady state anisotropy variation during the duplication cycle of k562 cells. Measurements were made at three different temperatures.

In whole cell membranes studies, the polarization or anisotropy value can give information on the variation of the ordering of the probe during physiological processes such as the cell duplication cycle (4) (Fig.1) and the differentiation (5) (Tab.I).

------------------------ Table I --------------------------------

DPH steady-state anisotropy variation in non differentiated K562 cells and after four days from the induction of erythroid differentiation.

	NON-DIFFERENTIATED	DIFFERENTIATED	Δr%
r	0.155 + 0.002	0.125 + 0.002	-19.4

More complex and interesting information can be obtained by fluorescence lifetime determinations, mainly using the powerful multifrequency phase and modulation technique (6). The technique employes sinusoidally modulated exciting light at variable frequencies in the MHz range. The fluorescent response is also sinusoidally modulated at the same frequency but phase shifted and demodulated with respect to the excitation. From the phase angle and from the relative demodulation apparent lifetime values can be derived, at each modulation frequency. As the excitation source, a lamp or a laser or synchrotron radiation can be used. Synchrotron radiation offers the advantage of excellent temporal properties and of the simultaneous presence of all wavelenghts. Phase and modulation data obtained at the various frequencies can be analysed by comparison with a model of mono-or multiexponential decay, and the goodness of the obtained fits is evaluated by the reduced chisquare value (7).

For studying biological membranes the choice of the proper fluorophore to be used as the probe is a preliminary condition. Chemical properties such as hydrophobicity or amphiphilicity allow the inference of the probe location in the bilayer. Spectroscopic properties such as the comparison of the quantum yield in water and in membranes, the lifetime value sensitivity to the environment variations are of relevance, too.

By using the multifrequency phase and modulation fluorometry technique relevant results have been recently obtained. The main reason should be attributed to the high detection and resolution power of the technique in the case of multiple exponential decays. Since the complex composition of biological materials and since the microenvironment multiplicity the probe can experience, the possibility of exactly determining and describing a complex decay in studies of natural samples is of maximum relevance.

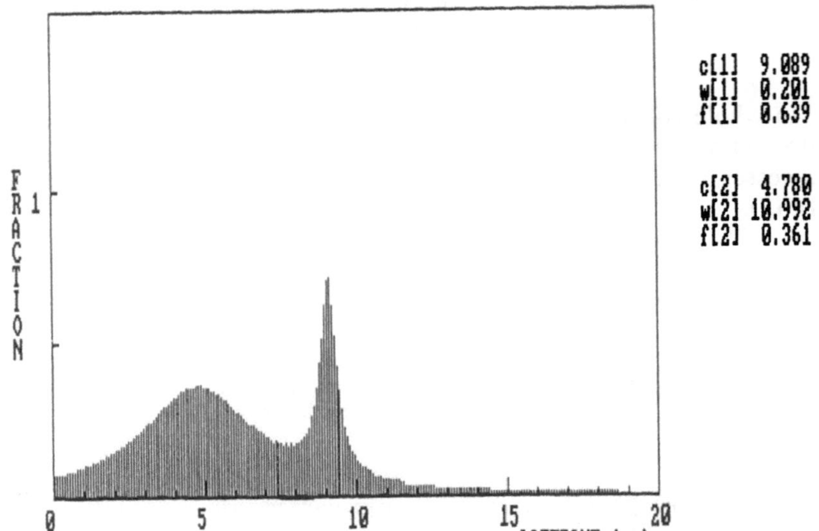

Fig. 2. Continuous lifetime distribution of cis-PnA in ethanol at -20°C. C1,2, center; W1,2, full width at half maximum; and f1,2, fractional intensity of the two distributional components.

The parinaric acid cis and trans isomers (cis, trans, trans, cis-9, 11, 13, 15-octadecatetraenoic acid, cis-PnA; all-trans-9, 11, 13, 15-octadecatetraenoic acid, trans-PnA) decay determined in an isotropic solvent by using the pulsed light technique was best described with a single exponential component (8). On the contrary, multifrequency phase and modulation determinations resulted in a more complex decay, best described by three exponential components (9). Moreover, the behaviour of the associated preexponential factors was strongly temperature dependent. Furthermore, in single-phase phospholipid vesicles the cis- and trans-PnA decay was satisfactorily described only using a model of four exponential components. Even if these results were better representative of the PnAs decay with respect to the results obtained using the pulsed light technique, nevertheless the attribution of a physical origin to the observed exponential components was dubious.

A new approach to the analysis of the phase and modulation data was then developed, assuming a continuous lifetime distribution. The distribution is characterized by a lorentzian shape, centered at a decay time, C, and having a width, W.

By using the continuous lifetime distribution analysis method interesting results on the PnAs decay in isotropic solvents have been obtained (Figs.2 and 3).

We observed that cis- and trans-PnA decays were best described by two lorentzian distributions at all temperatures, in the range from -60°C to 40°C. Moreover, the two isomers showed a different decay behaviour.

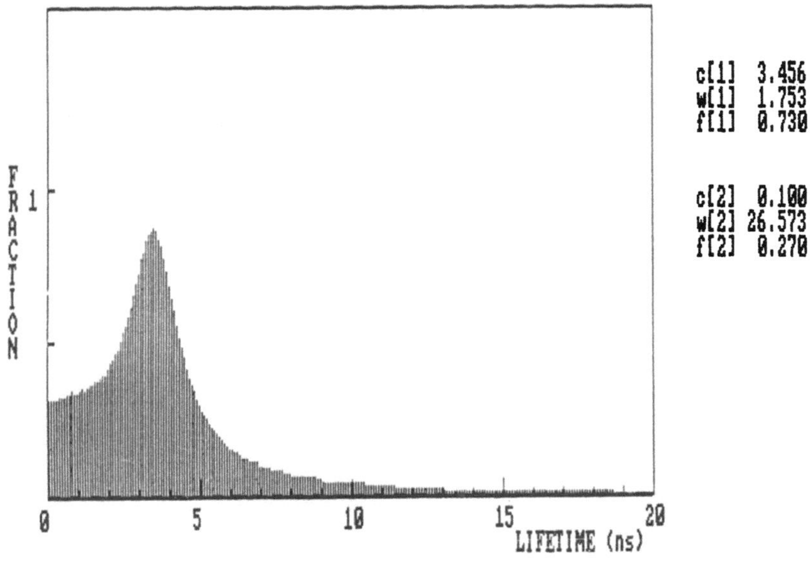

Fig. 3. Continuous lifetime distributions of trans-PnA in ethanol at -20°C. C1,2, center; W1,2, full width at half maximum; and f1,2, fractional intensity of the two distributional components.

The reported observations introduce new theoretical questions on the intrinsic distributional decay of the PnAs in isotropic solvents and on their decay dependence from the environment. Fluorescence lifetime values can be affected by the polarity or the dielectric constant of the solvent, and from the vibrational and rotational freedom of the fluorophore in the medium.

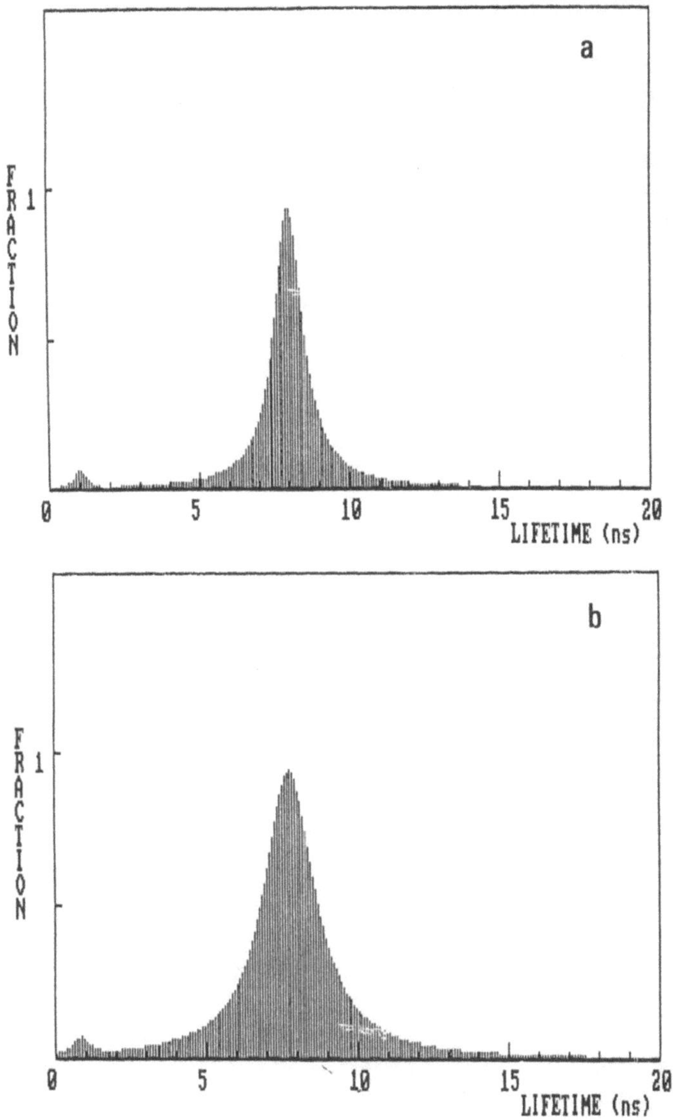

Fig. 4 DPH continuous lifetime distribution in K562 cell membranes; **a.** non differentiated and **b.** after four days from the induction of erythroid differentiation.

The probe interacts with the solvent molecules and the different exchange kinetics between the various solvating shells can influence the excited state lifetime value, giving rise to a continuous lifetime distribution.

Another largely used fluorescent membrane probe is DPH. The multifrequency phase and modulation technique allowed the exact determination of a monoexponential DPH decay in isotropic solvents (10).

Nevertheless, a more complex model was necessary to describe its decay mode in phospholipid vesicles and in natural membranes.

Using the distributional approach, gel to liquid-crystal transitions have been determined in dimyristoyl- and dipalmitoyl-phosphatidyl-choline membranes.

Investigating the cellular membranes physico - chemical characteristics, we have been able to observe a general reorganization of the cell membranes during the differentiation process (fig.4 a and b).

The K562 is a human proerythropoietic cell line that can be induced to differentiate along the erythroid pathway by various inductors (5) The cells were labelled by DPH. We observed that after four days from the induction of erythroid differentiation the width of the DPH lifetime distribution increased of more than a factor of two. From the observations discussed above, the width value increase should be considered to be representative of an increased heterogeneity in the probe environment, due to the membrane matrix reorganization during the differentiation process.

REFERENCES

(1) Lee, A.G., (1975) *Prog. Biophys. Molec. Biol.* **29**3-56.

(2) Jameson, D.M.,(1984) in *Fluorescein Hapten: An Immunological Probe*, Voss. E. Jr., ed., CRC Press, Boca Raton, 23-48.

(3) Parasassi, T., Rosato, N., de Felip, E., and Conti, F.,(1985) in *Physical Methods on Biological Membranes*, F. Conti, W. Blumberg, H. de Gier and F. Pocchiari, eds., Plenum Press, 251-260.

(4) Sapora, O., Parasassi, T., Padovani, L.M. and Conti, F., (1985) in *Physical Methods on Biological Membranes*, F. Conti, W. Blumberg, H. de Gier and F. Pocchiari, eds., Plenum Press , 291-296.

(5) Conti, F., Parasassi, T., Rosato, N., Sapora, O. and Gratton, (1984) E., *Biochim. Biophys. Acta* **805**, 117-122.

(6) Gratton, E. and Limkeman, M., (1983) *Biophys. J.* **44**, 315-324.

(7) Lakowicz, J.R., Gratton, E., Cherek, H. and Limkeman, M., (1984) *Biophys. J.* **46**, 463-477.

(8) Wolber, P.K. and Hudson, B.S., (1981) *Biochemistry* **20**, 2800-2819.

(9) Parasassi, T., Conti, F. and Gratton, E., (1984) *Biochemistry* **23**, 5660-5664.

(10) Parasassi, T., Conti, F., Glaser, M. and Gratton, E., (1984) *J. Biol. Chem.* **259**, 14011-14017.

MIX
Papier aus verantwortungsvollen Quellen
Paper from responsible sources
FSC® C105338

If you have any concerns about our products,
you can contact us on
ProductSafety@springernature.com

In case Publisher is established outside the EU,
the EU authorized representative is:
Springer Nature Customer Service Center GmbH
Europaplatz 3, 69115 Heidelberg, Germany

Printed by Libri Plureos GmbH
in Hamburg, Germany